INTRODUCTION TO TURBULENT TRANSPORT OF PARTICLES, TEMPERATURE AND MAGNETIC FIELDS

Turbulence and the associated turbulent transport of scalar and vector fields present a classical physics problem that has dazzled scientists for over a century, yet many fundamental questions remain. Igor Rogachevskii, in this concise book, systematically applies various analytical methods to the turbulent transport of temperature, particles and magnetic fields. Introducing key concepts in turbulent transport including essential physics principles and statistical tools, this interdisciplinary book is suited for a range of readers, such as theoretical physicists, astrophysicists, geophysicists, plasma physicists and researchers in fluid mechanics and related topics in engineering. With an overview to various analytical methods, such as mean-field approach, dimensional analysis, multi-scale approach, quasi-linear approach, spectral tau approach, path-integral approach and analysis based on budget equations, it is also an accessible reference tool for advanced graduates, PhD students and researchers.

IGOR ROGACHEVSKII is a full Professor at the Department of Mechanical Engineering of Ben-Gurion University of the Negev and visiting Professor at Nordita of KTH Royal Institute of Technology and Stockholm University. He has published more than 160 research articles in several peer-reviewed journals on theoretical physics; dynamo; turbulent transport of temperature, particles and magnetic fields; plasma physics; solar physics and atmospheric physics.

INTRODUCTION TO TURBULENT TRANSPORT OF PARTICLES, TEMPERATURE AND MAGNETIC FIELDS

Analytical Methods for Physicists and Engineers

IGOR ROGACHEVSKII
Ben-Gurion University of the Negev

CAMBRIDGE
UNIVERSITY PRESS

CAMBRIDGE
UNIVERSITY PRESS

University Printing House, Cambridge CB2 8BS, United Kingdom

One Liberty Plaza, 20th Floor, New York, NY 10006, USA

477 Williamstown Road, Port Melbourne, VIC 3207, Australia

314–321, 3rd Floor, Plot 3, Splendor Forum, Jasola District Centre, New Delhi – 110025, India

103 Penang Road, #05–06/07, Visioncrest Commercial, Singapore 238467

Cambridge University Press is part of the University of Cambridge.

It furthers the University's mission by disseminating knowledge in the pursuit of
education, learning, and research at the highest international levels of excellence.

www.cambridge.org
Information on this title: www.cambridge.org/9781316518601
DOI: 10.1017/9781009000918

First published 2021

A catalogue record for this publication is available from the British Library.

ISBN 978-1-316-51860-1 Hardback

Contents

Preface

Turbulence, and the associated turbulent transport of scalar and vector fields, is one of the classical problems of physics. It has been studied systematically for more than a hundred years. However, many fundamental questions related to the nature of turbulence remain. Turbulent transport is a part of the field related to turbulence. Many excellent books on turbulence that describe velocity fluctuations in detail have been published over the last hundred years. Some of these books include problems related to the turbulent transport of passive fields. However, there are no books that systematically apply different analytical methods to turbulent transport of temperature, particles and magnetic fields.

The current book is an introduction to the various analytical methods of theoretical physics and applied mathematics used to develop the mean-field theories for studying the turbulent transport of particles, temperature and magnetic fields. In particular, the following analytical methods are systematically applied in this book: the dimensional analysis, the multi-scale approach, the quasi-linear approach, the tau approach (the relaxation approach), the path-integral approach and analyses based on the budget equations. One-way and two-way couplings between turbulence and particles, or temperature, or magnetic fields are described.

This book is written for theoretical physicists, astrophysicists, geophysicists, plasma physicists, space science physicists and also for researchers working in fluid mechanics in engineering sciences. The current book can be useful for postgraduate students, specialist researchers of turbulence and turbulent transport and nonspecialist researchers from related fields. It can be presented as a source of advanced teaching material for specialized seminars, courses and schools. This book has appeared as a development and extension of the material of graduate and postgraduate lecture courses given by the author at Ben-Gurion University of the Negev. Some of these lectures have been given at the Nordic Institute of Theoretical Physics (Nordita) of KTH Royal Institute of Technology and Stockholm University.

The current book assumes prior knowledge of the basic equations and principles of fluid mechanics and magnetohydrodynamics, as well as initial knowledge about turbulence. For instance, the first part of the textbook *Turbulence* by Peter Davidson (Oxford University Press, 2015) provides a solid introduction to the field. For educational purposes, the current book is written with detailed analytical calculations and numerous practice problems and exercises. Every chapter ends with a "Further Reading" section containing a short review in the field. The book contains 60 exercises of varying difficulty with solutions.

I would like to express my warmest thanks to my friend and coauthor, Nathan Kleeorin, with whom I discussed various aspects related to turbulent transport during our joint research. I am grateful to my friends and colleagues with whom I collaborated in different areas of turbulent transport, magnetohydrodynamics and plasma physics over the past 40 years: Alexey Boyarsky, Axel Brandenburg, Steve Cowley, Oliver Gressel, Alexander Gurevich, Alex Eidelman, David Eichler, Tov Elperin (1949–2018), Jürg Fröhlich, Nils Haugen, Maarit Käpylä, Petri Käpylä, Alexander Khain, Kirill Kuzanyan, Avi Levy, Michael Liberman, Victor L'vov, Baruh Meerson, Dhruba Mitra, Michael Mond, David Moss (1943–2020), Karl-Heintz Rädler (1935–2020), Oleg Ruchayskiy, Alexander Ruzmaikin, Pavel Sasorov, Alex Schekochihin, Jennifer Schober, Nishant Singh, Dmitry Sokoloff, Andrew Soward, Jörn Warnecke and Sergej Zilitinkevich (1936–2021).

I appreciate valuable and illuminating discussions with Alexander Baklanov, Akshay Bhatnagar, Erick Blackman, Fritz Busse, Pathric Diamond, Emmanuel Dormy, Chris Jones, Gregory Falkovich, Joe Fernando, Peter Frick, Lazar Friedland, George Golitsyn, David Hughes, Leonid Kitchatinov, Dmitri Kharzeev, Eliezer Kitt, Alex Liberzon, Larry Mahrt, Pablo Mininni, Keith Moffatt, Sergey Nazarenko, Lev Ostrovsky, Eugene Parker, Jean-François Pinton, Valery Pipin, Annick Pouquet, Michael Proctor, Matthias Rheinhardt, Paul Roberts, Anvar Shukurov, Edward Spiegel (1932–2020), Kandu Subramanian, Steve Tobias, Arkady Tsinober (1937–2020), Victor Yakhot, Nobu Yokoi and Vladimir Zakharov.

I am grateful to my friend Michael Feldman, who convinced and encouraged me to write this book and helped me a lot. I would like to express my warmest thanks to Rimma Shekhtman for the painting of the cover image for the book. This book is dedicated to my family.

1

Turbulent Transport of Temperature Fields

In this chapter, we consider a turbulent transport of temperature field in an isotropic homogeneous and incompressible turbulence. We discuss the Kolmogorov theory of hydrodynamic turbulence and obtain spectrum of velocity fluctuations for fully developed turbulence using the dimensional analysis. We study isotropic and anisotropic spectra of temperature fluctuations in different subranges of turbulent scales and different Prandtl numbers applying the dimensional analysis. We derive mean-field equations for the temperature field and obtain expressions for turbulent heat flux, turbulent diffusion and level of temperature fluctuations for small and large Péclet numbers by means of various analytical methods, namely the dimensional analysis, the quasi-linear approach and the spectral tau approach (the relaxation approach).

1.1 Hydrodynamic Turbulence: Dimensional Analysis

In this section, we consider a theory of hydrodynamic isotropic homogeneous and incompressible turbulence using the dimensional analysis.

1.1.1 Governing Equations and Basic Parameters

The fluid velocity field in an incompressible flow is determined by the Navier[1]-Stokes[2] equation (Landau and Lifshits, 1987; Batchelor, 1967; Lighthill, 1986; Tritton, 1988; Faber, 1995; Falkovich, 2011):

$$\boxed{\frac{\partial \boldsymbol{U}}{\partial t} + (\boldsymbol{U} \cdot \boldsymbol{\nabla})\boldsymbol{U} = -\frac{\boldsymbol{\nabla} P}{\rho} + \nu\, \Delta \boldsymbol{U} + \boldsymbol{f}.} \tag{1.1}$$

[1] Claude-Louis Navier (1785–1836) was a French engineer and physicist well-known for his works in mechanics, fluid dynamics, theory of elasticity and structural analysis.

[2] Sir George Gabriel Stokes (1819–1903) was a mathematician and physicist (who was born in Ireland and worked at the University of Cambridge) well-known for his works in fluid dynamics, optics and mathematical physics.

Equation (1.1) is the second law of Newton[3] for a unit mass of a fluid:

$$\rho \frac{\mathrm{d}\boldsymbol{U}}{\mathrm{d}t} = -\nabla P + \nabla \cdot \left(2\nu\, \rho\, \mathbf{S}^{(\mathrm{U})}\right) + \rho \boldsymbol{f}, \tag{1.2}$$

where according to the chain rule of differentiation of the function $\boldsymbol{U}[t, \boldsymbol{r}(t)]$, the substantial time derivative $\mathrm{d}\boldsymbol{U}/\mathrm{d}t$ for the moving fluid element is the sum of a local time derivative $\partial \boldsymbol{U}/\partial t$ and convective derivative $(\boldsymbol{U}\cdot\nabla)\boldsymbol{U}$. We take into account here that most fluids obey Newton's law of viscosity [see the second term on the right-hand side of Eq. (1.2)], where $\mathsf{S}_{ij}^{(\mathrm{U})} = \frac{1}{2}(\nabla_j U_i + \nabla_i U_j)$ are the components of the rate-of-strain-tensor $\mathbf{S}^{(\mathrm{U})}$ for incompressible fluid, ν is the kinematic viscosity, $\rho\boldsymbol{f}$ is the external force, that, e.g., creates a turbulent random velocity field, and P and ρ are the fluid pressure and density, respectively. The operators ∇ and $\Delta = \nabla^2$ in the Cartesian coordinates are defined as

$$\nabla = \boldsymbol{e}_x \frac{\partial}{\partial x} + \boldsymbol{e}_y \frac{\partial}{\partial y} + \boldsymbol{e}_z \frac{\partial}{\partial z}, \qquad \Delta = \frac{\partial^2}{\partial x^2} + \frac{\partial^2}{\partial y^2} + \frac{\partial^2}{\partial z^2}, \tag{1.3}$$

and $\boldsymbol{e}_x, \boldsymbol{e}_y$ and $\boldsymbol{e}_z$ are unit vectors along the x-, y- and z-axes. When the viscosity ν tends to zero, Eq. (1.1) is reduced to the Euler[4] equation. The fluid pressure and density are the macroscopic variables that determine the internal state of the fluid, and they are related by the equation of state for the perfect gas, $P = (k_B/m_\mu)\,\rho\, T \equiv (R/\mu)\,\rho\, T$, where $k_B = R/N_A$ is the Boltzmann constant, R is the gas constant, N_A is the Avogadro number, $\mu = m_\mu N_A$ is the molar mass and m_μ is the mass of the molecules of the surrounding fluid. Generally for arbitrary fluid flows, the continuity equation which is the conservation law for the fluid mass reads

$$\boxed{\frac{\partial \rho}{\partial t} + \nabla \cdot (\rho\, \boldsymbol{U}) = 0\,.} \tag{1.4}$$

This equation implies that for any volume, the change of the fluid mass inside the volume is compensated by the fluid flux through this volume.

For an incompressible fluid flow, the continuity equation (1.4) is reduced to

$$\boxed{\operatorname{div} \boldsymbol{U} \equiv \nabla \cdot \boldsymbol{U} = 0\,,} \tag{1.5}$$

where the fluid density ρ is constant in time and space. The second term $(\boldsymbol{U}\cdot\nabla)\boldsymbol{U}$ on the left-hand side of Eq. (1.1) is a nonlinear term that describes inertia. The

[3] Sir Isaac Newton (1642–1726) was an English mathematician, physicist and astronomer who made key contributions to the foundations of classical mechanics, optics and the infinitesimal calculus and built the first practical reflecting telescope.

[4] Leonhard Euler (1707–1783) is a Swiss mathematician, physicist, astronomer, geographer and engineer who made influential discoveries in mathematics (infinitesimal calculus and graph theory, topology, analytic number theory and mathematical analysis), mechanics, fluid dynamics, optics, astronomy and music theory.

dimensionless ratio of the nonlinear term to the viscous term in Eq. (1.1) is the Reynolds[5] number, which is a key parameter in the system:

$$\mathrm{Re} = \frac{|(\boldsymbol{U} \cdot \nabla)\boldsymbol{U}|}{|\nu\, \Delta \boldsymbol{U}|}. \tag{1.6}$$

For very large Reynolds numbers, the fluid flow is turbulent. There are many examples of turbulent flows in nature, laboratory experiments and industrial applications (Landau and Lifshits, 1987; Batchelor, 1953; Monin and Yaglom, 1971, 2013; Tennekes and Lumley, 1972; Frisch, 1995; Pope, 2000; Bernard and Wallace, 2002; Lesieur, 2008; Davidson, 2015). For instance, turbulence in laboratory experiments is produced, e.g., by oscillating grids, propellers, shear flows, etc. The atmospheric turbulence is produced by convective motions and large-scale shear flow (a non-uniform wind). Turbulence inside the Sun is produced by convection in the solar convective zone located under the solar surface. Turbulence in galaxies is produced by random supernova explosions. In astrophysics, turbulence can be also produced by shear motions and various plasma instabilities. Various pictures of turbulent flows can be found in the book by Van Dyke (1982).

In turbulent flows, the fluid velocity is a random field. Large-scale effects caused by small-scale turbulence can be studied using a mean-field approach. In the framework of this approach velocity field can be decomposed into the mean velocity and fluctuations, $\boldsymbol{U} = \overline{\boldsymbol{U}} + \boldsymbol{u}$, where according to the Reynolds rule velocity fluctuations $\boldsymbol{u}$ have zero mean value, $\langle \boldsymbol{u} \rangle = 0$ and $\overline{\boldsymbol{U}} = \langle \boldsymbol{U} \rangle$ is the mean fluid velocity. The angular brackets $\langle \ldots \rangle$ denote an averaging. Different kinds of averaging procedures will be discussed in the next section.

The Reynolds number defined by Eq. (1.6) can be estimated using the dimensional analysis. In particular, in Eq. (1.6) we replace operators $|\nabla|$ by ℓ_0^{-1} and Δ by ℓ_0^{-2}. This yields

$$\boxed{\mathrm{Re} = \frac{\ell_0\, u_0}{\nu},} \tag{1.7}$$

where ℓ_0 is the integral (energy-containing or maximum) scale of turbulence and $u_0 = \left[\langle \boldsymbol{u}^2 \rangle_{\ell=\ell_0}\right]^{1/2}$ is the characteristic turbulent velocity in the integral scale of turbulence ℓ_0. For example, in Table 1.1 we give turbulence parameters for various flows, e.g., for laboratory experiments in air flows, industrial flows in a wind tunnel and a diesel engine, atmospheric turbulence in the low troposphere (about 1 or 2 kilometers height from the Earth surface), and astrophysical turbulence, e.g., in the solar convective zone located under the solar surface with the depth about 1/3 of the solar radius and inside a galactic disk with a high concentration of stars. Here

[5] Osborne Reynolds (1842–1912) was an engineer (who was born in Ireland and worked at Owens College in Manchester, now the University of Manchester), well-known for his works in fluid dynamics and heat transfer.

Table 1.1 *Parameters for engineering, geophysical and astrophysical turbulence*

	ℓ_0 (cm)	u_0 (cm/s)	τ_0 (s)	ν (cm^2/s)	Re
Laboratory experiments	1–10	10–10^2	10^{-2}–1	10^{-1} (air)	10^2–10^4
Diesel engine	0.3	3×10^2	10^{-3}	10^{-2}	10^4
Wind tunnel	(1–3) $\times 10^2$	(1–3) $\times 10^3$	0.03–0.3	10^{-1}	10^6–10^7
Atmospheric turbulence	10^4	10^2	10^2	10^{-1}	10^7
Sun ($r \approx R_\odot$)	3×10^7	10^5	3×10^2	3×10^{-2}	10^{14}
Sun $\left(r \approx \frac{2}{3} R_\odot\right)$	5×10^9	2×10^3	3×10^6	10^{-1}	10^{14}
Galactic disk	10^{20}	10^6	10^{14}	10^{18}	10^8

$\tau_0 = \ell_0/u_0$ is the characteristic turbulent time in the scale ℓ_0, the radius $r = \frac{2}{3} R_\odot$ corresponds to the bottom of the solar convective zone and $R_\odot = 6.96 \times 10^{10}$ cm is the solar radius.

A fully developed turbulence for very large Reynolds numbers can be qualitatively regarded as a sea of eddies, i.e., an ensemble of turbulent eddies of different scales varying from the integral energy-containing scale ℓ_0 to very small viscous scale ℓ_ν. Turbulent eddy can be considered as a blob of vorticity $\nabla \times \boldsymbol{u}$. In the scale ℓ_ν, the viscous dissipation of the turbulent kinetic energy becomes important. The dynamics of the turbulent eddies is as follows. The large eddies are unstable, and they break down into the small eddies. The new small eddies are also unstable and continue to breakdown into the very small eddies. This process is called the Richardson[6] energy cascade and implies the transfer of the turbulent kinetic energy from the integral scale to smaller ones (Richardson, 1920). The energy cascade stops when the size of the small eddies is of the order of the viscous scale of turbulence. At this scale, turbulent kinetic energy is dissipated into thermal energy. The rate of the dissipation of the turbulent kinetic energy ε can be estimated as

$$\varepsilon = \frac{u_0^2}{\tau_0} = \frac{u_0^3}{\ell_0}. \tag{1.8}$$

1.1.2 Kolmogorov Theory of Hydrodynamic Turbulence

In this section, we consider Kolmogorov[7] theory of hydrodynamic turbulence. Let us assume that

[6] Lewis F. Richardson (1881–1953) was an English mathematician, physicist and meteorologist, well-known for his works in turbulence, mathematical physics and mathematical techniques of weather forecasting.

[7] Andrey N. Kolmogorov (1903–1987) was a Russian mathematician, well-known for his works in theory of random processes and probability theory, theory of turbulence, topology, theory of differential equations, functional analysis and information theory.

- turbulence is homogeneous, i.e., $\nabla \langle \boldsymbol{u}^2 \rangle = 0$;
- turbulence is isotropic, i.e., there is no preferential direction;
- turbulent flow is incompressible, i.e., $\nabla \cdot \boldsymbol{u} = 0$ and the fluid density ρ is constant in time and in space;
- interactions in the turbulence are local, i.e., there are only interactions between turbulent eddies of the same size, and there are no interactions between the eddies of different sizes;
- in a subrange of turbulent scales $\ell_\nu \le \ell \le \ell_0$, the dissipation rate of the turbulent kinetic energy density is constant

$$\boxed{\varepsilon = \frac{u_0^3}{\ell_0} = \frac{u_\ell^3}{\ell} = \ldots = \frac{u_\nu^3}{\ell_\nu} = \text{const},} \tag{1.9}$$

where $u_\ell = \left[\langle \boldsymbol{u}^2 \rangle_\ell\right]^{1/2}$ is the characteristic turbulent velocity at the scale ℓ inside the inertial subrange of turbulence scales $\ell_\nu \le \ell \le \ell_0$ and $u_\nu = \left[\langle \boldsymbol{u}^2 \rangle_{\ell=\ell_\nu}\right]^{1/2}$ is the characteristic velocity at the viscous scale ℓ_ν. For the simplicity we assume here that the constant fluid density is unity. Equation (1.9) allows us to determine turbulent velocities in different scales,

$$u_0 = (\varepsilon\, \ell_0)^{1/3}, \quad u_\ell = (\varepsilon\, \ell)^{1/3}, \quad u_\nu = (\varepsilon\, \ell_\nu)^{1/3}. \tag{1.10}$$

Equation $u_\ell = (\varepsilon\, \ell)^{1/3}$ implies that the scaling for u_ℓ^2 in the inertial subrange of turbulent scales $\ell_\nu \ll \ell \ll \ell_0$ is given by

$$\boxed{u_\ell^2 = \varepsilon^{2/3}\, \ell^{2/3}} \tag{1.11}$$

[see Kolmogorov (1941), and its English translation in Kolmogorov (1991)], and the characteristic time $\tau_\ell = \ell / u_\ell$ in the inertial subrange of scales is

$$\boxed{\tau_\ell = \varepsilon^{-1/3}\, \ell^{2/3}.} \tag{1.12}$$

Using Eq. (1.10), we rewrite the Reynolds number as

$$\text{Re} = \frac{\ell_0\, u_0}{\nu} = \frac{\varepsilon^{1/3}\, \ell_0^{4/3}}{\nu}. \tag{1.13}$$

We introduce the local Reynolds number:

$$\text{Re}_\ell = \frac{\ell\, u_\ell}{\nu} = \frac{\varepsilon^{1/3}\, \ell^{4/3}}{\nu}. \tag{1.14}$$

Equations (1.13)–(1.14) allow us to determine the ratio Re_ℓ/Re as

$$\frac{\text{Re}_\ell}{\text{Re}} = \left(\frac{\ell}{\ell_0}\right)^{4/3}. \tag{1.15}$$

The viscous scale ℓ_ν (the Kolmogorov scale) is defined as the scale in which the local Reynolds number is 1. This implies that in the Kolmogorov scale, the nonlinear term in the Navier-Stokes equation is of the order of the viscous term. Therefore, Eq. (1.15) with the condition $\mathrm{Re}_{\ell=\ell_\nu} = 1$ allow us to relate the Kolmogorov scale ℓ_ν with the integral scale ℓ_0 of turbulence as

$$\boxed{\ell_\nu = \frac{\ell_0}{\mathrm{Re}^{3/4}}\,.} \tag{1.16}$$

Substituting the Kolmogorov scale ℓ_ν given by Eq. (1.16) into Eq. (1.10) for $u_\nu = (\varepsilon\,\ell_\nu)^{1/3}$, we obtain the characteristic velocity in the Kolmogorov scale as $u_\nu = (\varepsilon\,\ell_0)^{1/3}\,\mathrm{Re}^{-1/4}$, so that

$$\boxed{u_\nu = \frac{u_0}{\mathrm{Re}^{1/4}}\,,} \tag{1.17}$$

where $u_0 = (\varepsilon\,\ell_0)^{1/3}$. Therefore, the characteristic viscous time $\tau_\nu = \ell_\nu/u_\nu$ (the Kolmogorov time) is given by

$$\boxed{\tau_\nu = \frac{\tau_0}{\mathrm{Re}^{1/2}}\,.} \tag{1.18}$$

Next, we determine the spectrum of velocity fluctuations in the inertial subrange of scales (the Kolmogorov-Obukhov[8] spectrum). We define the energy spectrum function of the velocity field as

$$u_\ell^2 = \int_{k_0}^{k} E_{\mathrm{u}}(k')\,dk', \tag{1.19}$$

where wave numbers $k_0 = \ell_0^{-1}$ and $k = \ell^{-1}$. Using the dimensional analysis, we rewrite Eq. (1.19) as $u_\ell^2 = E_{\mathrm{u}}(k)\,k$. Therefore, the Kolmogorov-Obukhov spectrum $E_{\mathrm{u}}(k)$ in the inertial subrange of turbulent scales $k_0 \ll k \ll k_\nu$ is given by (Kolmogorov, 1941; Obukhov, 1941)

$$\boxed{E_{\mathrm{u}}(k) = \varepsilon^{2/3} k^{-5/3},} \tag{1.20}$$

where $k_\nu = \ell_\nu^{-1}$ and we take into account that $u_\ell = (\varepsilon/k)^{1/3}$. Equation (1.20) also directly follows from Eq. (1.11) using the relations $\ell = k^{-1}$ and $E_{\mathrm{u}}(k) = u_\ell^2/k$. Since $\varepsilon = u_\ell^2/\tau_\ell = E_{\mathrm{u}}(k)\,k/\tau(k)$, we obtain the scaling for the characteristic time $\tau(k)$ in the inertial subrange of turbulent scales as

$$\boxed{\tau(k) = \varepsilon^{-1/3} k^{-2/3}.} \tag{1.21}$$

[8] Alexander M. Obukhov (1918–1998) was a Russian geophysicist well-known for his works in atmospheric physics, meteorology, turbulence and mathematical statistics.

Equation (1.21) also directly follows from Eq. (1.12) using the relation $\ell = k^{-1}$. The Kolmogorov-Obukhov spectrum has been detected in many laboratory experiments where turbulence is produced by various sources. This spectrum also has been observed in atmospheric turbulence, space experiments with solar wind, and solar and galactic turbulence. The Kolmogorov-Obukhov spectrum can be considered as a universal spectrum since it is observed in various turbulent systems of different origins.

1.2 Spectra of Temperature Fluctuations: Dimensional Analysis

In this section, we obtain various spectra of temperature fluctuations in a hydrodynamic isotropic homogeneous and incompressible turbulence using the dimensional analysis.

1.2.1 Governing Equations, Averaging and Basic Parameters

The equation for the evolution of fluid temperature field $T(t, \boldsymbol{x})$ in an incompressible fluid velocity field $\boldsymbol{U}(t, \boldsymbol{x})$ reads (Landau and Lifshits, 1987; Batchelor, 1967)

$$\boxed{\frac{\partial T}{\partial t} + (\boldsymbol{U} \cdot \boldsymbol{\nabla})T = D^{(\theta)} \, \Delta T + I_T,} \tag{1.22}$$

where $D^{(\theta)}$ is the coefficient of the molecular diffusion of the temperature field and I_T is the heat source/sink that for simplicity is neglected below. Equation (1.22) is the convective diffusion equation. The continuity equation for the incompressible fluid velocity field is $\boldsymbol{\nabla} \cdot \boldsymbol{U} = 0$. We apply a mean-field approach, i.e., all quantities are decomposed into the mean and fluctuating parts, where the fluctuating parts have zero mean values. For example, the temperature field $T = \overline{T} + \theta$, where $\overline{T} = \langle T \rangle$ is the mean fluid temperature, θ are temperature fluctuations, and $\langle \theta \rangle = 0$. The angular brackets $\langle \ldots \rangle$ denote an averaging. Similarly, $\boldsymbol{U} = \overline{\boldsymbol{U}} + \boldsymbol{u}$, where $\overline{\boldsymbol{U}} = \langle \boldsymbol{U} \rangle$ is the mean fluid velocity, $\boldsymbol{u}$ are velocity fluctuations and $\langle \boldsymbol{u} \rangle = 0$. There are three main ways of averaging:

- The time averaging (i.e., the averaging over the time):

$$\overline{T} = \frac{1}{t_M} \int_0^{t_M} T(t, \boldsymbol{x}) \, dt, \tag{1.23}$$

  where t_M is the total time of measurements (e.g., in the case of laboratory or field experiments) or the total time of calculations (e.g., in the case of numerical simulations).

- The spatial (volume) averaging:

$$\overline{T} = \frac{1}{L_x L_y L_z} \int_0^{L_x} dx \int_0^{L_y} dy \int_0^{L_z} T(t, \boldsymbol{x})\, dz, \tag{1.24}$$

  where L_x, L_y, L_z are the sizes of the box along x, y, z directions. The plane averaging,

$$\overline{T} = \frac{1}{L_x L_y} \int_0^{L_x} dx \int_0^{L_y} T(t, \boldsymbol{x})\, dy, \tag{1.25}$$

  is used or even the averaging along one direction,

$$\overline{T} = \frac{1}{L_y} \int_0^{L_y} T(t, \boldsymbol{x})\, dy, \tag{1.26}$$

  is also used.
- The ensemble averaging (e.g., averaging over independent spatial distributions $T_n = T(t_n, \boldsymbol{x})$ of temperature fields taken in different times: $t_1, t_2, \ldots, t_N$):

$$\overline{T} = \frac{1}{N} \sum_{n=1}^{N} T_n(\boldsymbol{x}), \tag{1.27}$$

  where t_n are the instants of measurements and N is the total number of data points.

Averaging Eq. (1.22) over an ensemble of turbulent velocity field, we arrive at the mean-field equation for the mean temperature field:

$$\boxed{\frac{\partial \overline{T}}{\partial t} + \boldsymbol{\nabla} \cdot \left(\overline{T}\, \overline{\boldsymbol{U}} + \langle \theta\, \boldsymbol{u} \rangle\right) = D^{(\theta)}\, \Delta \overline{T},} \tag{1.28}$$

where $\langle \theta\, \boldsymbol{u} \rangle$ is the turbulent heat flux. In our derivation of Eq. (1.28), we take into account that

- various operators, like the averaging $\langle \ldots \rangle$, the partial derivative over time, the spatial partial derivatives, the operators $\boldsymbol{\nabla}$ and Δ, are linear commutative operators;
- $\langle \boldsymbol{u} \overline{T} \rangle = \overline{T} \langle \boldsymbol{u} \rangle = 0$ and $\langle \overline{\boldsymbol{U}} \theta \rangle = \overline{\boldsymbol{U}} \langle \theta \rangle = 0$.

Let us consider for simplicity the case $\overline{\boldsymbol{U}} = 0$. The obtained results will be the same for the constant mean fluid velocity due to the Galilean[9] invariance.

[9] Galileo Galilei (1564–1642) was an Italian astronomer, physicist, engineer, philosopher and mathematician, well-known for his works in physics, astronomy and applied science.

The equation for temperature fluctuations $\theta = T - \overline{T}$ is obtained by subtracting Eq. (1.28) from Eq. (1.22):

$$\boxed{\frac{\partial \theta}{\partial t} + \nabla \cdot [\theta\, \boldsymbol{u} - \langle \theta\, \boldsymbol{u} \rangle] - D^{(\theta)} \Delta \theta = -(\boldsymbol{u} \cdot \nabla) \overline{T}.} \tag{1.29}$$

The second term, $\nabla \cdot (\theta\, \boldsymbol{u} - \langle \theta\, \boldsymbol{u} \rangle)$, on the left-hand side of Eq. (1.29) is the nonlinear term, while the first term, $-(\boldsymbol{u} \cdot \nabla)\overline{T}$, on the right-hand side of Eq. (1.29) is the source of temperature fluctuations produced by the tangling of the gradient of the mean temperature $\nabla \overline{T}$ by random velocity fluctuations $\boldsymbol{u}$. The dimensionless ratio of the nonlinear term to the diffusion term in Eq. (1.29) is the Péclet[10] number which is a key parameter in the system:

$$\mathrm{Pe} = \frac{|\nabla \cdot (\theta\, \boldsymbol{u} - \langle \theta\, \boldsymbol{u} \rangle)\,|}{|D^{(\theta)} \Delta \theta|}. \tag{1.30}$$

The Péclet number, defined by Eq. (1.30), can be estimated using dimensional analysis as

$$\boxed{\mathrm{Pe} = \frac{\ell_0\, u_0}{D^{(\theta)}}.} \tag{1.31}$$

Using Eq. (1.10) for the turbulent velocity $u_0 = (\varepsilon\, \ell_0)^{1/3}$ at the integral scale, we rewrite the Péclet number as

$$\mathrm{Pe} = \frac{\varepsilon^{1/3} \ell_0^{4/3}}{D^{(\theta)}}. \tag{1.32}$$

Next, we introduce the local Péclet number $\mathrm{Pe}_\ell = \ell\, u_\ell / D^{(\theta)}$ at the scale ℓ and use Eq. (1.10) for the turbulent velocity $u_\ell = (\varepsilon\, \ell)^{1/3}$, so that the local Péclet number is

$$\mathrm{Pe}_\ell = \frac{\ell\, u_\ell}{D^{(\theta)}} = \frac{\varepsilon^{1/3} \ell^{4/3}}{D^{(\theta)}}. \tag{1.33}$$

We determine the ratio $\mathrm{Pe}_\ell / \mathrm{Pe}$ as

$$\frac{\mathrm{Pe}_\ell}{\mathrm{Pe}} = \left(\frac{\ell}{\ell_0} \right)^{4/3}. \tag{1.34}$$

We introduce a diffusion scale ℓ_D defined as the scale in which the local Péclet number is 1. This implies that at the scale ℓ_D, the nonlinear terms in the equation

10 Jean Claude Eugène Péclet (1793–1857) was a French physicist well-known for his works in fluid dynamics, heat transfer and theory of combustion.

for temperature fluctuations equal the diffusion term. Therefore, Eq. (1.34) yields the diffusion scale ℓ_D as

$$\boxed{\ell_D = \frac{\ell_0}{\mathrm{Pe}^{3/4}}.} \tag{1.35}$$

Let us consider the case when the diffusion scale is inside the inertial subrange of turbulent scales, $\ell_0 \geq \ell_D \geq \ell_\nu$. This implies that $u_D = (\varepsilon\, \ell_D)^{1/3}$ [see Eqs. (1.9)–(1.10)]. Substituting the diffusion scale (1.35) into the equation for u_D, we obtain the characteristic velocity at the diffusion scale as $u_D = (\varepsilon\, \ell_0)^{1/3}\ \mathrm{Pe}^{-1/4}$, so that

$$\boxed{u_D = \frac{u_0}{\mathrm{Pe}^{1/4}},} \tag{1.36}$$

where $u_0 = (\varepsilon\, \ell_0)^{1/3}$. Using Eqs. (1.35) and (1.36), we determine the characteristic diffusion time $\tau_D = \ell_D/u_D$ as

$$\boxed{\tau_D = \frac{\tau_0}{\mathrm{Pe}^{1/2}} \equiv \frac{\ell_D^2}{D^{(\theta)}},} \tag{1.37}$$

where we take into account that $\mathrm{Pe}_{\ell=\ell_D} = 1$, i.e., $u_D \ell_D = D^{(\theta)}$. Let us determine the ratio of the diffusion scale to the viscous scale ℓ_D/ℓ_ν:

$$\frac{\ell_D}{\ell_\nu} = \left(\frac{\mathrm{Re}}{\mathrm{Pe}}\right)^{3/4} = \left(\frac{D^{(\theta)}}{\nu}\right)^{3/4} = \mathrm{Pr}^{-3/4}, \tag{1.38}$$

where

$$\boxed{\mathrm{Pr} = \frac{\nu}{D^{(\theta)}}} \tag{1.39}$$

is the Prandtl[11] number. Small Prandtl numbers $\mathrm{Pr} \ll 1$ implies that $\ell_\nu \ll \ell_D$, i.e., the viscous scale ℓ_ν is the smallest scale. In the opposite case of large Prandtl numbers $\mathrm{Pr} \gg 1$, the diffusion scale $\ell_D \ll \ell_\nu$ is the smallest scale.

1.2.2 Isotropic Temperature Fluctuations

In this section, we consider the case of small Prandtl numbers ($\mathrm{Pr} \ll 1$) and study temperature fluctuations in the inertial subrange of turbulence, $\ell_\nu \ll \ell_D < \ell < \ell_0$. The energy spectrum function of the velocity field is defined as $u_\ell^2 = \int_{k_0}^{k} E_{\mathrm{u}}(k')\, dk'$, where $k_0 = \ell_0^{-1}$ and $k = \ell^{-1}$. The Kolmogorov-Obukhov spectrum of velocity fluctuations in the inertial subrange of turbulent scales is given by

$$E_{\mathrm{u}}(k) = \frac{u_\ell^2}{k} = \varepsilon^{2/3}\, k^{-5/3}, \tag{1.40}$$

[11] Ludwig Prandtl (1875–1953) was a German engineer and physicist well-known for his works in fluid dynamics, aerodynamics, shock waves, plasticity, structural mechanics and meteorology.

and the scaling for the turbulent time $\tau(k)$ is

$$\tau(k) = \frac{\ell}{u_\ell} = \frac{\ell}{(\varepsilon \ell)^{1/3}} = \varepsilon^{-1/3}\, k^{-2/3}. \tag{1.41}$$

Equations (1.40)–(1.41) are only valid when the rate of dissipation of the turbulent kinetic energy density is constant inside the inertial subrange of turbulent scales, i.e.,

$$\varepsilon \equiv \frac{u_\ell^2}{\tau_\ell} = \frac{E_{\rm u}(k)\, k}{\tau(k)} = {\rm const}, \tag{1.42}$$

and $u_\ell = (\varepsilon / k)^{1/3}$.

Spectrum function of temperature fluctuations is defined as

$$\langle \theta^2 \rangle_\ell = \int_{k_0}^{k} \tilde{E}_\theta(k')\, dk'. \tag{1.43}$$

Using the dimensional analysis, we rewrite this expression as $\langle \theta^2 \rangle_\ell = \tilde{E}_\theta(k)\, k$ and assume that the rate of dissipation of temperature fluctuations is constant inside the subrange of scales $\ell_D < \ell < \ell_0$, i.e.,

$$\varepsilon_\theta \equiv \frac{\langle \theta^2 \rangle_\ell}{\tau_\ell} = \frac{\tilde{E}_\theta(k)\, k}{\tau(k)} = {\rm const}. \tag{1.44}$$

The condition (1.44) for temperature fluctuations is analogous to condition (1.42) for velocity fluctuations in the inertial range of turbulence. Equations (1.42) and (1.44) yield the spectrum of isotropic temperature fluctuations inside the scale-dependent turbulent diffusion range of scales, $\ell_D < \ell < \ell_0$:

$$\boxed{\tilde{E}_\theta(k) \sim E_{\rm u}(k) \sim \varepsilon^{2/3}\, k^{-5/3}.} \tag{1.45}$$

This spectrum was obtained by Obukhov (1949) and Corrsin (1951).[12]

Let us consider the case, $\Pr > 1$, and study temperature fluctuations in the viscous subrange of scales, $\ell_D < \ell < \ell_\nu$. In this range of scales, Eq. (1.44) is valid, but the time $\tau(k)$ does not have a universal scaling. If $\tau(k) = {\rm const}$, the spectrum of temperature fluctuations is

$$\boxed{\tilde{E}_\theta(k) \sim k^{-1}.} \tag{1.46}$$

This spectrum was obtained by Batchelor (1959)[13] and Kraichnan (1968).[14]

12 Stanley Corrsin (1920–1986) was an American physicist, well-known for his works in experimental and theoretical fluid dynamics, turbulence and turbulent mixing.

13 George Keith Batchelor (1920–2000) was an applied mathematician (who was born in Australia and worked at the University of Cambridge) well-known for his works in fluid dynamics, theory of turbulence and turbulent transport.

14 Robert Harry Kraichnan (1928–2008) was an American theoretical physicist well-known for his works in the theory of turbulence, turbulent transport and magnetohydrodynamics.

1.2.3 Anisotropic Temperature Fluctuations in the Inertial-Diffusion Range

We consider anisotropic temperature fluctuations caused by the tangling of the mean temperature gradient by velocity fluctuations in the inertial-diffusion subrange of scales, $\ell_\nu < \ell < \ell_D$. This subrange of scales corresponds to the Prandtl numbers, $\mathrm{Pr} < 1$. We use dimensional analysis, taking into account that molecular diffusion is a key effect in this subrange of scales. This implies that the molecular diffusion term, $D^{(\theta)}\Delta\theta$, in Eq. (1.29) should be balanced by the source term, $(\boldsymbol{u}\cdot\nabla)\overline{T}$, for temperature fluctuations, i.e.,

$$|D^{(\theta)}\Delta\theta| \sim |(\boldsymbol{u}\cdot\nabla)\overline{T}|, \tag{1.47}$$

which yields

$$\left\langle\theta^2\right\rangle_\ell \sim u_\ell^2 \left(\frac{\ell^2 \nabla\overline{T}}{D^{(\theta)}}\right)^2 . \tag{1.48}$$

In the $\boldsymbol{k}$ space, Eq. (1.48) implies that

$$\tilde{E}_\theta(k) \sim E_\mathrm{u}(k)\, k^{-4} \left(\frac{\nabla\overline{T}}{D^{(\theta)}}\right)^2 , \tag{1.49}$$

where $E_\mathrm{u}(k)$ is the spectrum function of velocity fluctuations. Since the subrange of scales $\ell_\nu < \ell < \ell_D$ corresponds to the inertial range of scales, velocity fluctuations have the Kolmogorov-Obukhov spectrum (1.40). Therefore, the spectrum of the anisotropic temperature fluctuations in the inertial-diffusion range of scales is

$$\boxed{\tilde{E}_\theta(k) \sim \varepsilon^{2/3}\, k^{-17/3} \left(\frac{\nabla\overline{T}}{D^{(\theta)}}\right)^2 .} \tag{1.50}$$

This spectrum was obtained by G. Batchelor, I. Howells and A. Townsend[15] (Batchelor et al., 1959).

1.2.4 Anisotropic Temperature Fluctuations in the Inertial-Turbulent Diffusion Range

We consider anisotropic temperature fluctuations caused by the tangling of the mean temperature gradient by velocity fluctuations in the inertial-turbulent diffusion range of scales, $\ell_D < \ell < \ell_0$. This subrange of scales corresponds to the small Prandtl numbers, $\mathrm{Pr} < 1$. We take into account that the main effect of turbulence

[15] Albert Alan Townsend (1917–2010) was a physicist (who was born in Australia and worked at the University of Cambridge) well-known for his works in fluid dynamics, experimental study of turbulence and turbulent transport, meteorology and nuclear physics.

on temperature fluctuations in incompressible flow is the scale-dependent turbulent diffusion that is much larger than the molecular diffusion for large Péclet numbers. Let us average Eq. (1.29) over an ensemble up to the scale ℓ_* that is inside the interval: $\ell_D \ll \ell_* \ll \ell_0$. This yields the renormalized equation for temperature fluctuations:

$$\frac{\partial \theta}{\partial t} - D_{_T}(\ell)\, \Delta\theta = -(\boldsymbol{u} \cdot \nabla)\overline{T}, \tag{1.51}$$

where $D_{_T}(\ell)$ is the scale-dependent turbulent diffusion coefficient that can be estimated as

$$\boxed{D_{_T}(\ell) = \ell\, u_\ell.} \tag{1.52}$$

In the subrange of scales $\ell_D \ll \ell_*$ the turbulent diffusion term $D_{_T}(\ell)\Delta\theta$ in Eq. (1.51) should be balanced by the source term, $(\boldsymbol{u}\cdot\nabla)\overline{T}$, for temperature fluctuations, i.e.,

$$|D_{_T}(\ell)\, \Delta\theta| \sim |(\boldsymbol{u} \cdot \nabla)\overline{T}|. \tag{1.53}$$

This implies that

$$\left\langle\theta^2\right\rangle_\ell \sim u_\ell^2 \left(\frac{\ell^2\, \nabla\overline{T}}{D_{_T}(\ell)}\right)^2 \sim u_\ell^2 \left(\frac{\ell^2\, \nabla\overline{T}}{\ell\, u_\ell}\right)^2 \sim \left(\ell\, \nabla\overline{T}\right)^2, \tag{1.54}$$

where we used Eq. (1.52). Equation (1.54) written in the $\boldsymbol{k}$ space yields the spectrum of anisotropic temperature fluctuations in the inertial-turbulent diffusion range of scales $\ell_D \ll \ell \ll \ell_0$:

$$\boxed{\tilde{E}_\theta(k) \sim k^{-3} \left(\nabla\overline{T}\right)^2,} \tag{1.55}$$

where we take into account in Eq. (1.54) that according to dimensional analysis, $\langle\theta^2\rangle_\ell = \tilde{E}_\theta(k)\, k$ and $k = \ell^{-1}$. This spectrum is independent of the spectrum of the turbulent velocity field because u_ℓ^2 is canceled in Eq. (1.54). The spectrum (1.55) was obtained by A. Wheelon using the dimensional analysis (Wheelon, 1957) and by T. Elperin, N. Kleeorin and I. Rogachevskii, applying the renormalization approach (Elperin et al., 1996a).

1.3 Turbulent Transport of Temperature Fields: Dimensional Analysis

1.3.1 Governing Equations

We consider incompressible fluid velocity field $\boldsymbol{U}(t, \boldsymbol{x})$ satisfying the continuity equation: $\nabla \cdot \boldsymbol{U} = 0$. Since the velocity field is incompressible, Eq. (1.22) for the fluid temperature field $T(t, \boldsymbol{x})$ can be rewritten in the following form:

$$\boxed{\frac{\partial T}{\partial t} + \nabla \cdot (T\, \boldsymbol{U}) = D^{(\theta)}\, \Delta T.} \tag{1.56}$$

The velocity field is a random turbulent field created, e.g., by external forcing.

1.3.2 Mean-Field Approach

Our goal is to study the long-term evolution of the temperature field in the large scales, i.e., in spatial scales $L_T \gg \ell_0$, and the time scales $t_T \gg \tau_0$, where τ_0 is the characteristic turbulent time in the integral turbulent scale ℓ_0, L_T is the characteristic spatial scale of variations of the mean temperature field and t_T is the characteristic time-scale of variations of the mean temperature field. We use a mean-field approach in which all quantities are decomposed into the mean and fluctuating parts, where the fluctuating parts have zero mean values. In particular, the temperature field $T = \overline{T} + \theta$, where $\overline{T} = \langle T \rangle$ is the mean fluid temperature, θ are temperature fluctuations and $\langle \theta \rangle = 0$. The angular brackets $\langle \ldots \rangle$ denote ensemble averaging. In similar fashion, we decompose a velocity field, $\boldsymbol{U} = \overline{\boldsymbol{U}} + \boldsymbol{u}$, where $\overline{\boldsymbol{U}} = \langle \boldsymbol{U} \rangle$ is the mean fluid velocity, $\boldsymbol{u}$ are velocity fluctuations and $\langle \boldsymbol{u} \rangle = 0$. This decomposition corresponds to the Reynolds rules. Averaging Eq. (1.56) over an ensemble of turbulent velocity field, we arrive at the mean-field equation (1.28) for the mean temperature field.

1.3.3 Equation for Temperature Fluctuations

Equation (1.28) is not closed because we do not know the turbulent heat flux $\boldsymbol{F}^{(\theta)} = \langle \boldsymbol{u}\theta \rangle$. To determine the turbulent heat flux, we use the equation for temperature fluctuations that is obtained by subtracting Eq. (1.28) from Eq. (1.56):

$$\boxed{\frac{\partial \theta}{\partial t} + \nabla \cdot (\theta\, \boldsymbol{u} - \langle \theta\, \boldsymbol{u} \rangle) - D^{(\theta)} \Delta \theta = -(\boldsymbol{u} \cdot \nabla)\overline{T}.} \tag{1.57}$$

The terms, $\nabla \cdot (\theta\, \boldsymbol{u} - \langle \theta\, \boldsymbol{u} \rangle)$, on the left-hand side of Eq. (1.57) are the nonlinear terms, while the first term, $-(\boldsymbol{u} \cdot \nabla)\overline{T}$, on the right-hand side of Eq. (1.57) determines the source of temperature fluctuations produced by the tangling of the gradient of the mean temperature, $\nabla \overline{T}$, by random velocity fluctuations, $\boldsymbol{u}$. Since Eq. (1.57) is nonlinear equation for temperature fluctuations, it cannot be solved exactly for the arbitrary range of parameters and arbitrary velocity field. Therefore, we have to use different approximate methods for the solution of Eq. (1.57). First, we consider a one-way coupling, i.e., we take into account the effect of the turbulent velocity on the temperature field, but neglect the feedback effect of the temperature field on the turbulent fluid flow. This implies that the temperature field is a passive scalar.

1.3.4 Dimensional Analysis

The first method that we apply here is the dimensional analysis. The dimension of the left-hand side of Eq. (1.57) is the rate of change of temperature fluctuations, i.e., θ/τ_θ, where τ_θ is the characteristic time of temperature fluctuations. We replace the left-hand side of Eq. (1.57) by θ/τ_θ, that yields:

$$\theta = -\tau_\theta \, (\boldsymbol{u} \cdot \nabla)\overline{T}. \tag{1.58}$$

We consider two cases of large and small Péclet numbers:

- *Large Péclet numbers*, $\mathrm{Pe} = u_0 \, \ell_0 / D^{(\theta)} \gg 1$. We also consider the case of large Reynolds numbers, $\mathrm{Re} = u_0 \, \ell_0 / \nu \gg 1$. This implies that we consider fully developed turbulence. In this case, the characteristic time of temperature fluctuations, τ_θ, can be identified with the correlation time τ_0 of the turbulent velocity field. Therefore, in the framework of the dimensional analysis, we replace the left-hand side of Eq. (1.57) with θ/τ_0. This yields the expression for temperature fluctuations:

$$\theta = -\tau_0 \, (\boldsymbol{u} \cdot \nabla)\overline{T}. \tag{1.59}$$

- *Small Péclet numbers*, $\mathrm{Pe} \ll 1$. In this case, the nonlinear terms are much smaller than the diffusion terms. This implies that the molecular diffusion for $\mathrm{Pe} \ll 1$ is the main process, which determines the dynamics of temperature fluctuations. Therefore, we assume that the time τ_θ can be identified with the molecular diffusion time $\tau_D = \ell_0^2 / D^{(\theta)}$, and the solution of Eq. (1.57) for $\mathrm{Pe} \ll 1$ reads

$$\theta = -\frac{\ell_0^2}{D^{(\theta)}} \, (\boldsymbol{u} \cdot \nabla)\overline{T}. \tag{1.60}$$

1.3.5 Turbulent Heat Flux and Level of Temperature Fluctuations

Large Péclet Numbers

Let us consider the case of large Péclet numbers $\mathrm{Pe} \gg 1$ and determine the turbulent heat flux and level of temperature fluctuations. Multiplying Eq. (1.59) by velocity fluctuations, u_i, and averaging over an ensemble of turbulent velocity field, we obtain the turbulent heat flux:

$$\langle \theta \, u_i \rangle = -\tau_0 \, \langle u_i (\boldsymbol{u} \cdot \nabla)\overline{T} \rangle = -\tau_0 \, \langle u_i u_j \rangle \, \nabla_j \overline{T}, \tag{1.61}$$

where we took into account that $\boldsymbol{u} \cdot \nabla \equiv u_j \, \nabla_j = u_1 \, \nabla_1 + u_2 \, \nabla_2 + u_3 \, \nabla_3 \equiv u_x \, \nabla_x + u_y \, \nabla_y + u_z \, \nabla_z$ (i.e., there is summation in the repeating indexes). In isotropic turbulence, $\langle u_i u_j \rangle = \delta_{ij} \, \langle \boldsymbol{u}^2 \rangle/3$, where δ_{ij} is the Kronecker tensor (or the unit matrix), that is defined as $\delta_{ij} = 1$ for $i = j$, and $\delta_{ij} = 0$ for $i \neq j$. Therefore, for an isotropic turbulence, the turbulent heat flux reads

$$\boxed{\boldsymbol{F}^{(\theta)} \equiv \langle \theta \, \boldsymbol{u} \rangle = -D_{_T} \, \boldsymbol{\nabla} \overline{T},} \tag{1.62}$$

with the coefficient of turbulent diffusion of the temperature field for large Péclet numbers:

$$\boxed{D_{_T} = \frac{1}{3} \, \tau_0 \, \left\langle \boldsymbol{u}^2 \right\rangle .} \tag{1.63}$$

Using Eq. (1.59) for θ^2 and averaging over an ensemble of turbulent velocity field, we determine the level of temperature fluctuations $\langle \theta^2 \rangle$:

$$\left\langle \theta^2 \right\rangle = \tau_0^2 \, \left\langle \left[(\boldsymbol{u} \cdot \boldsymbol{\nabla}) \overline{T} \, \right]^2 \right\rangle = \tau_0^2 \, \langle u_i u_j \rangle \, \left(\nabla_i \overline{T} \right) \left(\nabla_j \overline{T} \right) . \tag{1.64}$$

Therefore, for an isotropic turbulence, $\langle u_i u_j \rangle = \delta_{ij} \, \langle \boldsymbol{u}^2 \rangle/3$, the level of temperature fluctuations for large Péclet numbers is given by

$$\boxed{\left\langle \theta^2 \right\rangle = \frac{1}{3} \ell_0^2 \, \left(\boldsymbol{\nabla} \overline{T} \right)^2 ,} \tag{1.65}$$

where $\ell_0 = \tau_0 \, u_0$ and $u_0 \equiv u_{\rm rms} = \sqrt{\langle \boldsymbol{u}^2 \rangle}$ is the r.m.s. velocity fluctuations (characteristic turbulent velocity).

Small Péclet Numbers

Now we consider the case of small Péclet numbers $\mathrm{Pe} \ll 1$ and determine the turbulent heat flux and level of temperature fluctuations. Multiplying Eq. (1.60) by u_i and averaging this equation over a statistics of a random velocity field, we obtain

$$\langle \theta \, u_i \rangle = -\frac{\ell_0^2}{D^{(\theta)}} \, \langle u_i u_j \rangle \, (\nabla_j \overline{T}). \tag{1.66}$$

Therefore, for isotropic turbulence, $\langle u_i u_j \rangle = \delta_{ij} \, \langle \boldsymbol{u}^2 \rangle/3$, the turbulent heat flux $F_i^{(\theta)}$ for small Péclet numbers is given by $\boldsymbol{F}^{(\theta)} \equiv \langle \theta \, \boldsymbol{u} \rangle = -D_{_T} \, \boldsymbol{\nabla} \overline{T}$, which coincides with Eq. (1.62) derived for large Péclet numbers, but with a different coefficient of turbulent diffusion:

$$\boxed{D_T = \frac{u_0 \, \ell_0}{3} \, \mathrm{Pe}.} \tag{1.67}$$

Since $\mathrm{Pe} \ll 1$, the coefficient of turbulent diffusion D_T is much smaller than the molecular diffusion coefficient $D^{(\theta)}$. Using Eq. (1.60) for θ^2 and averaging over an ensemble of turbulent velocity field, we determine the level of temperature fluctuations $\langle \theta^2 \rangle$:

$$\langle\theta^2\rangle = \left(\frac{\ell_0^2}{D^{(\theta)}}\right)^2 \left\langle\left[(\boldsymbol{u}\cdot\nabla)\overline{T}\,\right]^2\right\rangle = \left(\frac{\ell_0^2}{D^{(\theta)}}\right)^2 \langle u_i u_j\rangle\,(\nabla_i\overline{T})(\nabla_j\overline{T}). \quad (1.68)$$

Therefore, for isotropic turbulence the level of temperature fluctuations for small Péclet numbers is given by

$$\boxed{\langle\theta^2\rangle = \frac{1}{3}\mathrm{Pe}^2\ell_0^2\left(\nabla\overline{T}\right)^2.} \quad (1.69)$$

1.3.6 Mean-Field Equation

Substituting the turbulent heat flux (1.62) into Eq. (1.28), and taking into account that for homogeneous turbulence the coefficient of turbulent diffusion is independent of coordinate, so that $\nabla\cdot\left(D_{_T}\,\nabla\overline{T}\right) = D_{_T}\Delta\overline{T}$, we obtain the mean-field equation for temperature field for homogeneous, isotropic and incompressible turbulence:

$$\frac{\partial\overline{T}}{\partial t} = \left(D^{(\theta)} + D_{_T}\right)\Delta\overline{T}. \quad (1.70)$$

Since the coefficient of turbulent diffusion $D_{_T}$ is positive, Eq. (1.70) implies that *the main effect of turbulence is enhancement of the diffusion of the mean temperature field, i.e., turbulence enhances the mixing.*

1.3.7 Solving the Diffusion Equation

Let us solve the diffusion equation (1.70) with the initial condition $T_0(\boldsymbol{r}) = \overline{T}(t=0,\boldsymbol{r})$. We use the Fourier[16] transform in the $\boldsymbol{k}$ space:

$$\overline{T}(t,\boldsymbol{r}) = \frac{1}{(2\pi)^3}\int \overline{T}(t,\boldsymbol{k})\exp(\mathrm{i}\boldsymbol{k}\cdot\boldsymbol{r})\,d\boldsymbol{k}, \quad (1.71)$$

$$\overline{T}(t,\boldsymbol{k}) = \int \overline{T}(t,\boldsymbol{r})\exp(-\mathrm{i}\boldsymbol{k}\cdot\boldsymbol{r})\,d\boldsymbol{r}, \quad (1.72)$$

where $\boldsymbol{k}$ is the wave vector. We seek a solution for Eq. (1.70) in the form given by Eq. (1.71). Substituting solution (1.71) into Eq. (1.70), we obtain

$$\int\left(\frac{d\overline{T}(t,\boldsymbol{k})}{dt} + D_*\boldsymbol{k}^2\overline{T}(t,\boldsymbol{k})\right)\exp(\mathrm{i}\boldsymbol{k}\cdot\boldsymbol{r})\,d\boldsymbol{k} = 0, \quad (1.73)$$

[16] Jean Baptiste Joseph Fourier (1768–1830) was a French mathematician well-known for his works in mathematical physics, algebra, etc.

where $D_* = D^{(\theta)} + D_T$ is the total diffusion coefficient that is independent of the coordinate. Equation (1.73) yields the following ordinary differential equation where $\boldsymbol{k}$ is considered as a parameter, and the time t is a variable:

$$\frac{d\overline{T}(t,\boldsymbol{k})}{dt} = -D_* \boldsymbol{k}^2 \overline{T}(t,\boldsymbol{k}). \tag{1.74}$$

Equation (1.74) with the initial condition $\overline{T}(t=0,\boldsymbol{k}) = \overline{T}_0(\boldsymbol{k})$ has the following solution:

$$\overline{T}(t,\boldsymbol{k}) = \overline{T}_0(\boldsymbol{k}) \exp(-D_* \boldsymbol{k}^2 t). \tag{1.75}$$

The Fourier transform (1.72) for the initial temperature distribution $\overline{T}_0(\boldsymbol{k})$ reads:

$$\overline{T}_0(\boldsymbol{k}) = \int \overline{T}_0(\boldsymbol{r}') \exp(-\mathrm{i}\boldsymbol{k}\cdot\boldsymbol{r}')\, d\boldsymbol{r}'. \tag{1.76}$$

Substituting Eq. (1.76) into Eq. (1.75), we obtain

$$\overline{T}(t,\boldsymbol{k}) = \int \overline{T}_0(\boldsymbol{r}') \exp(-\mathrm{i}\boldsymbol{k}\cdot\boldsymbol{r}')\, \exp(-D_* \boldsymbol{k}^2 t)\, d\boldsymbol{r}'. \tag{1.77}$$

Now we substitute Eq. (1.77) for $\overline{T}(t,\boldsymbol{k})$ into Eq. (1.71), which yields

$$\overline{T}(t,\boldsymbol{r}) = \frac{1}{(2\pi)^3} \int d\boldsymbol{k} \int \overline{T}_0(\boldsymbol{r}') \exp\left[\mathrm{i}\boldsymbol{k}\cdot(\boldsymbol{r}-\boldsymbol{r}') - D_* \boldsymbol{k}^2 t\right] d\boldsymbol{r}'. \tag{1.78}$$

We use the following identity:

$$\begin{aligned}
\mathrm{i}\boldsymbol{k}\cdot(\boldsymbol{r}-\boldsymbol{r}') - D_* \boldsymbol{k}^2 t &= -D_* t \left[\boldsymbol{k}^2 - \frac{\mathrm{i}\boldsymbol{k}\cdot(\boldsymbol{r}-\boldsymbol{r}')}{D_* t} + \left(\frac{\mathrm{i}\,(\boldsymbol{r}-\boldsymbol{r}')}{2D_* t}\right)^2\right] - \frac{(\boldsymbol{r}-\boldsymbol{r}')^2}{4D_* t} \\
&= -D_* t \left(\boldsymbol{k} - \frac{\mathrm{i}\,(\boldsymbol{r}-\boldsymbol{r}')}{2D_* t}\right)^2 - \frac{(\boldsymbol{r}-\boldsymbol{r}')^2}{4D_* t}.
\end{aligned} \tag{1.79}$$

This identity allows us to rewrite solution (1.78) of the diffusion equation (1.70) in the final form:

$$\boxed{\overline{T}(t,\boldsymbol{r}) = \frac{1}{(4\pi\, D_* t)^{3/2}} \int \overline{T}_0(\boldsymbol{r}') \exp\left[-\frac{(\boldsymbol{r}-\boldsymbol{r}')^2}{4D_* t}\right] d\boldsymbol{r}',} \tag{1.80}$$

where we calculate the following integral,

$$\begin{aligned}
I &= \frac{1}{(2\pi)^3} \int \exp\left[-D_* t \left(\boldsymbol{k} - \frac{\mathrm{i}\,(\boldsymbol{r}-\boldsymbol{r}')}{2D_* t}\right)^2\right] d\boldsymbol{k} = \frac{1}{(2\pi)^3} \int \exp(-a^2 \tilde{\boldsymbol{k}}^2)\, d\tilde{\boldsymbol{k}} \\
&= \frac{1}{(2\pi)^3} \int_0^\infty \exp(-a^2 \tilde{k}^2)\, \tilde{k}^2\, d\tilde{k} \int_0^\pi \sin\theta\, d\theta \int_0^{2\pi} d\varphi = \frac{1}{(4\pi\, D_* t)^{3/2}},
\end{aligned} \tag{1.81}$$

by using the following new variables: $a = (D_* t)^{1/2}$ and

$$\tilde{\boldsymbol{k}} = \boldsymbol{k} - \frac{\mathrm{i}\,(\boldsymbol{r} - \boldsymbol{r}')}{2D_* t}. \tag{1.82}$$

When the initial condition is localized at $\boldsymbol{r} = 0$, i.e., $\overline{T}_0(\boldsymbol{r}) = T_* \delta(\boldsymbol{r})$, the solution (1.80) of the diffusion equation reads

$$\boxed{\overline{T}(t, \boldsymbol{r}) = \frac{T_*}{(4\pi\, D_* t)^{3/2}} \exp\left[-\frac{\boldsymbol{r}^2}{4D_* t}\right],} \tag{1.83}$$

where $\delta(\boldsymbol{r})$ is the Dirac delta function defined by the following properties: $\delta(\boldsymbol{r}) = 0$ when $\boldsymbol{r} \neq 0$; $\delta(\boldsymbol{r}) = \infty$ when $\boldsymbol{r} = 0$; the normalization condition for this function is $\int_{-\infty}^{\infty} \delta(x)\, dx = 1$ and $\int f(\boldsymbol{r}) \delta(\boldsymbol{r} - \boldsymbol{r}_0)\, d\boldsymbol{r} = f(\boldsymbol{r}_0)$.

When the initial condition is one-dimensional (i.e., it depends only on x), the solution (1.80) of the diffusion equation reads

$$\boxed{\overline{T}(t, x) = \frac{1}{(4\pi\, D_* t)^{1/2}} \int \overline{T}_0(x') \exp\left[-\frac{(x - x')^2}{4D_* t}\right] dx'.} \tag{1.84}$$

When the initial condition is one-dimensional and localized at $x = 0$, i.e., $\overline{T}_0(\boldsymbol{r}) = T_* \delta(x)$, the solution (1.84) of the diffusion equation reads

$$\boxed{\overline{T}(t, x) = \frac{T_*}{(4\pi\, D_* t)^{1/2}} \exp\left[-\frac{x^2}{4D_* t}\right].} \tag{1.85}$$

Equation (1.85) implies that the mean temperature $\overline{T}(t, x = 0)$ at $x = 0$ decreases with time as $\overline{T}(t, x = 0) \sim T_* / \sqrt{t}$ and the characteristic size of the mean temperature variations is $L_T \sim \sqrt{D_* t}$. This implies that the characteristic diffusion time is

$$\boxed{\tau_{\mathrm{diffusion}} = \frac{L_T^2}{D_*}.} \tag{1.86}$$

1.4 Multi-Scale Approach

To apply rigorous methods for solving Eq. (1.57) for temperature fluctuations, we consider multi-scale approach (Roberts and Soward, 1975), which allows us to introduce fast and slow variables, and separate small-scale effects corresponding to fluctuations and large-scale effects describing mean fields. In particular, this allows us to study direct effect of small-scale turbulence on the evolution of the mean temperature field.

1.4.1 Multi-Scale Approach: Spatial Scales

In the framework of the mean-field approach, we assume that there is a separation of spatial and temporal scales, i.e., $\ell_0 \ll L_T$ and $\tau_0 \ll t_T$, where L_T and t_T are the characteristic spatial and temporal scales characterizing the variations of the mean temperature field. The mean fields depend on "slow" variables, while fluctuations depend on "fast" variables. Separation into slow and fast variables is widely used in theoretical physics, and all calculations are reduced to Taylor expansions of all functions using small parameters ℓ_0/L_T and τ_0/t_T. The findings are further truncated to leading order terms.

Separation to slow and fast variables is performed by means of a standard multi-scale approach. In the framework of this approach, we consider the two-point second-order correlation function of velocity field:

$$f_{ij}(t, \boldsymbol{x}, \boldsymbol{y}) = \big\langle u_i(t, \boldsymbol{x})\, u_j(t, \boldsymbol{y}) \big\rangle . \tag{1.87}$$

We use the Fourier transform in the $\boldsymbol{k}$ space:

$$u_i(t, \boldsymbol{x}) = \int u_i(t, \boldsymbol{k}_1) \exp(\mathrm{i} \boldsymbol{k}_1 \cdot \boldsymbol{x})\, d\boldsymbol{k}_1, \tag{1.88}$$

where hereafter for simplicity of notations we omit the coefficient $(2\pi)^{-3}$, multiplying by the Fourier integral (1.88). We apply the Fourier transform in $\boldsymbol{k}$ for the two-point correlation function as

$$\big\langle u_i(t, \boldsymbol{x})\, u_j(t, \boldsymbol{y}) \big\rangle = \int \int \big\langle u_i(t, \boldsymbol{k}_1)\, u_j(t, \boldsymbol{k}_2) \big\rangle \exp\left(\mathrm{i} \boldsymbol{k}_1 \cdot \boldsymbol{x} + \mathrm{i} \boldsymbol{k}_2 \cdot \boldsymbol{y}\right)\, d\boldsymbol{k}_1\, d\boldsymbol{k}_2. \tag{1.89}$$

We use new variables in the physical space, $\boldsymbol{r} = \boldsymbol{x} - \boldsymbol{y}$ and $\boldsymbol{R} = (\boldsymbol{x} + \boldsymbol{y})/2$, which correspond to small-scale and large-scale spatial variables, respectively. Now we express the old variables $\boldsymbol{x}$ and $\boldsymbol{y}$ in terms of new variables: $\boldsymbol{x} = \boldsymbol{r}/2 + \boldsymbol{R}$ and $\boldsymbol{y} = -\boldsymbol{r}/2 + \boldsymbol{R}$. Using these expressions, we rewrite $\boldsymbol{k}_1 \cdot \boldsymbol{x} + \boldsymbol{k}_2 \cdot \boldsymbol{y}$ in the new variables:

$$\boldsymbol{k}_1 \cdot \boldsymbol{x} + \boldsymbol{k}_2 \cdot \boldsymbol{y} = \frac{1}{2}(\boldsymbol{k}_1 - \boldsymbol{k}_2) \cdot \boldsymbol{r} + (\boldsymbol{k}_1 + \boldsymbol{k}_2) \cdot \boldsymbol{R} = \boldsymbol{k} \cdot \boldsymbol{r} + \boldsymbol{K} \cdot \boldsymbol{R}, \tag{1.90}$$

where we also use new variables in the $\boldsymbol{k}$ space: $\boldsymbol{k} = (\boldsymbol{k}_1 - \boldsymbol{k}_2)/2$ that corresponds to the small scales, and $\boldsymbol{K} = \boldsymbol{k}_1 + \boldsymbol{k}_2$ that characterizes the large scales. In the integral (1.89) we change variables:

$$\begin{aligned}\big\langle u_i(t, \boldsymbol{x})\, u_j(t, \boldsymbol{y}) \big\rangle &= \int \int f_{ij}(t, \boldsymbol{k}, \boldsymbol{K}) \exp(\mathrm{i} \boldsymbol{k} \cdot \boldsymbol{r} + \mathrm{i} \boldsymbol{K} \cdot \boldsymbol{R})\, d\boldsymbol{k}\, d\boldsymbol{K} \\ &= \int f_{ij}(t, \boldsymbol{k}, \boldsymbol{R}) \exp(\mathrm{i} \boldsymbol{k} \cdot \boldsymbol{r})\, d\boldsymbol{k},\end{aligned} \tag{1.91}$$

where

$$f_{ij}(t, \boldsymbol{k}, \boldsymbol{K}) = \left\langle u_i(t, \boldsymbol{k} + \boldsymbol{K}/2) u_j(t, -\boldsymbol{k} + \boldsymbol{K}/2) \right\rangle, \tag{1.92}$$

$$f_{ij}(t, \boldsymbol{k}, \boldsymbol{R}) = \int f_{ij}(t, \boldsymbol{k}, \boldsymbol{K}) \exp(\mathrm{i}\, \boldsymbol{K} \cdot \boldsymbol{R})\, d\boldsymbol{K}, \tag{1.93}$$

and we take into account that the Jacobian of this transformation equals 1. For the one-point correlation function:

$$\begin{aligned} \left\langle u_i(t, \boldsymbol{x})\, u_j(t, \boldsymbol{x}) \right\rangle &\equiv \lim_{\boldsymbol{x} \to \boldsymbol{y}} \left\langle u_i(t, \boldsymbol{x})\, u_j(t, \boldsymbol{y}) \right\rangle \\ &= \int \left\langle u_i(t, \boldsymbol{k} + \boldsymbol{K}/2)\, u_j(t, -\boldsymbol{k} + \boldsymbol{K}/2) \right\rangle d\boldsymbol{k} \end{aligned} \tag{1.94}$$

[see Eq. (1.91) for $\boldsymbol{r} \to 0$]. For homogeneous turbulence, the correlation function $f_{ij}(t, \boldsymbol{k}, \boldsymbol{R})$ is independent of $\boldsymbol{R}$, i.e., $f_{ij}(t, \boldsymbol{k}, \boldsymbol{R}) = f_{ij}(t, \boldsymbol{k})$. This implies that for homogeneous turbulence Eq. (1.94) reads

$$\boxed{\left\langle u_i(t, \boldsymbol{x})\, u_j(t, \boldsymbol{x}) \right\rangle = \lim_{\boldsymbol{x} \to \boldsymbol{y}} \left\langle u_i(t, \boldsymbol{x})\, u_j(t, \boldsymbol{y}) \right\rangle = \int \left\langle u_i(t, \boldsymbol{k})\, u_j(t, -\boldsymbol{k}) \right\rangle d\boldsymbol{k}.} \tag{1.95}$$

Trace of the tensor in Eq. (1.95) yields

$$\boxed{\left\langle \boldsymbol{u}^2 \right\rangle = \int \langle u_i(t, \boldsymbol{k})\, u_i(t, -\boldsymbol{k}) \rangle\, d\boldsymbol{k}.} \tag{1.96}$$

Therefore, such transformation allows us to separate the integration over small-scales, i.e., over turbulence scales, see Eq. (1.91), and over large scales, see Eq. (1.93). This implies that we introduce fast and slow variables and separate the integration over the fast and slow variables. We remember that in this analysis, we assumed that there exists a separation of scales, i.e., this implies that $r \le l_0 \ll R$. Usually, in final results one should check that this assumption is indeed valid.

1.4.2 Multi-Scale Approach: Temporal Scales

Let us consider a one-point non-instantaneous correlation function:

$$f_{ij}(t_1, t_2, \boldsymbol{x}) = \left\langle u_i(t_1, \boldsymbol{x})\, u_j(t_2, \boldsymbol{x}) \right\rangle. \tag{1.97}$$

We use the Fourier transform in a spectral ω space,

$$u_i(t, \boldsymbol{x}) = \int u_i(\omega_1, \boldsymbol{x}) \exp(\mathrm{i}\, \omega_1 t_1)\, d\omega_1, \tag{1.98}$$

where ω_1 is the frequency. We use the Fourier transform in the ω space for the non-instantaneous one-point correlation function:

$$\langle u_i(t_1, \boldsymbol{x})\, u_j(t_2, \boldsymbol{x})\rangle = \int\int d\omega_1\, d\omega_2\, \langle u_i(\omega_1, \boldsymbol{x})\, u_j(\omega_2, \boldsymbol{x})\rangle \times \exp\left[\mathrm{i}\,(\omega_1 t_1 + \omega_2 t_2)\right]. \tag{1.99}$$

We use new variables, $\tau = t_1 - t_2$ and $t = (t_1 + t_2)/2$, which correspond to fast and slow variables, respectively. We express the old variables t_1 and t_2 in terms of new variables: $t_1 = \tau/2 + t$ and $t_2 = -\tau/2 + t$. Using these expressions, we rewrite $\omega_1 t_1 + \omega_2 t_2$ in the new variables,

$$\omega_1 t_1 + \omega_2 t_2 = \frac{1}{2}(\omega_1 - \omega_2)\tau + (\omega_1 + \omega_2)t = \omega\,\tau + \Omega\, t, \tag{1.100}$$

where we also use new variables in the spectral space: $\omega = (\omega_1 - \omega_2)/2$, which corresponds to turbulent time scales (the fast variables), and $\Omega = \omega_1 + \omega_2$, which corresponds to the long-term time scales (the slow variables) describing the mean fields. In the integral (1.99), we change variables:

$$\begin{aligned}\langle u_i(t_1, \boldsymbol{x})\, u_j(t_2, \boldsymbol{x})\rangle &= \int\int f_{ij}(\omega, \Omega, \boldsymbol{x}) \exp(\mathrm{i}\,\omega\,\tau + \mathrm{i}\,\Omega\, t)\, d\omega\, d\Omega \\ &= \int f_{ij}(\omega, \Omega, \boldsymbol{x}) \exp(\mathrm{i}\,\omega\,\tau)\, d\omega,\end{aligned} \tag{1.101}$$

where

$$f_{ij}(\omega, \Omega, \boldsymbol{x}) = \langle u_i(\omega + \Omega/2, \boldsymbol{x})\, u_j(-\omega + \Omega/2, \boldsymbol{x})\rangle, \tag{1.102}$$

$$f_{ij}(\omega, t, \boldsymbol{x}) = \int f_{ij}(\omega, \Omega, \boldsymbol{x}) \exp(\mathrm{i}\,\Omega\, t)\, d\Omega, \tag{1.103}$$

and we take into account that the Jacobian of this transformation equals 1. For the one-point correlation function,

$$\boxed{\langle u_i(t, \boldsymbol{x})\, u_j(t, \boldsymbol{x})\rangle \equiv \lim_{t_1 \to t_2} \langle u_i(t_1, \boldsymbol{x})\, u_j(t_2, \boldsymbol{x})\rangle = \int \langle u_i(\omega, \boldsymbol{x})\, u_j(-\omega, \boldsymbol{x})\rangle\, d\omega} \tag{1.104}$$

[see Eq. (1.101) for $\tau \to 0$]. For stationary (in statistical sense) turbulence, the correlation function $f_{ij}(\omega, t, \boldsymbol{x})$ is independent of t, i.e., $f_{ij}(\omega, t, \boldsymbol{x}) = f_{ij}(\omega, \boldsymbol{x})$. Trace of the tensor in Eq. (1.104) yields

$$\boxed{\langle \boldsymbol{u}^2\rangle = \int \langle u_i(\omega, \boldsymbol{x})\, u_i(-\omega, \boldsymbol{x})\rangle\, d\omega.} \tag{1.105}$$

Therefore, such transformation allows us to separate the integration over fast variables, i.e., the integration over turbulence time scales, see Eq. (1.104), and slow variables corresponding to mean fields, see Eq. (1.103). This implies we introduce fast and slow variables and separate the integration over the fast and slow variables. In this analysis, we assume that there is a separation of time scales. In particular, this implies that $\tau_0 \ll t_L$. In the final results, one should check that this assumption is indeed valid.

1.5 Turbulent Transport of Temperature Fields: Quasi-Linear Approach

In this section, we apply the multi-scale approach and the quasi-linear approach to study the turbulent transport of a temperature field in an incompressible homogeneous and isotropic random velocity field. The quasi-linear approach is valid for small Péclet numbers.

1.5.1 Governing Equations

Temperature field $T(t, \mathrm{x})$ is determined by Eq. (1.56). Averaging Eq. (1.56) over an ensemble of turbulent velocity field, we obtain the mean-field equation for the mean temperature $\overline{T}$:

$$\frac{\partial \overline{T}}{\partial t} + \nabla \cdot \left(\overline{T}\, \overline{\boldsymbol{U}} + \langle \theta\, \boldsymbol{u} \rangle \right) = D^{(\theta)} \Delta \overline{T}. \tag{1.106}$$

For simplicity, we consider the case $\overline{\boldsymbol{U}} = 0$. To determine the turbulent heat flux $\boldsymbol{F}^{(\theta)} \equiv \langle \theta\, \boldsymbol{u} \rangle$, we use the equation for temperature fluctuations obtained by subtracting Eq. (1.106) from Eq. (1.56):

$$\frac{\partial \theta}{\partial t} + \nabla \cdot (\theta\, \boldsymbol{u} - \langle \theta\, \boldsymbol{u} \rangle) - D^{(\theta)} \Delta \theta = -(\boldsymbol{u} \cdot \nabla) \overline{T}. \tag{1.107}$$

In the previous Section, we used the dimensional analysis to solve the nonlinear equation (1.107) and to determine the turbulent heat flux. In this section, we consider the case when the nonlinear terms in Eq. (1.107) are much smaller than the diffusion term, i.e., for small Péclet numbers $\mathrm{Pe} \ll 1$. We also consider a one-way coupling that implies that the temperature field is a passive scalar.

1.5.2 Quasi-Linear Approach

To study temperature fluctuations for small Péclet numbers, we use a quasi-linear approach. In the framework of this approach, we neglect the nonlinear terms but keep the molecular diffusion term in Eq. (1.107). For this reason, this approach is called a quasi-linear or a perturbation approach. In astrophysics, this approach is known as a first-order smoothing approximation or a second-order correlation approximation (Moffatt, 1978; Krause and Rädler, 1980). In scattering theory, this approach is similar to the Born approximation (Taylor, 1972). We rewrite Eq. (1.107) in a Fourier space, i.e., we use the Fourier transform in the $\boldsymbol{k} - \omega$ space:

$$\theta(t, \boldsymbol{x}) = \int \int \theta(\omega, \boldsymbol{k}) \exp(\mathrm{i}\boldsymbol{k} \cdot \boldsymbol{x} + \mathrm{i}\,\omega t)\, d\boldsymbol{k}\, d\omega, \tag{1.108}$$

$$u_i(t, \boldsymbol{x}) = \int \int u_i(\omega, \boldsymbol{k}) \exp(\mathrm{i}\boldsymbol{k} \cdot \boldsymbol{x} + \mathrm{i}\,\omega t)\, d\boldsymbol{k}\, d\omega. \tag{1.109}$$

Hereafter in all book, we do not use the factor $(2\pi)^{-4}$ multiplying the Fourier integrals (1.108) and (1.109); particularly, in the three-dimensional $\boldsymbol{k}$ space, the factor is $(2\pi)^{-3}$, and in the ω space, it is $(2\pi)^{-1}$.

We use the following identities:

$$\frac{\partial \theta(t, \boldsymbol{x})}{\partial t} = \int \int (\mathrm{i}\,\omega)\, \theta(\omega, \boldsymbol{k}) \exp(\mathrm{i}\boldsymbol{k} \cdot \boldsymbol{x} + \mathrm{i}\,\omega t)\, d\boldsymbol{k}\, d\omega, \tag{1.110}$$

$$\Delta\theta(t, \boldsymbol{x}) = \int \int (-k^2)\, \theta(\omega, \boldsymbol{k}) \exp(\mathrm{i}\boldsymbol{k} \cdot \boldsymbol{x} + \mathrm{i}\,\omega t)\, d\boldsymbol{k}\, d\omega. \tag{1.111}$$

Substituting Eqs. (1.109)–(1.111) into the linearized Eq. (1.107), we obtain

$$\int \int \left[\left(D^{(\theta)} k^2 + \mathrm{i}\,\omega\right) \theta(\omega, \boldsymbol{k}) + u_i(\omega, \boldsymbol{k}) \nabla_i \overline{T} \right] \exp(\mathrm{i}\boldsymbol{k} \cdot \boldsymbol{x} + \mathrm{i}\,\omega t)\, d\boldsymbol{k}\, d\omega = 0. \tag{1.112}$$

The integral in Eq. (1.112) vanishes only if

$$\left(D^{(\theta)} k^2 + \mathrm{i}\,\omega\right) \theta(\omega, \boldsymbol{k}) + u_i(\omega, \boldsymbol{k})\, \nabla_i \overline{T} = 0. \tag{1.113}$$

This equation yields the expression for temperature fluctuations $\theta(\omega, \boldsymbol{k})$ in the $\boldsymbol{k} - \omega$ space,

$$\theta(\omega, \boldsymbol{k}) = -G_D(\omega, \boldsymbol{k})\, u_i(\omega, \boldsymbol{k})\, \nabla_i \overline{T}, \tag{1.114}$$

where the function $G_D(\omega, \boldsymbol{k}) = \left(D^{(\theta)} k^2 + \mathrm{i}\,\omega\right)^{-1}$.

Note that Eq. (1.114) is not valid in a general case. However, since (i) we take into account in turbulent heat flux only the effects that are linear in the gradient of the mean temperature $\nabla_j \overline{T}$ and neglect higher-order spatial derivatives of the mean temperature field, and (ii) we do not consider in this section compressibility effects of a turbulent velocity field in the evolution of the mean temperature field (i.e., we consider an incompressible flow), we can use Eq. (1.114) to determine the turbulent heat flux in this case [see Section 2.2.6 for a rigorous derivation of an equation for $\theta(\omega, \boldsymbol{k})$ in a more general case].

We determine the turbulent heat flux using the multi-scale approach:

$$\left\langle \theta(t, \boldsymbol{x})\, u_j(t, \boldsymbol{x}) \right\rangle = \lim_{t_1 \to t_2, \boldsymbol{x} \to \boldsymbol{y}} \left\langle \theta(t_1, \boldsymbol{x})\, u_j(t_2, \boldsymbol{y}) \right\rangle$$

$$= \lim_{\tau \to 0, \boldsymbol{r} \to 0} \int \int \left\langle \theta(\omega, \boldsymbol{k})\, u_j(-\omega, -\boldsymbol{k}) \right\rangle \exp[\mathrm{i}\,\boldsymbol{k} \cdot \boldsymbol{r} + \mathrm{i}\,\omega\,\tau]\, d\omega\, d\boldsymbol{k}$$

$$= \int \int \left\langle \theta(\omega, \boldsymbol{k})\, u_j(-\omega, -\boldsymbol{k}) \right\rangle \, d\omega\, d\boldsymbol{k}, \tag{1.115}$$

where $\boldsymbol{r} = \boldsymbol{x} - \boldsymbol{y}$, $\boldsymbol{k} = (\boldsymbol{k}_1 - \boldsymbol{k}_2)/2$, $\tau = t_1 - t_2$, $\omega = (\omega_1 - \omega_2)/2$, and we took into account that turbulence is homogeneous and stationary in the statistical sense. The latter implies that the correlation function $\langle u_i(\omega, \boldsymbol{k})\, u_j(-\omega, -\boldsymbol{k}) \rangle$ is independent of the large-scale variables Ω, $\boldsymbol{K}$ and t, $\boldsymbol{R}$. We assume also here that there exists a separation of scales, i.e., the maximum scale of random motions ℓ_0 is much smaller than the characteristic scales of inhomogeneities of the mean temperature. In a similar way, we determine the level of temperature fluctuations $\langle \theta^2 \rangle$:

$$\langle \theta(t, \boldsymbol{x})\, \theta(t, \boldsymbol{x}) \rangle = \lim_{t_1 \to t_2, \boldsymbol{x} \to \boldsymbol{y}} \langle \theta(t_1, \boldsymbol{x})\, \theta(t_2, \boldsymbol{y}) \rangle$$

$$= \lim_{\tau \to 0, \boldsymbol{r} \to 0} \int \int \langle \theta(\omega, \boldsymbol{k})\, \theta(-\omega, -\boldsymbol{k}) \rangle \exp\left[\mathrm{i}\,\boldsymbol{k} \cdot \boldsymbol{r} + \mathrm{i}\,\omega\,\tau\right] \, d\omega\, d\boldsymbol{k}$$

$$= \int \int \langle \theta(\omega, \boldsymbol{k})\, \theta(-\omega, -\boldsymbol{k}) \rangle \, d\omega\, d\boldsymbol{k}. \tag{1.116}$$

Substituting Eq. (1.114) into Eqs. (1.115) and (1.116), we determine the turbulent heat flux $F_i^{(\theta)} = \langle \theta\, u_i \rangle$ and the level of temperature fluctuations $\langle \theta^2 \rangle$:

$$F_i^{(\theta)} = - \left(\nabla_j \overline{T}\right) \int \int \left\langle u_i(\omega, \boldsymbol{k})\, u_j(-\omega, -\boldsymbol{k}) \right\rangle \, G_D \, d\omega\, d\boldsymbol{k}, \tag{1.117}$$

$$\left\langle \theta^2 \right\rangle = \left(\nabla_i \overline{T}\right) \left(\nabla_j \overline{T}\right) \int \int \left\langle u_i(\omega, \boldsymbol{k})\, u_j(-\omega, -\boldsymbol{k}) \right\rangle \, G_D\, G_D^* \, d\omega\, d\boldsymbol{k}, \tag{1.118}$$

where the function G_D^* is given by $G_D^*(\omega, \boldsymbol{k}) = \left(D^{(\theta)} k^2 - \mathrm{i}\,\omega\right)^{-1}$.

1.5.3 Model of Isotropic Homogeneous and Incompressible Random Velocity Field

We use a simple model for the second-order moment $\langle u_i(\omega, \boldsymbol{k})\, u_j(-\omega, -\boldsymbol{k}) \rangle$ of an isotropic, homogeneous and incompressible random velocity field in a Fourier space:

$$\boxed{\left\langle u_i(\omega, \boldsymbol{k})\, u_j(-\omega, -\boldsymbol{k}) \right\rangle = \frac{\left\langle \boldsymbol{u}^2 \right\rangle \, E(k)\, \Phi(\omega)}{8\pi k^2} \left(\delta_{ij} - \frac{k_i\, k_j}{k^2}\right),} \tag{1.119}$$

where δ_{ij} is the Kronecker tensor, the energy spectrum function is

$$\boxed{E(k) = (q - 1)\, k_0^{-1} \left(\frac{k}{k_0}\right)^{-q},} \tag{1.120}$$

and the exponent q varies in the interval $1 < q < 3$, the interval in the wave numbers is $k_0 \le k \le k_d$, the wave number $k_0 = 1/\ell_0$, the length ℓ_0 is the maximum scale of random motions and $k_d \gg k_0$. We use the Lorentz profile for the frequency function $\Phi(\omega)$:

$$\boxed{\Phi(\omega) = \frac{1}{\pi \tau_0 \, (\omega^2 + \tau_0^{-2})},} \tag{1.121}$$

where τ_0 is the correlation time of a random velocity field. This model for the frequency function corresponds to the following non-instantaneous correlation function: $\langle u_i(t) u_j(t+\tau)\rangle \propto \exp(-\tau/\tau_0)$. Since the Péclet number is small, we consider the case $\tau_0 \gg (D^{(\theta)} k^2)^{-1}$.

Equation (1.119) is derived using symmetry arguments. In particular, since turbulence is homogeneous, the two-point correlation function of random velocity field is independent of the large-scale variables $\boldsymbol{K}$ and $\boldsymbol{R}$. For homogeneous and isotropic turbulence, there are only two symmetric tensors δ_{ij} and $k_i \, k_j / k^2$. Therefore, we can construct the tensor $\big\langle u_i(\omega, \boldsymbol{k}) \, u_j(-\omega, -\boldsymbol{k})\big\rangle$ as a linear combination of the tensors δ_{ij} and $k_i \, k_j / k^2$:

$$\big\langle u_i(\omega, \boldsymbol{k}) \, u_j(-\omega, -\boldsymbol{k})\big\rangle = A_1 \, \delta_{ij} + A_2 \, \frac{k_i \, k_j}{k^2}. \tag{1.122}$$

The velocity field is incompressible, i.e., div $\boldsymbol{u} = 0$. This condition in a Fourier space implies that $k_i \, u_i(\omega, \boldsymbol{k}) = 0$. Multiplying the latter equation by $u_j(-\omega, -\boldsymbol{k})$ and average over an ensemble of random velocity fluctuations, we arrive at the following condition for the second-order moment of the velocity field:

$$k_i \, \big\langle u_i(\omega, \boldsymbol{k}) \, u_j(-\omega, -\boldsymbol{k})\big\rangle = 0. \tag{1.123}$$

Multiplying Eq. (1.122) by k_i and using Eq. (1.123), we obtain:

$$k_i \, \big\langle u_i(\omega, \boldsymbol{k}) \, u_j(-\omega, -\boldsymbol{k})\big\rangle = A_1 \, k_i \, \delta_{ij} + A_2 \, k_i \, \frac{k_i \, k_j}{k^2} = (A_1 + A_2) k_j = 0. \tag{1.124}$$

Since Eq. (1.124) should be valid for any wave vector k_j, it yields the following relation $A_1 = -A_2$. Next, we can rewrite A_1 in the following form: $A_1 = A \, \Phi(\omega) \, E(k)$, where the normalization conditions for the functions $\Phi(\omega)$ and $E(k)$ are $\int_{-\infty}^{\infty} \Phi(\omega) \, d\omega = 1$ and $\int_{k_0}^{k_d} E(k) \, dk = 1$. This implies that the correlation function $\langle u_i(\omega, \boldsymbol{k}) \, u_j(-\omega, -\boldsymbol{k})\rangle$ is

$$\big\langle u_i(\omega, \boldsymbol{k}) \, u_j(-\omega, -\boldsymbol{k})\big\rangle = A \, \Phi(\omega) \, E(k) \, \left(\delta_{ij} - \frac{k_i \, k_j}{k^2}\right). \tag{1.125}$$

Let us show that

$$A = \frac{\langle \boldsymbol{u}^2 \rangle}{8\pi k^2}, \tag{1.126}$$

where $\langle \boldsymbol{u}^2 \rangle = \int \int \langle u_i(\omega, \boldsymbol{k}) \, u_i(-\omega, -\boldsymbol{k}) \rangle \, d\omega \, d\boldsymbol{k}$. The trace of the tensor $\langle u_i \, u_j \rangle$ in Eq. (1.125) is given by

$$\langle u_i(\omega, \boldsymbol{k}) \, u_i(-\omega, -\boldsymbol{k}) \rangle = \frac{\langle \boldsymbol{u}^2 \rangle \, E(k)}{4\pi k^2} \Phi(\omega), \tag{1.127}$$

where we take into account Eq. (1.126). Integrating Eq. (1.127) in the ω space and $\boldsymbol{k}$ space, we obtain

$$\int \int \langle u_i(\omega, \boldsymbol{k}) \, u_i(-\omega, -\boldsymbol{k}) \rangle \, d\omega \, d\boldsymbol{k} = \frac{\langle \boldsymbol{u}^2 \rangle}{4\pi} \int_{k_0}^{k_d} E(k) \, dk \\ \times \int_0^{2\pi} d\varphi \int_0^{\pi} \sin\vartheta \, d\vartheta \int_{-\infty}^{\infty} \Phi(\omega) \, d\omega, \tag{1.128}$$

where we use the spherical coordinates in the $\boldsymbol{k}$ space. Here ϑ, φ are the angles in the spherical coordinates, $0 \leq \varphi \leq 2\pi$ and $0 \leq \vartheta \leq \pi$, and $d\boldsymbol{k} = k^2 \, dk \sin\vartheta \, d\vartheta \, d\varphi$. We take into account the following identity:

$$\int_0^{2\pi} d\varphi \int_0^{\pi} \sin\vartheta \, d\vartheta = 4\pi, \tag{1.129}$$

so that Eq. (1.128) yields

$$\int \int \langle u_i(\omega, \boldsymbol{k}) \, u_i(-\omega, -\boldsymbol{k}) \rangle \, d\omega \, d\boldsymbol{k} = \langle \boldsymbol{u}^2 \rangle. \tag{1.130}$$

Equation (1.130) is identity, which shows that the coefficient A is indeed determined by Eq. (1.126).

1.5.4 Turbulent Heat Flux and Level of Temperature Fluctuations

Substituting Eq. (1.119) for the second-order moment $\langle u_i(\omega, \boldsymbol{k}) \, u_j(-\omega, -\boldsymbol{k}) \rangle$ into Eq. (1.117) and (1.118), we determine the turbulent heat flux $F_i^{(\theta)}$ and the level of temperature fluctuations $\langle \theta^2 \rangle$ as

$$F_i^{(\theta)} = -\frac{\langle \boldsymbol{u}^2 \rangle}{8\pi} \int_{k_0}^{k_d} E(k) \, dk \int_0^{2\pi} d\varphi \int_0^{\pi} \sin\vartheta \, d\vartheta \left[\delta_{ij} - \frac{k_i \, k_j}{k^2} \right] \\ \times \int_{-\infty}^{\infty} G_D \, \Phi(\omega) \, d\omega \, \left(\nabla_j \overline{T} \right), \tag{1.131}$$

$$\left\langle\theta^2\right\rangle = \frac{\left\langle\boldsymbol{u}^2\right\rangle}{8\pi} \int_{k_0}^{k_d} E(k)\,dk \int_0^{2\pi} d\varphi \int_0^{\pi} \sin\vartheta\,d\vartheta \left[\delta_{ij} - \frac{k_i\,k_j}{k^2}\right]$$
$$\times \int_{-\infty}^{\infty} G_D\,G_D^*\,\Phi(\omega)\,d\omega\;\left(\nabla_i\overline{T}\right). \tag{1.132}$$

For the integration over ω, we use the following integrals:

$$\int_{-\infty}^{\infty} \frac{d\omega}{(\pm\mathrm{i}\,\omega + D^{(\theta)}k^2)\,(\omega^2+\tau_0^{-2})} = \frac{\pi\,\tau_0}{\tau_0^{-1} + D^{(\theta)}\,k^2} \approx \frac{\pi\,\tau_0}{D^{(\theta)}\,k^2}, \tag{1.133}$$

$$\int_{-\infty}^{\infty} \frac{d\omega}{(\mathrm{i}\,\omega + D^{(\theta)}k^2)\,(-\mathrm{i}\,\omega + D^{(\theta)}k^2)\,(\omega^2+\tau_0^{-2})}$$
$$= \frac{\pi\,\tau_0}{D^{(\theta)}\,k^2\left(\tau_0^{-1} + D^{(\theta)}\,k^2\right)} \approx \frac{\pi\,\tau_0}{(D^{(\theta)}\,k^2)^2}, \tag{1.134}$$

which are determined in the limit when the correlation time $\tau_0 \gg (D^{(\theta)}k^2)^{-1}$. For the integration over angles in the $\boldsymbol{k}$ space, we use the following integral:

$$\int_0^{2\pi} d\varphi \int_0^{\pi} \sin\vartheta\,d\vartheta\,\frac{k_i\,k_j}{k^2} = \frac{4\pi}{3}\,\delta_{ij}. \tag{1.135}$$

For the integration over k we use the following integrals:

$$\int_{k_0}^{k_d} \frac{E(k)}{k^2}\,dk = \frac{q-1}{q+1}\,\ell_0^2, \tag{1.136}$$

$$\int_{k_0}^{k_d} \frac{E(k)}{k^4}\,dk = \frac{q-1}{q+3}\,\ell_0^4. \tag{1.137}$$

After integration in the ω space and in $\boldsymbol{k}$ space in Eqs. (1.131) and (1.132), we obtain the final formulae for the turbulent heat flux $F_i^{(\theta)}$ and the level of temperature fluctuations $\langle\theta^2\rangle$:

$$\boxed{\langle\theta\,u_i\rangle = -D_T\,\nabla_i\overline{T},} \tag{1.138}$$

$$\boxed{D_T = \frac{q-1}{3(q+1)}\,u_0\,\ell_0\,\mathrm{Pe},} \tag{1.139}$$

$$\boxed{\left\langle\theta^2\right\rangle = \frac{q-1}{3(q+3)}\,\mathrm{Pe}^2\,\ell_0^2\,\left(\boldsymbol{\nabla}\overline{T}\right)^2,} \tag{1.140}$$

where $u_0 = \sqrt{\langle\boldsymbol{u}^2\rangle}$, $\mathrm{Pe} = \ell_0 u_0/D^{(\theta)}$ is the Péclet number and D_T is the turbulent diffusion coefficient.

Let us analyze the obtained results determined by Eqs. (1.138)–(1.140). We compare the turbulent heat flux $\langle\theta\,u_i\rangle$, the turbulent diffusion coefficient D_T and the second moment $\langle\theta^2\rangle$ derived using the quasi-linear approach for Pe $\ll 1$ with the results obtained by the dimensional analysis for small Péclet numbers. In particular, using the dimensional analysis, we find that the turbulent diffusion coefficient is given by Eq. (1.67), and the level of temperature fluctuations is determined by Eq. (1.69). In both methods, the obtained turbulent heat flux has the same form, $\langle\theta\,\boldsymbol{u}\rangle = -D_T\nabla\overline{T}$; the turbulent diffusion coefficient is linear in the Péclet number, $D_T \propto$ Pe; while the second-order moment for temperature fluctuations is quadratic in the Péclet number, $\langle\theta^2\rangle \propto \mathrm{Pe}^2$. The minor difference in the results obtained by these methods is only in numerical factors. Therefore, Eqs. (1.67) and (1.69) obtained by the dimensional analysis are in agreement with Eqs. (1.138)–(1.140), derived using the quasi-linear approach.

1.6 Turbulent Transport of Temperature Fields: Spectral Tau Approach

In this section, we study temperature fluctuations in the case when nonlinear terms in Eq. (1.107) for θ are much larger than the molecular diffusion term, i.e., for large Péclet numbers. We also consider the case of large Reynolds numbers, so that turbulence is fully developed. In this case, the Strouhal number Sr, which is the ratio of the correlation time τ_0 of velocity fluctuations at the integral scale to the turn over time ℓ_0/u_0 of turbulent eddies, is of the order of unity, and the turbulent correlation time is scale-dependent in the inertial range. We apply the Fourier transform only in the $\boldsymbol{k}$ space but not in the ω space, because in a fully developed Kolmogorov-type turbulence, the turbulent time is universally related to spatial scales. We take into account the nonlinear terms in equations for velocity and temperature fluctuations and apply the spectral τ approach (Orszag, 1970; Pouquet et al., 1976; Kleeorin et al., 1990).

The spectral τ approach is a universal tool in turbulent transport for strongly nonlinear systems that allows us to obtain closed results and compare them with the results of laboratory experiments, observations and numerical simulations. The spectral τ approximation reproduces many well-known phenomena found by other methods in turbulent transport of particles and temperature field, in turbulent convection and stably stratified turbulent flows for large Reynolds and Péclet numbers.

1.6.1 Equations for Second Moments

To determine the turbulent heat flux and the level of temperature fluctuations, we use the Navier-Stokes equation for velocity fluctuations and Eq. (1.107) for temperature fluctuations. Equation (1.107) can be rewritten as

$$\frac{\partial\theta}{\partial t} = -Q - (\boldsymbol{u}\cdot\nabla)\overline{T}, \tag{1.141}$$

where $Q = \nabla \cdot (\theta\, \boldsymbol{u} - \langle \theta\, \boldsymbol{u} \rangle) - D^{(\theta)} \Delta\theta$. We consider a one-way coupling, when turbulence affects the temperature field, while the feedback of the temperature field on the turbulence is ignored. In such treatment, the temperature field is a passive scalar. Using Eq. (1.141), written in a Fourier space, we derive equations for the following second-order moments: $F_i^{(\theta)}(t, \boldsymbol{k}) = \langle u_i(t, \boldsymbol{k})\, \theta(t, -\boldsymbol{k}) \rangle$ and $E^{(\theta)}(t, \boldsymbol{k}) = \langle \theta(t, \boldsymbol{k})\, \theta(t, -\boldsymbol{k}) \rangle$:

$$\frac{\partial F_i^{(\theta)}}{\partial t} = - \big\langle u_i(t, \boldsymbol{k})\, u_j(t, -\boldsymbol{k}) \big\rangle\, \nabla_j \overline{T} + \hat{\mathcal{M}} F_i^{(III)}(\boldsymbol{k}), \tag{1.142}$$

$$\frac{\partial E^{(\theta)}}{\partial t} = - \big[\big\langle u_j(t, \boldsymbol{k})\, \theta(t, -\boldsymbol{k}) \big\rangle + \big\langle \theta(t, \boldsymbol{k})\, u_j(t, -\boldsymbol{k}) \big\rangle \big]\, \nabla_j \overline{T} + \hat{\mathcal{M}} E_\theta^{(III)}(\boldsymbol{k}), \tag{1.143}$$

where

$$\hat{\mathcal{M}} F_i^{(III)}(\boldsymbol{k}) = - \langle u_i(t, \boldsymbol{k})\, Q(t, -\boldsymbol{k}) \rangle + \left\langle \frac{\partial u_i(t, \boldsymbol{k})}{\partial t}\, \theta(t, -\boldsymbol{k}) \right\rangle, \tag{1.144}$$

$$\hat{\mathcal{M}} E_\theta^{(III)}(\boldsymbol{k}) = - \langle \theta(t, \boldsymbol{k})\, Q(t, -\boldsymbol{k}) \rangle - \langle Q(t, \boldsymbol{k})\, \theta(t, -\boldsymbol{k}) \rangle \tag{1.145}$$

are the third-order moments appearing due to the nonlinear terms in equations for velocity and temperature fluctuations. Small dissipative terms are also included in $\hat{\mathcal{M}} F_i^{(III)}(\boldsymbol{k})$ and $\hat{\mathcal{M}} E_\theta^{(III)}(\boldsymbol{k})$.

1.6.2 Spectral Tau Approach

Due to the nonlinearity of the governing equation for the temperature and velocity fields, it follows that equations for the second-order moments involve third-order moments. The equation for the third-order moments brings in fourth-order moments and so on. This leads to the famous closure problem of turbulence and turbulent transport – how to close the system to retain only a finite number of terms (Monin and Yaglom, 1971, 2013; McComb, 1990). In particular, the equations for the second-order moments include the first-order spatial differential operators $\hat{\mathcal{M}}$ applied to the third-order moments $F^{(III)}$. A problem arises how to close the system, i.e., how to express the third-order moments $\hat{\mathcal{M}} F^{(III)}$ through the lower-order moments $F^{(II)}$.

We apply here the spectral τ approach (or the spectral relaxation approach) that is valid only for large Péclet and Reynolds numbers. Let us first formulate the main assumption in this method which is a sort of a turbulence closure. The nature of the spectral tau approach and ranges of validity of this method are discussed in Section 1.7. The spectral τ approach postulates that the deviations of the third-order moments, $\hat{\mathcal{M}} F^{(III)}(\boldsymbol{k})$, from the contributions to these terms afforded by the background turbulence, $\hat{\mathcal{M}}F^{(III,0)}(\boldsymbol{k})$, can be expressed through the similar deviations of the second-order moments, $F^{(II)}(\boldsymbol{k}) - F^{(II,0)}(\boldsymbol{k})$:

$$\boxed{\hat{\mathcal{M}}F^{(III)}(\boldsymbol{k}) - \hat{\mathcal{M}}F^{(III,0)}(\boldsymbol{k}) = -\frac{1}{\tau_r(k)}\left[F^{(II)}(\boldsymbol{k}) - F^{(II,0)}(\boldsymbol{k})\right]} \tag{1.146}$$

(Orszag, 1970; Pouquet et al., 1976; Kleeorin et al., 1990), where $\tau_r(k)$ is the scale-dependent relaxation time, which can be identified with the correlation time $\tau(k)$ of the turbulent velocity field for large Reynolds and Péclet numbers. The functions with the superscript (0) correspond to the background turbulence with a zero gradient of the mean temperature. Note that for incompressible turbulence, the contributions of the terms with the superscript (0) vanish because when the gradient of the mean temperature is zero, the turbulent heat flux vanishes. Consequently, Eq. (1.146) for $\hat{\mathcal{M}}F_i^{(III)}(\boldsymbol{k})$ reduces to $\hat{\mathcal{M}}F_i^{(III)}(\boldsymbol{k}) = -F_i(\boldsymbol{k})/\tau(k)$ and for $\hat{\mathcal{M}}E_\theta^{(III)}(\boldsymbol{k}) = -E^{(\theta)}(\boldsymbol{k})/\tau(k)$. Validations of the τ approach for different situations have been performed in various direct numerical simulations (see Sections 1.8, 2.6 and 3.8).

1.6.3 Turbulent Heat Flux and Level of Temperature Fluctuations

We assume that the characteristic time of variation of the second moments $F_i(\boldsymbol{k})$ and $E_\theta(\boldsymbol{k})$ are substantially larger than the correlation time $\tau(k)$ for all turbulence scales. This allows us to use a steady-state solution of Eqs. (1.142) and (1.143). Therefore, applying the spectral τ approximation discussed in the previous Section and using the steady-state solution of Eqs. (1.142) and (1.143), we obtain the following expressions for the turbulent heat flux $F_i^{(\theta)}$ and the level of temperature fluctuations:

$$F_i^{(\theta)} = -\left(\nabla_j \overline{T}\right) \int \tau(k)\, \left\langle u_i(t,\boldsymbol{k})\, u_j(t,-\boldsymbol{k})\right\rangle\, d\boldsymbol{k}, \tag{1.147}$$

$$\left\langle \theta^2\right\rangle = -\left(\nabla_j \overline{T}\right) \int \tau(k)\, \left[\left\langle u_j(t,\boldsymbol{k})\, \theta(t,-\boldsymbol{k})\right\rangle + \left\langle \theta(t,\boldsymbol{k})\, u_j(t,-\boldsymbol{k})\right\rangle\right]\, d\boldsymbol{k}. \tag{1.148}$$

Equation (1.147) implies that

$$\langle u_i(t,\boldsymbol{k})\, \theta(t,-\boldsymbol{k})\rangle = -\tau(k)\, \left\langle u_i(t,\boldsymbol{k})\, u_j(t,-\boldsymbol{k})\right\rangle\, \nabla_j \overline{T}. \tag{1.149}$$

Substituting Eq. (1.149) into Eq. (1.148), we obtain:

$$\left\langle \theta^2\right\rangle = 2\, \left(\nabla_i \overline{T}\right) \left(\nabla_j \overline{T}\right) \int \tau^2(k)\, \left\langle u_i(t,\boldsymbol{k})\, u_j(t,-\boldsymbol{k})\right\rangle\, d\boldsymbol{k}. \tag{1.150}$$

We use the following model for the second moments of a turbulent velocity field, $\langle u_i(t,\boldsymbol{k})\, u_j(t,-\boldsymbol{k})\rangle$, of an isotropic, homogeneous and incompressible turbulence in a Fourier space:

$$\boxed{\left\langle u_i(t,\boldsymbol{k})\, u_j(t,-\boldsymbol{k})\right\rangle = \frac{\langle \boldsymbol{u}^2\rangle\, E(k)}{8\pi k^2}\left(\delta_{ij} - \frac{k_i\, k_j}{k^2}\right).} \tag{1.151}$$

Equation (1.151) is different from Eq. (1.119), because Eq. (1.151) is independent of ω. Since the fluid Reynolds number is very large, i.e., the background turbulence is fully developed, we can use the spectrum function of the turbulent kinetic energy density as $E(k) = k_0^{-1}\,(2/3)\,(k/k_0)^{-5/3}$, where the wave number k varies within the interval: $k_0 \le k \le k_\nu$, the smallest wave number $k_0 = 1/\ell_0$ with the length ℓ_0 being the maximum scale of random motions, the largest wave number $k_\nu = \ell_\nu^{-1}$ with the length $\ell_\nu = \ell_0\,\mathrm{Re}^{-3/4}$ being the Kolmogorov (viscous) scale. In the Kolmogorov turbulence, the expression for the turbulent correlation time is $\tau(k) = 2\,\tau_0\,(k/k_0)^{-2/3}$, where $\tau_0 = \ell_0/u_0$ is the characteristic turbulent time at the integral scale. Substituting Eq. (1.151) into Eqs. (1.147) and (1.150), we determine the turbulent heat flux $F_i^{(\theta)}$ and the level of temperature fluctuations $\langle\theta^2\rangle$:

$$F_i^{(\theta)} = -\frac{\langle \boldsymbol{u}^2\rangle}{8\pi}\left(\nabla_j \overline{T}\right)\int_{k_0}^{k_\nu}\tau(k)\,E(k)\,dk\int_0^{2\pi}d\varphi\int_0^{\pi}\sin\vartheta\,d\vartheta\left[\delta_{ij} - \frac{k_i\,k_j}{k^2}\right], \tag{1.152}$$

$$\begin{aligned}\langle\theta^2\rangle = \frac{\langle \boldsymbol{u}^2\rangle}{4\pi}\left(\nabla_i \overline{T}\right)\left(\nabla_j \overline{T}\right)\int_{k_0}^{k_\nu}\tau^2(k)\,E(k)\,dk \\ \times\int_0^{2\pi}d\varphi\int_0^{\pi}\sin\vartheta\,d\vartheta\left[\delta_{ij} - \frac{k_i\,k_j}{k^2}\right].\end{aligned} \tag{1.153}$$

For the integration over angels in the $\boldsymbol{k}$ space, we use identities (1.129) and (1.135), and for the integration over k, we use the following integrals for large Reynolds numbers, Re$= u_0\ell_0/\nu \gg 1$:

$$\int_{k_0}^{k_\nu}\tau(k)\,E(k)\,dk = \tau_0, \tag{1.154}$$

$$\int_{k_0}^{k_\nu}\tau^2(k)\,E(k)\,dk = \frac{4}{3}\tau_0^2. \tag{1.155}$$

After integration in the $\boldsymbol{k}$ space in Eqs. (1.152) and (1.153), we arrive at the final expressions for the turbulent heat flux $\langle\theta\, u_i\rangle$, and the level of temperature fluctuations, $\langle\theta^2\rangle$, in an isotropic homogeneous and incompressible turbulence for large Péclet and Reynolds numbers:

$$\boxed{\langle\theta\,\boldsymbol{u}\rangle = -D_T\,\boldsymbol{\nabla}\overline{T},} \tag{1.156}$$

$$\boxed{D_T = \frac{1}{3}\, u_0\, \ell_0,} \tag{1.157}$$

$$\boxed{\langle\theta^2\rangle = \frac{8}{9}\, \ell_0^2\, \left(\nabla \overline{T}\right)^2,} \tag{1.158}$$

where $u_0 = \sqrt{\langle \boldsymbol{u}^2 \rangle}$. Note that Eqs. (1.156)–(1.158), obtained using the tau approach, are in agreement with Eqs. (1.62)–(1.63) and (1.65), obtained by the dimensional analysis for large Péclet and Reynolds numbers. The form of the turbulent heat flux is similar for small and large Péclet numbers. However, for large Péclet numbers, the turbulent diffusion coefficient D_T and the level of temperature fluctuations $\langle\theta^2\rangle$ are independent of the Péclet number, while for small Péclet numbers, $D_T \propto$ Pe and $\langle\theta^2\rangle \propto \mathrm{Pe}^2$.

1.7 Ranges of Validity of Various Analytical Methods

In this section, we discuss ranges of validity for the various analytical methods applied in this book. The dimensional analysis, the quasi-linear approach and the spectral tau approach are applied in Chapters 1–3. The analysis based on the budget equations is applied in Chapter 4, and the path-integral approach is applied in Chapter 5.

In Table 1.2, we present a summary of the ranges of validity of various analytical methods applied in this book, as well as one-way and two-way couplings in the system. Here $\mathrm{Pe} = u_0\ell_0/D^{(\theta)}$ is the Péclet number used for temperature field, $\mathrm{Pe}^{(n)} = u_0\ell_0/D^{(n)}$ is the Péclet number for particles, $\mathrm{Re} = u_0\ell_0/\nu$ is the fluid Reynolds number, $\mathrm{Rm} = u_0\ell_0/\eta$ is the magnetic Reynolds number, $D^{(n)}$ is the coefficient of Brownian diffusion of particles, η is the coefficient of magnetic diffusion due to the electrical conductivity of the fluid and Sr is the Strouhal number that is the ratio of the correlation time τ_0 of velocity fluctuations at the integral scale to the turn over time ℓ_0/u_0 of turbulent eddies.

Both one-way and two-way couplings between turbulence and temperature, or particles, or magnetic field can be taken into account only in the spectral tau approach. The one-way coupling implies only the direct effect of turbulence on temperature, or particles, or magnetic field, but the feedback effect of temperature, or particles, or magnetic field on turbulence is ignored. On the other hand, the two-way coupling implies that both ways of coupling between turbulence and temperature, or particles, or magnetic field exist. The two-way coupling is taking into account when the dynamic equation for the Reynolds stress (including both diagonal and off-diagonal terms) is used explicitly, e.g., to determine nonlinear turbulent electromotive force (Rogachevskii and Kleeorin, 2000, 2001, 2006; Kleeorin and

Table 1.2 *Ranges of validity of analytical methods*

Method	Range of validity	Coupling
Dimensional analysis	$\mathrm{Pe} \ll 1$ or $\mathrm{Pe} \gg 1$ $\mathrm{Pe}^{(n)} \ll 1$ or $\mathrm{Pe}^{(n)} \gg 1$ $\mathrm{Rm} \ll 1$ or $\mathrm{Rm} \gg 1$	one-way
Quasi-linear approach	$\mathrm{Pe} \ll 1$ or $\mathrm{Pe} \gg 1$ but $\mathrm{Sr} \ll 1$ $\mathrm{Pe}^{(n)} \ll 1$ or $\mathrm{Pe}^{(n)} \gg 1$ but $\mathrm{Sr} \ll 1$ $\mathrm{Rm} \ll 1$ or $\mathrm{Rm} \gg 1$ but $\mathrm{Sr} \ll 1$	one-way
Spectral tau approach	$\mathrm{Re} \gg 1$ $\mathrm{Pe} \gg 1$, $\mathrm{Pe}^{(n)} \gg 1$ $\mathrm{Rm} \gg 1$	one-way (or two-way)
Budget equations approach	$\mathrm{Re} \gg 1$ $\mathrm{Pe} \gg 1$, $\mathrm{Pe}^{(n)} \gg 1$ $\mathrm{Rm} \gg 1$	two-way
Path-integral approach	Short correlation time of turbulent velocity field	one-way

Rogachevskii, 2007), or in the theory of formation of inhomogeneous magnetic structures (Kleeorin et al., 1990; Kleeorin and Rogachevskii, 1994a; Kleeorin et al., 1996; Rogachevskii and Kleeorin, 2007).

The analysis based on the budget equations takes into account the two-way coupling, where velocity fluctuations affect the turbulent transport of temperature, while temperature fluctuations and turbulent heat flux affect velocity fluctuations (Monin and Yaglom, 1971, 2013). This approach is a kind of energetic analysis, and it is effective for the study of turbulent transport in homogeneous turbulence when the divergence of third-order moments caused by nonlinear terms vanishes.

The quasi-linear approach (Krause and Rädler, 1980; Moffatt, 1978) and the path-integral approach (Feynman and Hibbs, 1965; Kac, 1959; Zeldovich et al., 1988, 1990; Piterbarg and Ostrovskii, 2013) are exact methods and can be considered as perturbation approaches. These two methods are applied for study of kinematic problems, i.e., for investigations of the evolution of passive or vector fields in a given prescribed random or turbulent velocity field. This implies that formal conditions on the fluid Reynolds number are not required for validity of these approaches. However, for dynamical problems when the velocity field is determined by the Navier-Stokes equation, an additional condition for validity of the quasi-linear approach is small fluid Reynolds numbers. A condition of very short correlation time of random velocity field in comparison with the characteristic time of the mean-field variations is necessary for validity of the path-integral approach.

The tau approach can be applied in physical space, referred to as the minimal tau approach (Field et al., 1999; Blackman and Field, 2003), as well as in the Fourier space referred to as the spectral tau approach (Orszag, 1970; Pouquet et al., 1976; Kleeorin et al., 1990). The spectral tau approach allows one to take into account the spectrum of turbulence and the scale-dependence of the turbulent correlation time. The spectral tau approach is a sort of high-order closure and in general is similar to Eddy Damped Quasi Normal Markovian (EDQNM) approximation. However, some principle differences exist between these two approaches (Orszag, 1970). The EDQNM closures do not relax to equilibrium (the background turbulence), and the EDQNM approach does not describe properly the motions in the equilibrium state in contrast to the spectral τ approach. Within the EDQNM theory, there is no dynamically determined relaxation time, and no slightly perturbed steady state can be approached. In the spectral tau approach, the relaxation time for small departures from equilibrium is determined by the turbulent motions in the equilibrium state, i.e., it is determined by the background turbulence, rather than by the departure from the equilibrium. As follows from the analysis (Orszag, 1970), the spectral tau approach describes the relaxation to the background turbulence much more accurately than the EDQNM approach.

Despite the fact that the spectral tau approach is a form of closure, this method works well enough if there is a small parameter. For instance, in turbulent transport of particles or temperature field, the small parameter can be the ratio of the integral turbulence scale to the characteristic scale of the mean particle number density variations (or the ratio of the integral turbulence scale to the mean fluid temperature variation scale). In the case of turbulent magnetohydrodynamics, the small parameter can be the ratio of the energy of the mean magnetic field to turbulent kinetic energy.

The spectral tau approach works well enough for developing the mean-field theories. This approach allows one to determine one-point second-order correlation functions, e.g., the turbulent heat flux, the turbulent flux of particles, the turbulent electromotive force. However, the spectral tau approach cannot be used for the study of intermittency of passive or vector fields, which require one to determine two-point or multi-point non-instantaneous second-order and high-order moments (Zeldovich et al., 1990; Frisch, 1995; Falkovich et al., 2001).

There are also other methods applied for large Péclet and Reynolds numbers. One of these methods is the renormalization procedure (Moffatt, 1981, 1983; Kleeorin and Rogachevskii, 1994a; Elperin et al., 1996a), which is applied for investigation of passive scalar fluctuations and magnetic fluctuations. The averaging in the renormalization procedure is performed over fluctuations of scales from the viscous scale ℓ_ν to scale ℓ_* inside the inertial range of turbulence $\ell_\nu < \ell_* < \ell_0$.

After this averaging, the range of scales $\ell > \ell_*$ in the renormalization procedure corresponds to "mean" fields, whereas fluctuations are considered in the range $\ell < \ell_*$. The effect of fluctuations on the "mean" fields is described by the turbulent transport coefficients, e.g., the turbulent viscosity and turbulent diffusivity which are functions of the scale of the averaging ℓ_*.

The next step of the renormalization procedure comprises a step-by-step increase of the scale of the averaging. This procedure allows one to derive equations for the turbulent transport coefficients. In the framework of the renormalization procedure, an interaction between the weakly inhomogeneous mean fields and background turbulence (with zero gradients of the mean passive scalar field) is studied. In particular, this procedure considers small deviations from the background turbulence under the influence of weak gradients of mean fields, while the spectrum and statistical properties of the background turbulence are assumed to be known.

This renormalization procedure is essentially different from the renormalization group method (RNG) (Yakhot and Orszag, 1986; McComb, 1990), which provides a closure for the dynamical problem of turbulence. Common in the renormalization procedure and in the renormalization group method is a step-by-step averaging. However, a renormalization of the background hydrodynamic turbulence as it is done in the RNG method is not performed in the renormalization procedure. In the renormalization procedure, a parameter of nonlinear interaction is not used contrary to the RNG method (Yakhot and Orszag, 1986; McComb, 1990).

1.8 Further Reading

Turbulence, and the associated transport of scalar and vector fields, is one of the classical problems of physics. Although it has been investigated systematically for more than a hundred years in theoretical, experimental and numerical studies, fundamental questions remain. This is particularly true in applications such as astrophysics, where the governing parameter values are too extreme to be modeled either experimentally or numerically. The large variety of interesting phenomena related to the nature of turbulent flows have been discussed in various books (Batchelor, 1953; Hinze, 1959; Monin and Yaglom, 1971, 2013; Bradshaw, 1971; Tennekes and Lumley, 1972; Townsend, 1980; Kraichnan and Montgomery, 1980; McComb, 1990; Chorin, 1994; Frisch, 1995; Libby, 1996; Pope, 2000; Mathieu and Scott, 2000; Bernard and Wallace, 2002; Biskamp, 2003; Jovanovic, 2004; Lumley, 2007; Lesieur, 2008; Sagaut and Cambon, 2008; Tsinober, 2009, 2014; Nazarenko, 2011; Zakharov et al., 2012; Piquet, 2013; Davidson, 2013, 2015; Beresnyak and Lazarian, 2019; Landau and Lifshits, 1987).

Turbulent transports of passive scalar and vector fields have various applications in geophysical, astrophysical and industrial flows. Different aspects of passive scalar transport in turbulent flows have been investigated theoretically, experimentally and using numerical simulations in a number of publications (Csanady, 2012; Pasquill and Smith, 1983; Piterbarg and Ostrovskii, 2013; Blackadar, 2012; Majda and Kramer, 1999; Warhaft, 2000). Interesting effects in turbulent transport include anomalous turbulent diffusion, intermittency and anomalous scalings (the violation of the dimensional analysis predictions for the scaling laws), fractal structure of a concentration field and formation of large-scale coherent structures in a small-scale turbulence (Kolmogorov, 1962; Zeldovich et al., 1988, 1990; Kraichnan, 1994; Frisch, 1995; Gawedzki and Kupiainen, 1995; Shraiman and Siggia, 1995, 2000; Chertkov and Falkovich, 1996; Falkovich et al., 2001; Cardy et al., 2008).

Compressibility of turbulent flow plays an essential role in a passive scalar dynamics and causes qualitative changes in the properties of both the mean passive scalar field and fluctuations (Elperin et al., 1995, 1997b; Rädler et al., 2011; Rogachevskii et al., 2018; Rogachevskii and Kleeorin, 2021). Compressibility of a surrounding fluid flow results in a slow scale-dependent turbulent diffusion of small-scale passive scalar fluctuations for large Pèclet numbers (Elperin et al., 1995). In addition, the level of passive scalar fluctuations in the presence of a gradient of the mean passive scalar field in compressible flow of fluid can be fairly strong. On the other hand, passive scalar transport in a density-stratified turbulent fluid flow [i.e., $\operatorname{div} \boldsymbol{u} = -(\boldsymbol{u} \cdot \nabla\overline{\rho})/\overline{\rho} \neq 0$] is accompanied by formation of large-scale structures due to instability of the mean passive scalar field in inhomogeneous turbulent velocity field (Elperin et al., 1995).

Compressibility of a fluid flow also reduces the turbulent diffusivity of the mean temperature field (Rogachevskii and Kleeorin, 2021) similarly to that for particle number density and magnetic field (Rädler et al., 2011; Rogachevskii et al., 2018). However, expression for turbulent diffusion coefficient for the mean temperature field in a compressible turbulence is different from that for particle number density and magnetic field. The combined effect of compressibility and inhomogeneity of turbulence for small Péclet numbers causes an increase of the mean temperature in the regions with more intense velocity fluctuations due to turbulent pumping. Formally, this effect is similar to a phenomenon of compressible turbophoresis found for non-inertial particles or gaseous admixtures (Rogachevskii et al., 2018). The gradient of the mean fluid pressure results in an additional turbulent pumping of the mean temperature field. The later effect is similar to turbulent barodiffusion of particles and gaseous admixtures (Elperin et al., 1997c; Rogachevskii et al., 2018). Compressibility of a fluid flow also causes a turbulent cooling of the surrounding fluid due to an additional sink term in the equation for the mean temperature field. There is no analog of this effect for particles (Rogachevskii and Kleeorin, 2021).

Another interesting feature for a stratified turbulence is that turbulent flux of entropy in compressible temperature stratified turbulence is different from turbulent convective flux of fluid internal energy (Braginsky and Roberts, 1995; Rogachevskii and Kleeorin, 2015). In particular, turbulent flux of entropy is given by $\boldsymbol{F}_s = \overline{\rho}\,\langle s'\,\boldsymbol{u}\rangle$, where $\overline{\rho}$ is the mean fluid density and s' are fluctuations of entropy. On the other hand, the turbulent convective flux of fluid internal energy is $\boldsymbol{F}_c = \overline{T}\,\overline{\rho}\,\langle s'\,\boldsymbol{u}\rangle$, where $\overline{T}$ is the mean fluid temperature. This turbulent convective flux is well-known in the astrophysical and geophysical literature, and it cannot be used as a turbulent flux in the equation for the mean entropy. This is the exact result for low-Mach-number temperature stratified turbulence and is independent of the turbulence model used. The velocity-entropy correlation is given by $\langle s'\,\boldsymbol{u}\rangle = D_T \boldsymbol{\nabla}\overline{s}$, where the turbulent diffusion coefficient D_T of the mean entropy depends on the Péclet number, i.e., $D_T \propto u_0\,\ell_0\,\mathrm{Pe}$ for small Péclet numbers, while $D_T \propto u_0\,\ell_0$ for large Péclet numbers (Rogachevskii and Kleeorin, 2015), where $\overline{s}$ is the mean entropy.

Numerous experiments demonstrate that the mean temperature distribution in turbulent convection is strongly inhomogeneous and anisotropic due to the large-scale circulations (Ahlers et al., 2009; Bukai et al., 2009; Barel et al., 2020; Ching, 2014), and the mean temperature gradient in the horizontal direction inside the large-scale circulations can be significantly larger than in the vertical direction, i.e., the vertical turbulent heat flux inside the large-scale circulations is very small. Moreover, even in convective turbulence, stably stratified regions can be formed, where the mean temperature gradient in the flow core inside the large-scale circulation is directed upward (Barel et al., 2020; Toppaladoddi et al., 2017).

Temperature fluctuations in turbulent flow can be passive in the case of a one-way coupling when there is no feedback of temperature fluctuations on turbulent flow. In this case, there is only an effect of turbulence on the temperature field. Such regime has been studied experimentally and theoretically in unstably stratified turbulent flow in externally forced Rayleigh-Bénard convection without large-scale circulation (Bukai et al., 2011). In these experiments, an additional source of turbulence has been produced by two oscillating grids located nearby the side walls of the chamber. When the frequency of the grid oscillations is larger than 2 Hz, the large-scale circulation in turbulent convection is destructed, and it is accompanied by a strong change in the mean temperature distribution. However, in all regimes of the unstably stratified turbulent flow the ratio $\sigma_* \equiv \left[(\ell_x \nabla_x \overline{T})^2 + (\ell_y \nabla_y \overline{T})^2 + (\ell_z \nabla_z \overline{T})^2\right]/\langle\theta^2\rangle$ varies slightly. This result is valid even in the range of parameters whereby the behavior of the temperature field is different from that of the passive scalar. Here ℓ_i are the integral scales of turbulence along x, y, z directions, $\overline{T}$ and θ are the mean and fluctuating parts of the fluid temperature. The ratio σ_* is also found to be constant (independent of the frequency of the grid oscillations) in the experiments with stably stratified forced turbulence in air flow produced by two oscillating grids, and it has the same magnitude for both stably and unstably stratified turbulent flows (Eidelman et al., 2013).

Many studies of turbulent transport of a passive scalar use a model of linear velocity field (Chertkov and Falkovich, 1996; Falkovich et al., 2001). Employment of this simple model allows one to perform calculations in a closed form for a passive scalar. A model of a linear velocity field can be viewed as: (i) an expansion in Taylor series of the velocity field in a local frame of reference moving with a fluid element and (ii) a real flow field in an infinite space (e.g., like the Hubble flow in cosmology). Some peculiar aspects of a transport of a passive scalar in a linear velocity field have been discussed by Elperin et al. (2001b), where the Cauchy problem for a passive scalar in both laminar and random linear velocity fields has been considered. It is shown that a spatial distribution of a passive scalar evolves either into a thin infinite "pancake" or into an infinite "rope" structures. The properties of a passive scalar are similar for laminar and random incompressible linear velocity fields. A compressibility of a fluid flow results in a non-zero flux of particles from infinity. For strong compressibility a spatial distribution of a passive scalar evolves into a ball (or ellipsoid) with a very small radius that is of the order of a diffusion scale. The higher statistical moments can grow exponentially in spite of a decay of a typical realization of a passive scalar. Such "strange" behavior of a passive scalar demonstrates that transport of a passive scalar in linear velocity field cannot be considered as a general and universal phenomenon. Remarkably, the dynamics of a passive scalar and a magnetic field (Zeldovich et al., 1984; Moffatt, 1963; Gvaramadze et al., 1988) in linear velocity fields are similar.

2
Particles and Gases in Density-Stratified Turbulence

In this chapter, we consider a turbulent transport of small particles and gaseous admixtures for incompressible and compressible flows as well as for a density-stratified flow in anelastic approximation. We derive mean-field equations for the particle number density and obtain expressions for level of particle number density fluctuations, turbulent flux of particles, turbulent diffusion and effective pumping velocity of non-inertial and inertial particles for small and large Péclet numbers by means of various analytical methods, namely the dimensional analysis, the quasi-linear approach and the spectral tau approach (the relaxation approach). We discuss physics of turbulent diffusion, turbulent thermal diffusion and turbophores of particles.

2.1 Turbulent Transport of Particles: Dimensional Analysis

2.1.1 Governing Equations

We consider the dynamics of small (non-inertial) particles or gaseous admixtures in a turbulent fluid flow. The equation for the evolution of the particle number density $n(t, \boldsymbol{x})$ in a compressible fluid velocity field $\boldsymbol{U}(t, \boldsymbol{x})$ reads (Chandrasekhar, 1943; Akhiezer and Peletminsky, 1981)

$$\boxed{\frac{\partial n}{\partial t} + \nabla \cdot \left(n\, \boldsymbol{U} - D^{(n)} \nabla n\right) = 0,} \tag{2.1}$$

where $D^{(n)} = k_B\, T/(6\pi \rho\, \nu\, a_p)$ is the coefficient of the molecular (Brownian) diffusion of particles having the radius a_p. Here ν is the kinematic viscosity of the fluid, T and ρ are the fluid temperature and density and k_B is the Boltzmann constant. The fluid velocity is a turbulent field produced by, e.g., an external steering force. Equation (2.1) is a conservation law for the total number of particles that implies that the total number of particles in a closed volume is conserved.

Here we do not consider a coagulation of particles or chemical reactions as well as condensation or evaporation of droplets which change the total number of particles or droplets in a closed volume (Pruppacher et al., 1998; Seinfeld and Pandis, 2016; Lamb and Verlinde, 2011). If one assumes for simplicity that $D^{(n)}$ is independent of coordinates, Eq. (2.1) can be rewritten as

$$\boxed{\frac{\partial n}{\partial t} + \nabla \cdot (n\, \boldsymbol{U}) = D^{(n)}\, \Delta n.} \tag{2.2}$$

Very often, instead of the particle number density, a mass concentration $C_{\rm p} = n\, m_{\rm p}/\rho$ of particles or gaseous admixtures is used, where $m_{\rm p}$ is the particle mass. The equation for the mass concentration $C_{\rm p}(t, \boldsymbol{x})$ is derived using Eq. (2.1) for the particle number density $n(t, \boldsymbol{x})$ and the continuity equation (1.4) for the fluid density $\rho(t, \boldsymbol{x})$, which yield

$$\boxed{\frac{\partial C_{\rm p}}{\partial t} + (\boldsymbol{U} \cdot \nabla) C_{\rm p} = \rho^{-1}\, \nabla \cdot \big(D^{(n)}\, \nabla (n\, m_{\rm p})\big).} \tag{2.3}$$

For an incompressible flow, Eq. (2.3) reads

$$\boxed{\frac{\partial C_{\rm p}}{\partial t} + (\boldsymbol{U} \cdot \nabla) C_{\rm p} = D^{(n)}\, \Delta C_{\rm p}.} \tag{2.4}$$

We use a point-particle approximation that implies that the size of particles is very small in comparison with all possible scales of fluid motions.

2.1.2 Anelastic Approximation

The continuity equation for the fluid density $\rho(t, \boldsymbol{x})$ in a compressible flow reads

$$\boxed{\frac{\partial \rho}{\partial t} + \nabla \cdot (\rho\, \boldsymbol{U}) = 0.} \tag{2.5}$$

We consider fluid flows with a low Mach number, ${\rm Ma} = |\boldsymbol{U}|/c_{\rm s} \ll 1$, which implies that fluid velocity is much less than the sound speed, $c_{\rm s}$. The first term on the left-hand side of Eq. (2.5), $|\partial \rho / \partial t| \sim {\rm Ma}^2$, i.e., the time derivative of the fluid density is very small for a low-Mach-number fluid flow. In this case, the continuity equation for the fluid density can be used in an anelastic approximation:

$$\boxed{\nabla \cdot (\rho\, \boldsymbol{U}) = 0.} \tag{2.6}$$

This equation implies that

$$\boxed{\nabla \cdot \boldsymbol{U} = -\frac{(\boldsymbol{U} \cdot \nabla)\rho}{\rho},} \tag{2.7}$$

i.e., the anelastic approximation takes into account an inhomogeneous fluid density. In this section, we apply the anelastic approximation (2.7).

2.1.3 Mean-Field Approach

In this section, we study a long-term evolution of the particle number density in spatial scales L_n, which are much larger than the integral scale of turbulence ℓ_0, and during the time scales t_n, which are much larger than the turbulent time scales τ_0. We use here a mean-field approach in which all quantities are decomposed into the mean and fluctuating parts, where the fluctuating parts have zero mean values, i.e., we use the Reynolds rules. In particular, the particle number density $n = \overline{n} + n'$, where $\overline{n} = \langle n \rangle$ is the mean particle number density, and n' are particle number density fluctuations and $\langle n' \rangle = 0$. The angular brackets $\langle \ldots \rangle$ denote ensemble averaging. Averaging Eq. (2.2) over an ensemble of turbulent velocity field, we arrive at the mean-field equation for the particle number density:

$$\boxed{\frac{\partial \overline{n}}{\partial t} + \nabla \cdot \left(\overline{n}\, \overline{\boldsymbol{U}} + \langle n' \boldsymbol{u} \rangle \right) = D^{(n)} \, \Delta \overline{n},} \tag{2.8}$$

where $\boldsymbol{F}^{(n)} \equiv \langle n' \boldsymbol{u} \rangle$ is the turbulent flux of particles. We consider for simplicity the case $\overline{\boldsymbol{U}} = 0$.

2.1.4 Equation for Particle Number Density Fluctuations

Equation (2.8) is not yet closed because we do not know how the particle turbulent flux $\langle \boldsymbol{u} n' \rangle$ depends on the mean particle number density $\overline{n}$. To determine the particle turbulent flux, we derive an equation for particle number density fluctuations n', that is obtained by subtracting Eq. (2.8) from Eq. (2.2):

$$\boxed{\frac{\partial n'}{\partial t} + \nabla \cdot \left(n' \boldsymbol{u} - \langle n' \boldsymbol{u} \rangle \right) - D^{(n)} \Delta n' = -(\boldsymbol{u} \cdot \nabla) \overline{n} - \overline{n} (\nabla \cdot \boldsymbol{u}).} \tag{2.9}$$

The term $\mathcal{Q}^{(n)} = \nabla \cdot \left(n' \boldsymbol{u} - \langle n' \boldsymbol{u} \rangle \right)$, on the left-hand side of Eq. (2.9), is the nonlinear term, while the terms $-(\boldsymbol{u} \cdot \nabla)\overline{n} - \overline{n}(\nabla \cdot \boldsymbol{u})$, on the right-hand side of Eq. (2.9), are the source terms for particle number density fluctuations. The first source term, $-(\boldsymbol{u} \cdot \nabla)\overline{n}$, in Eq. (2.9) causes a production of particle number density fluctuations via tangling of the gradient of the mean particle number density $\nabla \overline{n}$ by velocity fluctuations. The second source term, $-\overline{n}(\nabla \cdot \boldsymbol{u}) = (\overline{n}/\overline{\rho})(\boldsymbol{u} \cdot \nabla)\overline{\rho}$, results in production of particle number density fluctuations by the tangling of the gradient of the mean fluid density $\nabla \overline{\rho}$ by velocity fluctuations, where $\overline{\rho}$ is the mean fluid density. Here we take into account the anelastic approximation,

$\nabla \cdot \boldsymbol{u} = -(1/\overline{\rho})\,(\boldsymbol{u} \cdot \nabla)\overline{\rho}$. The ratio of the absolute values of the nonlinear term $|\mathcal{Q}^{(n)}|$ to the diffusion term $|D^{(n)} \Delta n'|$ is the Péclet number for particles:

$$\boxed{\mathrm{Pe}^{(n)} = \frac{|\mathcal{Q}^{(n)}|}{|D^{(n)} \Delta n'|} = \frac{u_0 \ell_0}{D^{(n)}}.} \tag{2.10}$$

Equation (2.9) is a nonlinear equation for particle number density fluctuations. Since this nonlinear equation cannot be solved exactly for arbitrary Péclet numbers, we have to use different approximate methods for the solution of Eq. (2.9). We consider a one-way coupling, i.e., we take into account the effect of the turbulent velocity on the particle number density, but we neglect the feedback effect of the particle number density on the turbulent fluid flow. This approximation is valid when the spatial density of particles $n\, m_p$ is much smaller than the fluid density ρ, where m_p is the particle mass. We also consider non-inertial particles or gaseous admixtures. In this case, the particles move with the fluid velocity. These assumptions imply that the particle number density is a passive scalar.

2.1.5 Dimensional Analysis

First, we use the dimensional analysis to solve Eq. (2.9). The dimension of the left-hand side of Eq. (2.9) is the rate of change of particle number density fluctuations n'/τ_n, where τ_n is the characteristic time of particle number density fluctuations. We consider two cases:

- *Large Péclet numbers*, $\mathrm{Pe}^{(n)} \gg 1$. In this case the characteristic time of particle number density fluctuations τ_n can be identified with the correlation time τ_0 of the turbulent velocity field for large Reynolds numbers. Therefore, in the framework of the dimensional analysis, we replace the left-hand side of Eq. (2.9) by n'/τ_0. This yields:

$$n' = -\tau_0\, [(\boldsymbol{u} \cdot \nabla)\overline{n} + \overline{n}\, (\nabla \cdot \boldsymbol{u})]\,. \tag{2.11}$$

- *Small Péclet numbers*, $\mathrm{Pe}^{(n)} \ll 1$. This implies that the nonlinear terms are much smaller than the diffusion terms. Since the molecular diffusion is the dominant effect here, we assume that the time τ_n can be identified with the diffusion time $\tau_D = \ell_0^2/D^{(n)}$. Therefore the solution for Eq. (2.9) reads:

$$n' = -\frac{\ell_0^2}{D^{(n)}}\, [(\boldsymbol{u} \cdot \nabla)\overline{n} + \overline{n}\, (\nabla \cdot \boldsymbol{u})]\,. \tag{2.12}$$

2.1.6 Turbulent Diffusion, Turbulent Thermal Diffusion and Level of Fluctuations

Using Eqs. (2.11) and (2.12), we determine the particle turbulent flux $\langle n' \boldsymbol{u} \rangle$ and the level of particle number density fluctuations $\langle n'^2 \rangle$ for large and small Péclet numbers.

Large Péclet Numbers

We start with the case of large Péclet numbers. Multiplying Eq. (2.11) by velocity fluctuations, u_i, and averaging over an ensemble of turbulent velocity field, we arrive at the expression for the turbulent flux of particles:

$$\left\langle n'\, u_i \right\rangle = -\tau_0 \left\langle u_i u_j \right\rangle \nabla_j \overline{n} - \tau_0 \overline{n} \left\langle u_i (\boldsymbol{\nabla} \cdot \boldsymbol{u}) \right\rangle \equiv V_i^{\rm eff}\, \overline{n} - D_{ij}^{(n)} \nabla_j \overline{n}, \tag{2.13}$$

where the last term, $-D_{ij}^{(n)} \nabla_j \overline{n}$, in Eq. (2.13) determines the contribution to the flux of particles caused by turbulent diffusion. Here

$$D_{ij}^{(n)} = \tau_0 \left\langle u_i u_j \right\rangle \tag{2.14}$$

is the turbulent diffusion tensor. The first term on the right-hand side of Eq. (2.13), $\boldsymbol{V}^{\rm eff}\, \overline{n}$, determines the contribution to the turbulent flux of particles caused by the effective pumping velocity:

$$\boxed{\boldsymbol{V}^{\rm eff} = -\tau_0 \left\langle \boldsymbol{u} (\boldsymbol{\nabla} \cdot \boldsymbol{u}) \right\rangle.} \tag{2.15}$$

Let us decompose the fluid density to the mean fluid density $\overline{\rho}$ and fluctuations ρ', i.e., $\rho = \overline{\rho} + \rho'$, where for low Mach numbers $|\rho'| \ll \overline{\rho}$ (Chassaing et al., 2002). The anelastic approximation yields $\boldsymbol{\nabla} \cdot (\rho\, \boldsymbol{u}) \approx \boldsymbol{\nabla} \cdot (\overline{\rho}\, \boldsymbol{u}) = 0$, so that

$$\boldsymbol{\nabla} \cdot \boldsymbol{u} \approx -\boldsymbol{u} \cdot \frac{\boldsymbol{\nabla}\, \overline{\rho}}{\overline{\rho}} = \boldsymbol{\lambda} \cdot \boldsymbol{u}, \tag{2.16}$$

where $\boldsymbol{\lambda} = -\boldsymbol{\nabla} \ln \overline{\rho}$. Therefore, the effective pumping velocity is given by

$$V_i^{\rm eff} = -\tau_0 \left\langle u_i u_j \right\rangle \lambda_j. \tag{2.17}$$

For an isotropic turbulence $\langle u_i u_j \rangle = \delta_{ij} \langle \boldsymbol{u}^2 \rangle / 3$, so that the turbulent diffusion tensor for large Péclet numbers is given by $D_{ij}^{(n)} = D_T \delta_{ij}$, where the turbulent diffusion coefficient is

$$\boxed{D_T = \frac{\tau_0 \left\langle \boldsymbol{u}^2 \right\rangle}{3}.} \tag{2.18}$$

The turbulent diffusion of particles was predicted by G. I. Taylor[1] (Taylor, 1922). The effective pumping velocity for an isotropic turbulence and large Péclet numbers is given by

[1] Sir Geoffrey Ingram Taylor (1886–1975) was a British physicist and mathematician, well-known for his works in fluid dynamics, turbulence and wave theory.

$$\boxed{\boldsymbol{V}^{\rm eff} = -D_T \boldsymbol{\lambda} = D_T \, \frac{\boldsymbol{\nabla} \overline{\rho}}{\overline{\rho}},} \tag{2.19}$$

and the particle turbulent flux $\langle n' \boldsymbol{u} \rangle$ is

$$\boxed{\langle n' \boldsymbol{u} \rangle = \boldsymbol{V}^{\rm eff} \, \overline{n} - D_T \, \boldsymbol{\nabla} \overline{n}.} \tag{2.20}$$

To determine the level of particle number density fluctuations $\langle n'^2 \rangle$ for large Péclet numbers, we use Eq. (2.11), so that

$$n'^2 = \tau_0^2 \, \left\{ [(\boldsymbol{u} \cdot \boldsymbol{\nabla})\overline{n}]^2 + 2 \, [(\boldsymbol{u} \cdot \boldsymbol{\nabla})\overline{n}] \, [\overline{n} \, (\boldsymbol{\nabla} \cdot \boldsymbol{u})] + \overline{n}^2 (\boldsymbol{\nabla} \cdot \boldsymbol{u})^2 \right\}. \tag{2.21}$$

Ensemble averaging of Eq. (2.21) yields:

$$\langle n'^2 \rangle = \tau_0^2 \, \left[\langle u_i u_j \rangle \, (\nabla_i \overline{n}) \, \left(\nabla_j \overline{n}\right) + 2 \, \langle u_i (\boldsymbol{\nabla} \cdot \boldsymbol{u}) \rangle \, \overline{n} \, \nabla_i \overline{n} + \overline{n}^2 \, \langle (\boldsymbol{\nabla} \cdot \boldsymbol{u})^2 \rangle \right]. \tag{2.22}$$

In the anelastic approximation, $\boldsymbol{\nabla} \cdot \boldsymbol{u} = \boldsymbol{\lambda} \cdot \boldsymbol{u}$ and $\langle (\boldsymbol{\nabla} \cdot \boldsymbol{u})^2 \rangle = \langle u_i u_j \rangle \, \lambda_i \lambda_j$, so that

$$\langle n'^2 \rangle = \tau_0^2 \, \langle u_i u_j \rangle \, \left[(\nabla_i \overline{n}) \, \left(\nabla_j \overline{n}\right) + (\lambda_j \nabla_i) \, \overline{n}^2 + \overline{n}^2 \, \lambda_i \, \lambda_j \right]. \tag{2.23}$$

For an isotropic turbulence, $\langle u_i u_j \rangle = \delta_{ij} \, \langle \boldsymbol{u}^2 \rangle / 3$ and $\langle (\boldsymbol{\nabla} \cdot \boldsymbol{u})^2 \rangle = \boldsymbol{\lambda}^2 \, \langle \boldsymbol{u}^2 \rangle / 3$. Therefore, the level of particle number density fluctuations $\langle n'^2 \rangle$ for the isotropic turbulence and large Péclet numbers is given by

$$\boxed{\langle n'^2 \rangle = \frac{\ell_0^2}{3} \, \left[(\boldsymbol{\nabla} \overline{n})^2 + (\boldsymbol{\lambda} \cdot \boldsymbol{\nabla}) \, \overline{n}^2 + \overline{n}^2 \, \boldsymbol{\lambda}^2 \right],} \tag{2.24}$$

where $\ell_0^2 = \tau_0^2 \, \langle \boldsymbol{u}^2 \rangle$.

Small Péclet Numbers

Now let us consider the case of small Péclet numbers. Multiplying Eq. (2.12) by u_i and averaging this equation over a statistics of random velocity field, we obtain the particle turbulent flux $\langle n' \boldsymbol{u} \rangle$ as

$$\langle n' \, u_i \rangle = -\frac{\ell_0^2}{D^{(n)}} \, \left[\langle u_i u_j \rangle (\nabla_j \overline{n}) + \langle u_i (\boldsymbol{\nabla} \cdot \boldsymbol{u}) \rangle \, \overline{n} \right]. \tag{2.25}$$

In the anelastic approximation, $\boldsymbol{\nabla} \cdot \boldsymbol{u} = \boldsymbol{\lambda} \cdot \boldsymbol{u}$, and for isotropic turbulence, $\langle u_i u_j \rangle = \langle \boldsymbol{u}^2 \rangle \, \delta_{ij} / 3$. Substituting these equations into Eq. (2.25), we arrive at the following expressions for the turbulent particle flux, the turbulent diffusion coefficient and the effective pumping velocity of non-inertial particles for small Péclet numbers:

$$\boxed{\langle n' \boldsymbol{u} \rangle = \boldsymbol{V}^{\rm eff} \, \overline{n} - D_T \, \boldsymbol{\nabla} \overline{n},} \tag{2.26}$$

$$\boxed{D_T = \frac{u_0\,\ell_0}{3}\,\mathrm{Pe}^{(n)},} \tag{2.27}$$

$$\boxed{\boldsymbol{V}^{\mathrm{eff}} = -D_T \boldsymbol{\lambda}.} \tag{2.28}$$

To determine the level of particle number density fluctuations for small Péclet numbers, we use Eq. (2.12), which yields

$$n'^2 = \left(\frac{\ell_0^2}{D^{(n)}}\right)^2 \left([(\boldsymbol{u}\cdot\nabla)\overline{n}]^2 + 2\,[(\boldsymbol{u}\cdot\nabla)\overline{n}]\,[\overline{n}\,(\nabla\cdot\boldsymbol{u})] + \overline{n}^2(\nabla\cdot\boldsymbol{u})^2\right). \tag{2.29}$$

Averaging Eq. (2.29) over a statistics of random velocity field, we determine the level of particle number density fluctuations $\langle n'^2\rangle$:

$$\langle n'^2\rangle = \left(\frac{\ell_0^2}{D^{(n)}}\right)^2 \left[\langle u_i u_j\rangle\,(\nabla_i\overline{n})\,\left(\nabla_j\overline{n}\right) + 2\,\langle u_i(\nabla\cdot\boldsymbol{u})\rangle\,\overline{n}\,\nabla_i\overline{n} + \overline{n}^2\left\langle(\nabla\cdot\boldsymbol{u})^2\right\rangle\right]. \tag{2.30}$$

In the anelastic approximation, $\nabla\cdot\boldsymbol{u} = \boldsymbol{\lambda}\cdot\boldsymbol{u}$, so that

$$\left\langle n'^2\right\rangle = \left(\frac{\ell_0^2}{D^{(n)}}\right)^2 \left\langle u_i u_j\right\rangle \left[(\nabla_i\overline{n})\,\left(\nabla_j\overline{n}\right) + (\lambda_j\nabla_i)\,\overline{n}^2 + \overline{n}^2\,\lambda_i\,\lambda_j\right]. \tag{2.31}$$

For an isotropic turbulence, the level of particle number density fluctuations $\langle n'^2\rangle$ for small Péclet numbers is given by

$$\boxed{\langle n'^2\rangle = \frac{1}{3}\,\left(\mathrm{Pe}^{(n)}\right)^2\,\ell_0^2\,\left[(\nabla\overline{n})^2 + (\boldsymbol{\lambda}\cdot\nabla)\,\overline{n}^2 + \overline{n}^2\,\boldsymbol{\lambda}^2\right].} \tag{2.32}$$

2.1.7 Physics of the Effective Pumping Velocity

To understand the physics related to the effective pumping velocity $\boldsymbol{V}^{\mathrm{eff}}$, let us first express the effective pumping velocity via physical parameters. We use the equation of state for a perfect gas,

$$P = \frac{k_B}{m_\mu}\,\rho\,T \equiv \frac{R}{\mu}\,\rho\,T, \tag{2.33}$$

which can be also rewritten for the mean fields as

$$\overline{P} = \frac{k_B}{m_\mu}\,\overline{\rho}\,\overline{T}, \tag{2.34}$$

where $k_B = R/N_A$ is the Boltzmann constant, R is the gas constant, N_A is the Avogadro number, $\mu = m_\mu N_A$ is the molar mass, $\overline{P}$ is the mean fluid pressure and $\overline{T}$ is the mean fluid temperature. In Eq. (2.34), we assume that $\overline{\rho}\,\overline{T} \gg \langle\rho'\,\theta\rangle$.

Using Eq. (2.34), we express the gradient of the mean fluid density in terms of the gradients of the mean fluid pressure $\nabla \overline{P}$ and mean fluid temperature $\nabla \overline{T}$ as

$$\frac{\nabla \overline{\rho}}{\overline{\rho}} = \frac{\nabla \overline{P}}{\overline{P}} - \frac{\nabla \overline{T}}{\overline{T}}. \tag{2.35}$$

Substituting Eq. (2.35) into Eq. (2.19), we obtain the final expression for the effective pumping velocity of non-inertial particles as

$$\boxed{V^{\rm eff} = D_T \left(\frac{\nabla \overline{P}}{\overline{P}} - \frac{\nabla \overline{T}}{\overline{T}}\right).} \tag{2.36}$$

To understand different terms in Eq. (2.36), we compare the molecular and turbulent fluxes of particles (or gaseous admixtures). The equation for the number density of particles reads

$$\frac{\partial n}{\partial t} + \nabla \cdot (n\, \boldsymbol{U}) = -\nabla \cdot \boldsymbol{F}_{\rm M}, \tag{2.37}$$

where the molecular flux of particles $\boldsymbol{F}_{\rm M}$ is given by

$$\boxed{\boldsymbol{F}_M = -D^{(n)} \left(\nabla n + k_{\rm t} \frac{\nabla T}{T} + k_{\rm p} \frac{\nabla P}{P}\right),} \tag{2.38}$$

which comprises three terms: molecular diffusion ($\propto \nabla n$), molecular thermal diffusion for gases or thermophoresis for particles ($\propto k_{\rm t} \nabla T$) and molecular barodiffusion ($\propto k_{\rm p} \nabla P$), where $k_{\rm t}$ is the molecular thermal diffusion ratio and $k_{\rm p}$ is the molecular barodiffusion ratio (Landau and Lifshits, 1987). The phenomenon of molecular thermal diffusion in gases was predicted by D. Enskog[2] and S. Chapman[3] (Enskog, 1911, 1912; Chapman, 1912, 1916; Chapman and Dootson, 1917; Chapman and Cowling, 1970).

In turbulent flows, the turbulent flux of particles (or gaseous admixtures) can be rewritten as

$$\boldsymbol{F}_T \equiv \langle n' \boldsymbol{u} \rangle = -D_T \left(\nabla \overline{n} + \overline{n} \frac{\nabla \overline{T}}{\overline{T}} - \overline{n} \frac{\nabla \overline{P}}{\overline{P}}\right), \tag{2.39}$$

which is obtained by substitution of Eq. (2.36) to Eq. (2.20). Comparing the molecular flux of particles (2.38) and the turbulent flux of particles (2.39), we can

[2] David Enskog (1884–1947) was a Swedish mathematical physicist well-known for his works in the kinetic theory of gases.

[3] Sydney Chapman (1888–1970) was a British mathematician and geophysicist well-known for his works in the kinetic theory of gases and solar-terrestrial physics.

interpret the new additional turbulent fluxes as fluxes caused by the effects of turbulent thermal diffusion [$\propto k_T(\nabla\overline{T})/\overline{T}$] and turbulent barodiffusion [$\propto k_P(\nabla\overline{P})/\overline{P}$], where

$$\boxed{\boldsymbol{F}_T \equiv \langle n'\,\boldsymbol{u}\rangle = -D_T\left(\nabla\overline{n} + k_T\,\frac{\nabla\overline{T}}{\overline{T}} + k_P\,\frac{\nabla\overline{P}}{\overline{P}}\right),} \tag{2.40}$$

and $k_T = \overline{n}$ is the turbulent thermal diffusion ratio and $k_P = -\overline{n}$ is the turbulent barodiffusion ratio. These phenomena have been predicted by T. Elperin, N. Kleeorin and I. Rogachevskii (Elperin et al., 1996c, 1997c; see Section 2.6.1 for more references). The physics of these phenomena is discussed in more detail in Sections 2.2.5 and 2.4.4.

Note that the turbulent flux of the mass concentration $C_\mathrm{p} = n\,m_\mathrm{p}/\rho$ of particles or gaseous admixtures,

$$\boxed{\left\langle C'_\mathrm{p}\,\boldsymbol{u}\right\rangle = -D_T\,\nabla\overline{C}_\mathrm{p},} \tag{2.41}$$

does not contain the turbulent pumping term proportional to the effective pumping velocity $\boldsymbol{V}^\mathrm{eff}$. This is due to that Eq. (2.3) for the mass concentration C_p does not have the term proportional to $\nabla\cdot\boldsymbol{U}$, while Eq. (2.1) for the particle number density n does contain the term proportional to $\nabla\cdot\boldsymbol{U}$. Here the mass concentration $C_\mathrm{p} = \overline{C}_\mathrm{p} + C'_\mathrm{p}$ is characterized by the mean mass concentration $\overline{C}_\mathrm{p}$ and fluctuations C'_p.

2.2 Turbulent Transport of Particles: Quasi-linear Approach

In Section 2.1, we used the dimensional analysis to get a solution of Eq. (2.9) for particle number density fluctuations. In this section, we use a rigorous approach to solve Eq. (2.9), i.e., we apply the quasi-linear approach that is valid for small Péclet numbers.

2.2.1 Assumptions

We consider the following assumptions:

- We consider a low-Mach-number flow ($\mathrm{Ma} = U/c_\mathrm{s} \ll 1$), which means that fluid velocity is much less than the sound speed, and we use the anelastic approximation (2.6), where the mean fluid density depends on the coordinates. We consider a fluid flow with a zero mean fluid velocity.
- We apply the mean-field approach in which all quantities are decomposed into the mean and fluctuating parts, where the fluctuating parts have zero mean

values, i.e., we use the Reynolds rules and over-bars denote averaging over an ensemble of turbulent velocity field.

- We apply the point-particle approximation that implies that the size of particles is very small in comparison with all possible scales of fluid motions (we consider particles of the same size).
- We do not consider coagulation, condensation and evaporation of particles as well as chemical reactions, so that the total number of particles in a closed volume is conserved.
- We take into account only a one-way coupling, i.e., we take into account the effect of the turbulent velocity on the particle number density, and neglect the feedback effect of the particle number density on the turbulent fluid flow. We also consider non-inertial particles or gaseous admixtures. Therefore, the particle number density is a passive scalar field.
- We study the case when the nonlinear terms in the equation for the particle number density are much smaller than the diffusion terms, i.e., the Péclet number is small, $\mathrm{Pe}^{(n)} \equiv u_0 \ell_0 / D^{(n)} \ll 1$.

2.2.2 Quasi-linear Approach

To study particle number density fluctuations for small Péclet numbers we use a quasi-linear approach. We neglect the nonlinear term but keep the molecular diffusion term in Eq. (2.9). We rewrite Eq. (2.9) in a Fourier space, i.e., we use the Fourier transform in the $\boldsymbol{k} - \omega$ space:

$$n'(t,\boldsymbol{x}) = \int\int n'(\omega,\boldsymbol{k}) \exp(\mathrm{i}\,\boldsymbol{k}\cdot\boldsymbol{x} + \mathrm{i}\,\omega\,t)\, d\boldsymbol{k}\, d\omega, \tag{2.42}$$

$$u_i(t,\boldsymbol{x}) = \int\int u_i(\omega,\boldsymbol{k}) \exp(\mathrm{i}\,\boldsymbol{k}\cdot\boldsymbol{x} + \mathrm{i}\,\omega\,t)\, d\boldsymbol{k}\, d\omega. \tag{2.43}$$

We use the following identities:

$$\frac{\partial n'(t,\boldsymbol{x})}{\partial t} = \int\int (\mathrm{i}\,\omega)\, n'(\omega,\boldsymbol{k}) \exp(\mathrm{i}\,\boldsymbol{k}\cdot\boldsymbol{x} + \mathrm{i}\,\omega\,t)\, d\boldsymbol{k}\, d\omega, \tag{2.44}$$

$$\Delta n'(t,\boldsymbol{x}) = \int\int (-\boldsymbol{k}^2)\, n'(\omega,\boldsymbol{k}) \exp(\mathrm{i}\,\boldsymbol{k}\cdot\boldsymbol{x} + \mathrm{i}\,\omega\,t)\, d\boldsymbol{k}\, d\omega, \tag{2.45}$$

$$\nabla\cdot\boldsymbol{u} = \int\int (\mathrm{i}\,k_j)\, u_j(\omega,\boldsymbol{k}) \exp(\mathrm{i}\,\boldsymbol{k}\cdot\boldsymbol{x} + \mathrm{i}\,\omega\,t)\, d\boldsymbol{k}\, d\omega. \tag{2.46}$$

Substituting Eqs. (2.42)–(2.46) into the linearized equation (2.9), we obtain

$$\int\int \left[\left(D^{(n)}k^2 + \mathrm{i}\,\omega\right) n'(\omega,\boldsymbol{k}) + \left(\nabla_j \overline{n} + \mathrm{i}\,k_j\,\overline{n}\right) u_j(\omega,\boldsymbol{k})\right] \times \exp(\mathrm{i}\,\boldsymbol{k}\cdot\boldsymbol{x} + \mathrm{i}\,\omega\,t)\, d\boldsymbol{k}\, d\omega = 0. \tag{2.47}$$

Therefore, the solution of the linearized equation (2.9) in the Fourier space is given by

$$n'(\omega,\boldsymbol{k}) = -G_D(\omega,\boldsymbol{k})\left(\nabla_j \overline{n} + \mathrm{i}\, k_j\, \overline{n}\right) u_j(\omega,\boldsymbol{k}), \tag{2.48}$$

where the function $G_D(\omega,\boldsymbol{k}) = \left(D^{(n)}k^2 + \mathrm{i}\,\omega\right)^{-1}$.

Note that Eqs. (2.47) and (2.48) are not valid in a general case. However, we use here these equations because: (i) we take into account in the turbulent particle flux only the effects which are linear in $\nabla_j \overline{n}$ and neglect higher-order spatial derivatives of the mean particle number density, and (ii) we do not study in this section the compressibility effects of a turbulent velocity field in the evolution of the mean particle number density (i.e., we do not consider the finite Mach numbers effects). We will show in Section 2.2.6 that in this case we can use Eq. (2.48) to determine the turbulent particle flux for small Péclet numbers. The expression for particle number density fluctuations $n'(\omega,\boldsymbol{k})$ in a more general case in the Fourier space for small Péclet numbers is given by Eq. (2.74) in Section 2.2.6.

We determine the turbulent particle flux using the multi-scale approach (see Section 1.4), i.e.,

$$\begin{aligned}
\left\langle n'(t,\boldsymbol{x})\, u_j(t,\boldsymbol{x})\right\rangle &= \lim_{t_1\to t_2,\boldsymbol{x}\to\boldsymbol{y}} \left\langle n'(t_1,\boldsymbol{x})\, u_j(t_2,\boldsymbol{y})\right\rangle \\
&= \lim_{\tau\to 0,\boldsymbol{r}\to 0} \int\!\!\int \left\langle n'(\omega,\boldsymbol{k})\, u_j(-\omega,-\boldsymbol{k})\right\rangle \exp[\mathrm{i}\,\boldsymbol{k}\cdot\boldsymbol{r} + \mathrm{i}\,\omega\,\tau]\, d\omega\, d\boldsymbol{k} \\
&= \int\!\!\int \left\langle n'(\omega,\boldsymbol{k})\, u_j(-\omega,-\boldsymbol{k})\right\rangle d\omega\, d\boldsymbol{k},
\end{aligned} \tag{2.49}$$

where $\boldsymbol{r} = \boldsymbol{x} - \boldsymbol{y}$, $\tau = t_1 - t_2$, and we consider a homogeneous and stationary (in statistical sense) turbulence. The latter implies that the correlation function $\langle n'(\omega,\boldsymbol{k})\, u_j(-\omega,-\boldsymbol{k})\rangle$ is independent of the large-scale variables. We also assume that there exists a separation of scales, i.e., the maximum scale of random motions ℓ_0 is much smaller than the characteristic scales of inhomogeneities of the mean particle number density. In a similar way, we determine the level of particle number density fluctuations $\langle n'^2\rangle$:

$$\begin{aligned}
\left\langle n'(t,\boldsymbol{x})\, n'(t,\boldsymbol{x})\right\rangle &= \lim_{t_1\to t_2,\boldsymbol{x}\to\boldsymbol{y}} \left\langle n'(t_1,\boldsymbol{x})\, n'(t_2,\boldsymbol{y})\right\rangle \\
&= \lim_{\tau\to 0,\boldsymbol{r}\to 0} \int\!\!\int \left\langle n'(\omega,\boldsymbol{k})\, n'(-\omega,-\boldsymbol{k})\right\rangle \exp[\mathrm{i}\,\boldsymbol{k}\cdot\boldsymbol{r} + \mathrm{i}\,\omega\,\tau]\, d\omega\, d\boldsymbol{k} \\
&= \int\!\!\int \left\langle n'(\omega,\boldsymbol{k})\, n'(-\omega,-\boldsymbol{k})\right\rangle d\omega\, d\boldsymbol{k}.
\end{aligned} \tag{2.50}$$

Substituting Eq. (2.48) into Eqs. (2.49) and (2.50), we determine the particle turbulent flux $F_i^{(n)} = \langle u_i n' \rangle$ and the level of particle number density fluctuations $\langle n'^2 \rangle$:

$$F_i^{(n)} = -\int\int \langle u_i(\omega, \boldsymbol{k})\, u_j(-\omega, -\boldsymbol{k})\rangle\, G_D^* \left[\nabla_j \overline{n} - \mathrm{i}\, k_j \overline{n}\right] d\omega\, d\boldsymbol{k}, \tag{2.51}$$

$$\langle n'^2 \rangle = \int\int \langle u_i(\omega, \boldsymbol{k})\, u_j(-\omega, -\boldsymbol{k})\rangle\, G_D G_D^* \times \left\{(\nabla_i \overline{n})\; (\nabla_j \overline{n}) + \left[k_i k_j + \frac{\mathrm{i}}{2}(k_i \nabla_j - k_j \nabla_i)\right] \overline{n}^2\right\} d\omega\, d\boldsymbol{k}. \tag{2.52}$$

2.2.3 Model of Isotropic and Homogeneous Turbulence in Anelastic Approximation

We use the simple model for the second moments, $\langle u_i(\omega, \boldsymbol{k})\, u_j(-\omega, -\boldsymbol{k})\rangle$, of an isotropic and homogeneous turbulence in the anelastic approximation in a Fourier space:

$$\boxed{\langle u_i(\omega, \boldsymbol{k})\, u_j(-\omega, -\boldsymbol{k})\rangle = \frac{\langle \boldsymbol{u}^2 \rangle\, E(k)\, \Phi(\omega)}{8\pi k^2} \left[\delta_{ij} - \frac{k_i k_j}{k^2} + \frac{\mathrm{i}}{k^2}\left(\lambda_i k_j - \lambda_j k_i\right)\right],} \tag{2.53}$$

where δ_{ij} is the Kronecker unit tensor, the spectrum function $E(k)$ for the turbulent kinetic energy is $E(k) = (q - 1)\, k_0^{-1}\, (k/k_0)^{-q}$ with the exponent q varying in the interval $1 < q < 3$, the interval in the wave numbers is $k_0 \le k \le k_d$, the wave number $k_0 = 1/\ell_0$, the length ℓ_0 is the maximum scale of random motions and $k_d \gg k_0$. The frequency function $\Phi(\omega)$ is chosen in the form of the Lorentz profile: $\Phi(\omega) = [\pi \tau_0\, (\omega^2 + \tau_0^{-2})]^{-1}$, where τ_0 is the correlation time of a random velocity field. This model for the frequency function corresponds to the following non-instantaneous correlation function: $\langle u_i(t) u_j(t + \tau)\rangle \propto \exp(-\tau/\tau_0)$. For small Péclet numbers, $\tau_0 \gg (D^{(n)} k^2)^{-1}$. The velocity field for very small Mach numbers satisfies the continuity equation in the anelastic approximation

$$\mathrm{div}\, \boldsymbol{u} = \boldsymbol{u} \cdot \boldsymbol{\lambda}, \tag{2.54}$$

where $\boldsymbol{\lambda} = -\boldsymbol{\nabla}\overline{\rho}/\overline{\rho}$ and $\overline{\rho}$ is the fluid density.

Equation (2.53) is derived using symmetry arguments. The incompressible part for the second-order moment of velocity fluctuations in a Fourier space, $\langle u_i(\omega, \boldsymbol{k})\, u_j(-\omega, -\boldsymbol{k})\rangle$, was derived in Section 1.5.3. Density stratification in anelastic approximation is introduced in the model by perturbations using a small parameter $\ell_0/H_\rho \ll 1$, where $H_\rho = 1/|\boldsymbol{\lambda}|$ is the density stratification length. We take into account only linear in $\boldsymbol{\lambda}$ terms in Eq. (2.53), i.e., the model (2.53) has two additional tensors, $\lambda_i k_j$ and $\lambda_j k_i$, caused by the density stratification. Thus the model for the second-order moment of velocity fluctuations in a Fourier space is

$$\langle u_i(\omega, \boldsymbol{k})\, u_j(-\omega, -\boldsymbol{k})\rangle = \frac{\langle \boldsymbol{u}^2\rangle\, E(k)}{8\pi k^2}\left[\delta_{ij} - \frac{k_i\, k_j}{k^2} + A_3\, \lambda_i\, k_j + A_4 \lambda_j\, k_i\right]\Phi(\omega), \tag{2.55}$$

To determine unknown coefficients A_3 and A_4 we use the continuity equation in the anelastic approximation $\operatorname{div} \boldsymbol{u} - u_i\, \lambda_i = 0$. In the $\boldsymbol{k}$ space, this condition reads

$$\int\int (\mathrm{i}\, k_i - \lambda_i)\, u_i(\omega, \boldsymbol{k}) \exp(\mathrm{i}\, \boldsymbol{k} \cdot \boldsymbol{x} + \mathrm{i}\, \omega\, t)\, d\boldsymbol{k}\, d\omega = 0. \tag{2.56}$$

This implies that $(\mathrm{i}\, k_i - \lambda_i)\, u_i(\omega, \boldsymbol{k}) = 0$. Multiplying this equation by $u_j(-\omega, -\boldsymbol{k})$ and averaging over an ensemble, we arrive at the following condition:

$$(\mathrm{i}\, k_i - \lambda_i)\, \langle u_i(\omega, \boldsymbol{k})\, u_j(-\omega, -\boldsymbol{k})\rangle = 0. \tag{2.57}$$

Substituting the tensor $\langle u_i(\omega, \boldsymbol{k})\, u_j(-\omega, -\boldsymbol{k})\rangle$ determined by Eq. (2.55) into Eq. (2.57) we obtain the following equation:

$$\lambda_j \left(\mathrm{i}\, k^2 A_4 - 1\right) + \lambda_i \frac{k_i\, k_j}{k^2}\left(\mathrm{i}\, k^2 A_3 + 1\right) = 0. \tag{2.58}$$

Equation (2.58) is valid for an arbitrary vector $\boldsymbol{\lambda}$, which implies that $\mathrm{i}\, k^2 A_4 - 1 = 0$ and $\mathrm{i}\, k^2 A_3 + 1 = 0$. These conditions yield $A_3 = -A_4 = \mathrm{i}/k^2$. Therefore, the model (2.53) correctly describes the density-stratified turbulence in the anelastic approximation when $\ell_0 \ll H_\rho$.

2.2.4 Turbulent Diffusion, Turbulent Thermal Diffusion and Level of Fluctuations

Substituting Eq. (2.53) into Eq. (2.51) and (2.52), we determine the turbulent heat flux $F_i^{(n)} = \langle u_i\, n'\rangle$ and the level of particle number density fluctuations $\langle n'^2\rangle$ as

$$\begin{aligned} F_i^{(n)} &= -\frac{\langle \boldsymbol{u}^2\rangle}{8\pi}\int_{k_0}^{k_\nu} E(k)\, dk \int_0^{2\pi} d\varphi \int_0^{\pi} \sin\vartheta\, d\vartheta \left[\delta_{ij} - \frac{k_i\, k_j}{k^2}\right. \\ &\quad \left. + \frac{\mathrm{i}}{k^2}\left(\lambda_i\, k_j - \lambda_j\, k_i\right)\right]\left[\nabla_j \overline{n} - \mathrm{i}\, k_j\, \overline{n}\right] \int G_D^* \Phi(\omega)\, d\omega \\ &= -\frac{\langle \boldsymbol{u}^2\rangle}{3 D^{(n)}}\left(\nabla_i \overline{n} + \lambda_i\, \overline{n}\right) \int_{k_0}^{k_\nu} \frac{E(k)}{k^2}\, dk, \end{aligned} \tag{2.59}$$

$$\langle n'^2\rangle = \frac{\langle \boldsymbol{u}^2\rangle}{8\pi}\int_{k_0}^{k_\nu} E(k)\, dk \int_0^{2\pi} d\varphi \int_0^{\pi} \sin\vartheta\, d\vartheta \left[\delta_{ij} - \frac{k_i\, k_j}{k^2} + \frac{\mathrm{i}}{k^2}\left(\lambda_i\, k_j - \lambda_j\, k_i\right)\right]$$

$$\left\{(\nabla_i \overline{n})\,(\nabla_j \overline{n}) + \left[k_i k_j + \frac{\mathrm{i}}{2}(k_i \nabla_j - k_j \nabla_i)\right]\overline{n}^2\right\} \int G_D\, G_D^*\, \Phi(\omega)\, d\omega$$
$$= \frac{\langle \boldsymbol{u}^2 \rangle}{3\left(D^{(n)}\right)^2}\, \left[(\nabla \overline{n})^2 + (\boldsymbol{\lambda}\cdot\nabla)\,\overline{n}^2\right] \int_{k_0}^{k_\nu} \frac{E(k)}{k^4}\, dk. \tag{2.60}$$

For the integration over ω and $\boldsymbol{k}$, we use the integrals given by Eqs. (1.133)–(1.137). After integration over ω and in the $\boldsymbol{k}$ space in Eqs. (2.59) and (2.60), we obtain formulae for the turbulent particle flux $F_i^{(n)}$ and the level of particle number density fluctuations $\langle n'^2 \rangle$:

$$\boxed{\langle n' \boldsymbol{u} \rangle = \boldsymbol{V}^{\mathrm{eff}}\,\overline{n} - D_T\, \nabla \overline{n},} \tag{2.61}$$

$$\boxed{\langle n'^2 \rangle = \frac{q-1}{3(q+3)}\, \left(\mathrm{Pe}^{(n)}\right)^2\, \ell_0^2\, \left[(\nabla \overline{n})^2 + (\boldsymbol{\lambda}\cdot\nabla)\overline{n}^2\right],} \tag{2.62}$$

where

$$\boxed{D_T = \frac{q-1}{3(q+1)}\, u_0\, \ell_0\, \mathrm{Pe}^{(n)},} \tag{2.63}$$

$$\boxed{\boldsymbol{V}^{\mathrm{eff}} = -D_T \boldsymbol{\lambda} = D_T\, \frac{\nabla \overline{\rho}}{\overline{\rho}},} \tag{2.64}$$

$u_0 = \sqrt{\langle \boldsymbol{u}^2 \rangle}$. Let us compare these results with those obtained by dimensional analysis. This comparison shows that Eqs. (2.61)–(2.64) obtained by the quasi-linear approach are in agreement with Eqs. (2.26)–(2.28) and (2.32) obtained by the dimensional analysis for small Péclet numbers.

2.2.5 Physics of Turbulent Thermal Diffusion and Turbulent Barodiffusion

The effective pumping velocity $\boldsymbol{V}^{\mathrm{eff}}$ describes the phenomena of turbulent thermal diffusion and turbulent barodiffusion. Below we discuss the physics of these effects. Substituting the particle turbulent flux $\langle n' \boldsymbol{u} \rangle = \boldsymbol{V}^{\mathrm{eff}}\,\overline{n} - D_T \nabla \overline{n}$ into Eq. (2.8) for the particle mean number density $\overline{n}$, we obtain

$$\boxed{\frac{\partial \overline{n}}{\partial t} + \nabla \cdot \left[\overline{n}\, \boldsymbol{V}^{\mathrm{eff}} - \left(D^{(n)} + D_T\right)\, \nabla \overline{n}\right] = 0.} \tag{2.65}$$

Table 2.1 *Mechanism of particle transport $\langle n'\boldsymbol{u}\rangle$ for non-inertial particles along the direction of the mean fluid density gradient*

$\Longrightarrow$	$\nabla\overline{\rho}$
a	b
$u_x > 0$	$u_x < 0$
$\nabla \cdot \boldsymbol{u} < 0$	$\nabla \cdot \boldsymbol{u} > 0$
$n' > 0$	$n' < 0$
$n' u_x > 0$	$n' u_x > 0$
$\Longrightarrow$	$\langle n'\boldsymbol{u}\rangle$

The steady-state solution of Eq. (2.65) for the mean number density of non-inertial particles is given by

$$\boxed{\frac{\overline{n}}{\overline{n}_*} = \left(\frac{\overline{\rho}}{\overline{\rho}_*}\right)^{\frac{D_T}{D^{(n)}+D_T}}}, \tag{2.66}$$

where the subscripts $*$ represent the values of the mean particle number density $\overline{n}_*$ and the mean fluid density $\overline{\rho}_*$ at the boundary. The solution (2.66) is obtained for the boundary condition of a zero total (molecular + turbulent) particle flux at the boundary. Equation (2.66) implies that small particles are accumulated in the vicinity of the maximum of the fluid density.

Indeed, the non-diffusive flux of particles, $\overline{n}\,\boldsymbol{V}^{\rm eff}$, is directed toward the maximum of the fluid density [see Eq. (2.64)]. The physics of the accumulation of non-inertial particles in the vicinity of the maximum of the mean fluid density can be explained as follows (see Table 2.1). Let us assume that the mean fluid density $\overline{\rho}_2$ at point 2 is larger than the mean fluid density $\overline{\rho}_1$ at point 1. Consider two small control volumes "a" and "b" located between these two points, and let the direction of the local turbulent velocity in volume "a" at some instant be the same as the direction of the mean fluid density gradient $\nabla\,\overline{\rho}$ (i.e., along the x-axis toward point 2). Let the local turbulent velocity in volume "b" at this instant be directed opposite to the mean fluid density gradient (i.e., toward point 1).

In a fluid flow with a non-zero mean fluid density gradient, one of the sources of particle number density fluctuations, $n' \propto -\tau_0\,\overline{n}\,(\nabla \cdot \boldsymbol{u})$, is caused by a non-zero $\nabla \cdot \boldsymbol{u} \approx -\boldsymbol{u} \cdot \nabla \ln \rho \neq 0$ [see the second term on the right-hand side of Eq. (2.11)].

Since fluctuations of the fluid velocity $\boldsymbol{u}$ are positive in volume "a" and negative in volume "b", we have the negative divergence of the fluid velocity, $\nabla \cdot \boldsymbol{u} < 0$, in volume "a", and the positive divergence of the fluid velocity, $\nabla \cdot \boldsymbol{u} > 0$, in volume "b". Therefore, fluctuations of the particle number density $n' \propto -\tau_0 \overline{n} \, (\nabla \cdot \boldsymbol{u})$ are positive in volume "a" and negative in volume "b". However, the flux of particles $n' u_x$ is positive in volume "a" (i.e., it is directed toward point 2), and it is also positive in volume "b" (because both fluctuations of fluid velocity and number density of particles are negative in volume "b"). Therefore, the mean flux of particles $\langle n' \boldsymbol{u} \rangle$ is directed, as is the mean fluid density gradient $\nabla \overline{\rho}$, toward point 2. This results in formation large-scale heterogeneous structures of non-inertial particles in regions with a mean fluid density maximum. When the gradient $\nabla \overline{P}$ of the mean fluid pressure vanishes, $(\nabla \overline{\rho})/\overline{\rho} = -(\nabla \overline{T})/\overline{T}$. This implies that particles are accumulated at the vicinity of the minimum of the fluid temperature. This phenomenon is called turbulent thermal diffusion (Elperin et al., 1997c).

When the gradient $\nabla \overline{P}$ of the mean fluid pressure does not vanish, $(\nabla \overline{\rho})/\overline{\rho} = (\nabla \overline{P})/\overline{P} - (\nabla \overline{T})/\overline{T}$. The latter equation follows from the equation of state for the perfect gas. This implies that the effective pumping velocity has two contributions,

$$\boxed{\boldsymbol{V}^{\rm eff} = D_T \, \frac{\nabla \overline{P}}{\overline{P}} - D_T \, \frac{\nabla \overline{T}}{\overline{T}}.} \tag{2.67}$$

The second term on the right-hand side of Eq. (2.67) determines turbulent thermal diffusion, while the first term in Eq. (2.67) describes turbulent barodiffusion, which causes the accumulation of particles in the vicinity of the maximum of the fluid pressure. The physics of this effect is similar to that of turbulent thermal diffusion, and it is related to the non-diffusive flux of particles, $\overline{n} \, \boldsymbol{V}^{\rm eff}$, in the direction of the maximum of the fluid density.

2.2.6 Rigorous Derivation of Turbulent Particle Flux for Small Péclet Numbers

In Section 2.2.2, we used the approximate solution (2.48) of Eq. (2.9) for particle number density fluctuations $n'(\omega, \boldsymbol{k})$ in the Fourier space to derive an equation for the turbulent particle flux for small Péclet numbers. In this section, we give a rigorous derivation of the turbulent particle flux. To this end we use the following identities:

$$[(\boldsymbol{u} \cdot \nabla)\overline{n}]_{\boldsymbol{k}} = {\rm i} \int Q_p \, \overline{n}(\boldsymbol{Q}) \, u_p(\boldsymbol{k} - \boldsymbol{Q}) \, d\boldsymbol{Q}, \tag{2.68}$$

$$[\overline{n} \, (\nabla \cdot \boldsymbol{u})]_{\boldsymbol{k}} = {\rm i} \int \big(k_p - Q_p\big) \, \overline{n}(\boldsymbol{Q}) \, u_p(\boldsymbol{k} - \boldsymbol{Q}) \, d\boldsymbol{Q}, \tag{2.69}$$

where $[\ldots]_k$ denotes the Fourier transform. To derive Eq. (2.68), we rewrite the source term $(\boldsymbol{u}\cdot\nabla)\overline{n}$, using the Fourier transform in spatial variables for both functions $\boldsymbol{u}(t,\boldsymbol{x})$ and $\overline{n}(t,\boldsymbol{x})$:

$$(\boldsymbol{u}\cdot\nabla)\overline{n} = \int u_p(\boldsymbol{k}')\,\exp(\mathrm{i}\,\boldsymbol{k}'\cdot\boldsymbol{x})\,d\boldsymbol{k}' \int \mathrm{i}\,Q_p\overline{n}(\boldsymbol{Q})\,\exp(\mathrm{i}\,\boldsymbol{Q}\cdot\boldsymbol{x})\,d\boldsymbol{Q}. \quad (2.70)$$

For brevity of notation, we omit hereafter the arguments t and $\boldsymbol{x}$ in the functions $\overline{n}(t,\boldsymbol{x})$ and $\boldsymbol{u}(t,\boldsymbol{x})$, and the variable t in the functions $\overline{n}(t,\boldsymbol{Q})$ and $u_i(t,\boldsymbol{k})$. We introduce a new variable $\boldsymbol{k}=\boldsymbol{k}'+\boldsymbol{Q}$, so that after simple manipulations we obtain

$$\begin{aligned}(\boldsymbol{u}\cdot\nabla)\overline{n} &= \int \left(\mathrm{i}\int Q_p\,\overline{n}(\boldsymbol{Q})\,u_p(\boldsymbol{k}-\boldsymbol{Q})\,d\boldsymbol{Q}\right)\exp(\mathrm{i}\,\boldsymbol{k}\cdot\boldsymbol{x})\,d\boldsymbol{k} \\ &\equiv \int [(\boldsymbol{u}\cdot\nabla)\overline{n}]_k \exp(\mathrm{i}\,\boldsymbol{k}\cdot\boldsymbol{x})\,d\boldsymbol{k}, \end{aligned} \quad (2.71)$$

which yields Eq. (2.68). A similar derivation is performed for Eq. (2.69).

Substituting Eqs. (2.42)–(2.46) and (2.68)–(2.69) into linearized equation (2.9), we obtain

$$\begin{aligned}\int\!\!\int &\left[\left(D^{(n)}k^2+\mathrm{i}\,\omega\right)n'(\omega,\boldsymbol{k})+\mathrm{i}\,k_i\int \overline{n}(\boldsymbol{Q})\,u_i(\omega,\boldsymbol{k}-\boldsymbol{Q})\,d\boldsymbol{Q}\right] \\ &\times \exp(\mathrm{i}\boldsymbol{k}\cdot\boldsymbol{x}+\mathrm{i}\,\omega\,t)\,d\boldsymbol{k}\,d\omega = 0. \end{aligned} \quad (2.72)$$

Equation (2.72) implies that

$$\left(D^{(n)}k^2+\mathrm{i}\,\omega\right)n'(\omega,\boldsymbol{k})+\mathrm{i}\,k_i\int \overline{n}(\boldsymbol{Q})\,u_i(\omega,\boldsymbol{k}-\boldsymbol{Q})\,d\boldsymbol{Q}=0, \quad (2.73)$$

which yields the expression for particle number density fluctuations $n'(\omega,\boldsymbol{k})$ in the Fourier space at small Péclet numbers:

$$n'(\omega,\boldsymbol{k}) = -\mathrm{i}\;k_i\,G_D(\omega,\boldsymbol{k})\int \overline{n}(\boldsymbol{Q})\,u_i(\omega,\boldsymbol{k}-\boldsymbol{Q})\,d\boldsymbol{Q}. \quad (2.74)$$

We determine the turbulent particle flux using the multi-scale approach, i.e.,

$$\begin{aligned}\left\langle n'(t,\boldsymbol{x})\,u_j(t,\boldsymbol{x})\right\rangle &= \lim_{t_1\to t_2,\boldsymbol{x}\to\boldsymbol{y}} \left\langle n'(t_1,\boldsymbol{x})\,u_j(t_2,\boldsymbol{y})\right\rangle \\ &= \int\!\!\int \left\langle n'(\omega,\boldsymbol{k})\,u_j(-\omega,-\boldsymbol{k})\right\rangle d\omega\,d\boldsymbol{k} \end{aligned} \quad (2.75)$$

[see Eq. (2.49)]. Substituting Eq. (2.74) into Eqs. (2.75), we rewrite the turbulent particle flux $\left\langle n'(t,\boldsymbol{x})\,u_j(t,\boldsymbol{x})\right\rangle$ as

$$\begin{aligned}\left\langle n'(t,\boldsymbol{x})\,u_j(t,\boldsymbol{x})\right\rangle = -\mathrm{i}\int (k_i+K_i/2)\,&\left\langle u_i(\boldsymbol{k}+\boldsymbol{K}/2-\boldsymbol{Q})u_j(-\boldsymbol{k}+\boldsymbol{K}/2)\right\rangle \\ &\times G_D(\boldsymbol{k}+\boldsymbol{K}/2)\,\overline{n}(\boldsymbol{Q})\exp(\mathrm{i}\,\boldsymbol{K}\cdot\boldsymbol{R})\,d\boldsymbol{K}\,d\boldsymbol{Q}, \end{aligned} \quad (2.76)$$

where the functions G_D and u_i depend also on ω, and $\overline{n}$ depend on t as well. To simplify the notation, we do not show these dependencies here. We use new variables:

$$\tilde{\boldsymbol{k}} = (\tilde{\boldsymbol{k}}_1 - \tilde{\boldsymbol{k}}_2)/2 = \boldsymbol{k} - \boldsymbol{Q}/2, \;\; \tilde{\boldsymbol{K}} = \tilde{\boldsymbol{k}}_1 + \tilde{\boldsymbol{k}}_2 = \boldsymbol{K} - \boldsymbol{Q}, \tag{2.77}$$

where

$$\tilde{\boldsymbol{k}}_1 = \boldsymbol{k} + \boldsymbol{K}/2 - \boldsymbol{Q}, \quad \tilde{\boldsymbol{k}}_2 = -\boldsymbol{k} + \boldsymbol{K}/2. \tag{2.78}$$

Since $|\boldsymbol{Q}| \ll |\boldsymbol{k}|$ and $|\boldsymbol{K}| \ll |\boldsymbol{k}|$, we use the Taylor expansion of the functions:

$$f_{ij}(\boldsymbol{k} - \boldsymbol{Q}/2, \boldsymbol{K} - \boldsymbol{Q}) = f_{ij}(\boldsymbol{k}, \boldsymbol{K} - \boldsymbol{Q}) - \frac{1}{2}\frac{\partial f_{ij}}{\partial k_s} Q_s + O(\boldsymbol{Q}^2), \tag{2.79}$$

$$G_D(\boldsymbol{k} + \boldsymbol{K}/2) = G_D(\boldsymbol{k})\left[1 - D^{(n)}(\boldsymbol{k} \cdot \boldsymbol{K}) G_D(\boldsymbol{k})\right] + O(\boldsymbol{K}^2), \tag{2.80}$$

where $f_{ij}(\tilde{\boldsymbol{k}}_1, \tilde{\boldsymbol{k}}_2) \equiv \left\langle u_i(\tilde{\boldsymbol{k}}_1) u_j(\tilde{\boldsymbol{k}}_2)\right\rangle = f_{ij}(\boldsymbol{k} - \boldsymbol{Q}/2, \boldsymbol{K} - \boldsymbol{Q})$. Using Eqs. (2.76)–(2.80) and considering a homogeneous turbulence, we arrive at an expression for the turbulent flux of particles for small Péclet numbers:

$$\langle n' \, u_j \rangle = -\int G_D \left[\mathrm{i} k_i \, f_{ij} \, \overline{n} + \frac{1}{2}\left(f_{mj} - k_i \frac{\partial f_{ij}}{\partial k_m} - 2D^{(n)} k^2 G_D \, k_{im} \, f_{ij} \right) \nabla_m \overline{n} \right] d\boldsymbol{k} \, d\omega, \tag{2.81}$$

where $G_D \equiv G_D(\boldsymbol{k}, \omega)$ and $f_{ij} = f_{ij}(\boldsymbol{k}, \omega, \boldsymbol{R}, t)$. Since we consider a one-way coupling, the correlation function f_{ij} in Eq. (2.81) should be replaced by $f_{ij}^{(0)}$ for the background random flow with zero turbulent flux of particles. Substituting the model of turbulence (2.53) for $f_{ij}^{(0)}$ in Eq. (2.81), we obtain the turbulent particle flux $\boldsymbol{F}^{(n)}$, the turbulent diffusion coefficient D_T and the particle effective pumping velocity $\boldsymbol{V}^{\rm eff}$ which coincides with Eqs. (2.61)–(2.64) [see Section 2.2.6]. We remind that Eqs. (2.61)–(2.64) have been derived using an approximate solution (2.48) for particle number density fluctuations.

2.3 Turbulent Transport of Particles: Spectral Tau Approach

In Sections 2.1 and 2.2, we used the dimensional analysis and the quasi-linear approach to obtain a solution for Eq. (2.9) for particle number density fluctuations. In this section, we apply the tau approach (the relaxation approach) that is valid for large Péclet and Reynolds numbers.

We use the equation for particle number density fluctuations by subtracting the mean-field equation from the equation for the total particle number density:

$$\frac{\partial n'}{\partial t} = -\mathcal{Q}^{(n)} - (\boldsymbol{u} \cdot \nabla)\overline{n} - \overline{n}(\nabla \cdot \boldsymbol{u}), \tag{2.82}$$

where $\mathcal{Q}^{(n)} = \nabla \cdot \left(n' \boldsymbol{u} - \langle n' \boldsymbol{u}\rangle\right) - D^{(n)} \Delta n'$.

2.3.1 Equations for Second Moments

Using Eq. (2.82) written in a Fourier space we derive an equation for the two-point second-order correlation functions $F_j^{(n)}(t, \boldsymbol{k}) = \langle u_j(t, \boldsymbol{k})\, n'(t, -\boldsymbol{k})\rangle$ and $E^{(n)}(t, \boldsymbol{k}) = \langle n'(t, \boldsymbol{k})\, n'(t, -\boldsymbol{k})\rangle$:

$$\frac{\partial F_j^{(n)}(\boldsymbol{k})}{\partial t} = -\left(\nabla_i \overline{n} - \mathrm{i}\, k_i\, \overline{n}\right)\, f_{ij}(-\boldsymbol{k}) + \hat{\mathcal{M}}^{(n)} F_j^{(III)}(\boldsymbol{k}), \tag{2.83}$$

$$\frac{\partial E^{(n)}(\boldsymbol{k})}{\partial t} = -\left[F_j^{(n)}(\boldsymbol{k}) + F_j^{(n)}(-\boldsymbol{k})\right] \nabla_j \overline{n} - \left[F_j^{(n)}(\boldsymbol{k}) - F_j^{(n)}(-\boldsymbol{k})\right] \mathrm{i}\, k_j\, \overline{n} + \hat{\mathcal{M}}^{(n)} E_n^{(III)}(\boldsymbol{k}), \tag{2.84}$$

where

$$f_{ij}(\boldsymbol{k}) = \langle u_i(t, \boldsymbol{k})\, u_j(t, -\boldsymbol{k})\rangle, \tag{2.85}$$

and

$$\hat{\mathcal{M}}^{(n)} F_i^{(III)}(\boldsymbol{k}) = -\langle u_i(t, \boldsymbol{k})\, \mathcal{Q}^{(n)}(t, -\boldsymbol{k})\rangle + \left\langle \frac{\partial u_i(t, \boldsymbol{k})}{\partial t}\, n'(t, -\boldsymbol{k})\right\rangle, \tag{2.86}$$

$$\hat{\mathcal{M}}^{(n)} E_n^{(III)}(\boldsymbol{k}) = -\langle n'(t, \boldsymbol{k})\, \mathcal{Q}^{(n)}(t, -\boldsymbol{k})\rangle - \langle \mathcal{Q}^{(n)}(t, \boldsymbol{k})\, n'(t, -\boldsymbol{k})\rangle \tag{2.87}$$

are the third-order moments appearing due to the nonlinear terms $\mathcal{Q}^{(n)}$ and the nonlinear Navier-Stokes equation. They also include small molecular diffusion and viscosity terms. For the derivation of Eq. (2.84) we take into account that

$$\frac{\partial E^{(n)}(t, \boldsymbol{k})}{\partial t} = \left\langle \frac{\partial n'(t, \boldsymbol{k})}{\partial t}\, n'(t, -\boldsymbol{k})\right\rangle + \left\langle n'(t, \boldsymbol{k})\, \frac{\partial n'(t, -\boldsymbol{k})}{\partial t}\right\rangle. \tag{2.88}$$

We use the spectral τ approach which postulates that the deviations of the third-moment terms, $\hat{\mathcal{M}} F^{(III)}(\boldsymbol{k})$, from the contributions to these terms afforded by the background turbulence, $\hat{\mathcal{M}} F^{(III,0)}(\boldsymbol{k})$, can be expressed through the similar deviations of the second moments, $F^{(II)}(\boldsymbol{k}) - F^{(II,0)}(\boldsymbol{k})$,

$$\boxed{\hat{\mathcal{M}} F^{(III)}(\boldsymbol{k}) - \hat{\mathcal{M}} F^{(III,0)}(\boldsymbol{k}) = -\frac{1}{\tau_r(k)}\left[F^{(II)}(\boldsymbol{k}) - F^{(II,0)}(\boldsymbol{k})\right]} \tag{2.89}$$

(Orszag, 1970; Pouquet et al., 1976; Kleeorin et al., 1990), where $\tau_r(k)$ is the scale-dependent relaxation time, which can be identified with the correlation time $\tau(k)$ of the turbulent velocity field for large Reynolds and Péclet numbers. An equation similar to Eq. (2.89) is also applied for the third-moment terms $\hat{\mathcal{M}} E_n^{(III)}(\boldsymbol{k})$. The functions with the superscript (0) correspond to the background turbulence with a zero turbulent particle flux and a zero level of particle number density fluctuations. Consequently, Eq. (2.89) reduces to $\hat{\mathcal{M}}^{(n)} F_i^{(III)}(\boldsymbol{k}) = -F_i^{(n)}(\boldsymbol{k})/\tau(k)$ and for $\hat{\mathcal{M}}^{(n)} E_n^{(III)}(\boldsymbol{k}) = -E^{(n)}(\boldsymbol{k})/\tau(k)$.

2.3.2 Turbulent Flux of Particles and Level of Fluctuations

We assume that the characteristic time of variation of the second moments $F_i^{(n)}(\boldsymbol{k})$ and $E^{(n)}(\boldsymbol{k})$ are substantially larger than the correlation time $\tau(k)$ for all turbulence scales. This allows us to use a steady-state solution of Eqs. (2.83) and (2.84). Applying the tau approximation and using the steady-state solution of Eqs. (2.83) and (2.84), we obtain the following formulae for the turbulent flux of particles, $F_j^{(n)}$, and the function $E^{(n)}$:

$$F_j^{(n)} = \lim_{\boldsymbol{x}\to\boldsymbol{y}} \langle u_j(t,\boldsymbol{x})\, n'(t,-\boldsymbol{y})\rangle = -\int \tau(k)\; (\nabla_i \overline{n} - \mathrm{i}\, k_i\, \overline{n})\; f_{ij}(-\boldsymbol{k})\, d\boldsymbol{k}, \tag{2.90}$$

$$E^{(n)} = \lim_{\boldsymbol{x}\to\boldsymbol{y}} \langle n'(t,\boldsymbol{x})\, n'(t,-\boldsymbol{y})\rangle = -\int \tau(k) \Big\{ \Big[F_j^{(n)}(\boldsymbol{k}) + F_j^{(n)}(-\boldsymbol{k})\Big]\; \nabla_j \overline{n} + \Big[F_j^{(n)}(\boldsymbol{k}) - F_j^{(n)}(-\boldsymbol{k})\Big]\, (\mathrm{i}\, k_j\, \overline{n}) \Big\}\, d\boldsymbol{k}. \tag{2.91}$$

We take into account that $F_j^{(n)} = \int F_j^{(n)}(\boldsymbol{k})\, d\boldsymbol{k}$, which implies that

$$F_j^{(n)}(\boldsymbol{k}) = -\tau(k)\; (\nabla_i \overline{n} - \mathrm{i}\, k_i\, \overline{n})\; f_{ij}(-\boldsymbol{k}), \tag{2.92}$$

[see Eq. (2.90)]. Substituting Eq. (2.92) into Eq. (2.91), we obtain

$$E^{(n)} = \int \tau^2(k) \Big\{ \big[f_{ij}(\boldsymbol{k}) + f_{ij}(-\boldsymbol{k})\big] \big[(\nabla_i \overline{n})\,(\nabla_j \overline{n}) + k_i k_j\, \overline{n}^2\big] + \frac{i}{2} \big[f_{ij}(\boldsymbol{k}) - f_{ij}(-\boldsymbol{k})\big]\; \big[(k_i \nabla_j - k_j \nabla_i)\, \overline{n}^2\big] \Big\}\, d\boldsymbol{k}. \tag{2.93}$$

We use the following model for the second moments of the turbulent velocity field $\langle u_i(\boldsymbol{k})\, u_j(-\boldsymbol{k})\rangle$ of an isotropic and homogeneous turbulence in anelastic approximation in a Fourier space:

$$\boxed{f_{ij}^{(0)}(\boldsymbol{k}) \equiv \langle u_i(\boldsymbol{k})\, u_j(-\boldsymbol{k})\rangle = \frac{\langle \boldsymbol{u}^2\rangle\, E(k)}{8\pi k^2} \left(\delta_{ij} - \frac{k_i\, k_j}{k^2} + \frac{\mathrm{i}}{k^2}\, (\lambda_i\, k_j - \lambda_j\, k_i)\right),} \tag{2.94}$$

where δ_{ij} is the Kronecker unit tensor, the spectrum function of the turbulent kinetic energy density is $E(k) = (2/3)\, k_0^{-1}\, (k/k_0)^{-5/3}$. Here the wave number k varies within the interval $k_0 \le k \le k_\nu$ corresponding to the inertial range of scales, the wave number $k_0 = 1/\ell_0$, the length ℓ_0 is the integral scale of turbulence, the wave number $k_\nu = \ell_\nu^{-1}$, where $\ell_\nu = \ell_0 \mathrm{Re}^{-3/4}$ is the Kolmogorov (viscous) scale, and the turbulent correlation time is given by $\tau(k) = 2\,\tau_0\, (k/k_0)^{-2/3}$, where $\tau_0 = \ell_0/u_0$ is the characteristic turbulent time. The functions $E(k)$ and $\tau(k)$ correspond to fully developed turbulence with the Kolmogorov scalings.

Substituting Eq. (2.94) into Eq. (2.90) and (2.93), we determine the turbulent flux of particles $F_i^{(n)} = \langle n'\, u_i \rangle$ and the level of particle number density fluctuations $\langle n'^2 \rangle$:

$$F_i^{(n)} = -\frac{\langle \boldsymbol{u}^2 \rangle}{8\pi} \int_{k_0}^{k_\nu} \tau(k)\, E(k)\, dk \int_0^{2\pi} d\varphi \int_0^{\pi} \sin\vartheta\, d\vartheta \times \left[\left(\delta_{ij} - \frac{k_i\, k_j}{k^2} \right) \nabla_i \overline{n} + \left(\lambda_j - \lambda_i \frac{k_i\, k_j}{k^2} \right) \overline{n} \right], \tag{2.95}$$

$$\langle n'^2 \rangle = \frac{\langle \boldsymbol{u}^2 \rangle}{4\pi} \int_{k_0}^{k_\nu} \tau^2(k)\, E(k)\, dk \int_0^{2\pi} d\varphi \int_0^{\pi} \sin\vartheta\, d\vartheta \times \left[\left(\delta_{ij} - \frac{k_i\, k_j}{k^2} \right) (\nabla_i \overline{n})\, (\nabla_j \overline{n}) + \left(\lambda_j - \lambda_i \frac{k_i\, k_j}{k^2} \right) \nabla_j \overline{n}^2 \right], \tag{2.96}$$

where we took into account that

$$f_{ij}(\boldsymbol{k}) + f_{ij}(-\boldsymbol{k}) = \frac{\langle \boldsymbol{u}^2 \rangle\, E(k)}{4\pi k^2} \left(\delta_{ij} - \frac{k_i\, k_j}{k^2} \right), \tag{2.97}$$

$$f_{ij}(\boldsymbol{k}) - f_{ij}(-\boldsymbol{k}) = \frac{\mathrm{i}\, \langle \boldsymbol{u}^2 \rangle\, E(k)}{4\pi k^4} \left(\lambda_i\, k_j - \lambda_j\, k_i \right), \tag{2.98}$$

$$\left[f_{ij}(\boldsymbol{k}) - f_{ij}(-\boldsymbol{k}) \right] \mathrm{i}\, k_i \nabla_j = \frac{\langle \boldsymbol{u}^2 \rangle\, E(k)}{4\pi k^4} \left(\lambda_j - \lambda_i \frac{k_i\, k_j}{k^2} \right) \nabla_j, \tag{2.99}$$

$$\left[f_{ij}(\boldsymbol{k}) - f_{ij}(-\boldsymbol{k}) \right] \mathrm{i}\, k_j \nabla_i = \frac{\langle \boldsymbol{u}^2 \rangle\, E(k)}{4\pi k^4} \left(\lambda_j \frac{k_i\, k_j}{k^2} - \lambda_i \right) \nabla_i. \tag{2.100}$$

For the integration over $\boldsymbol{k}$, we use the integral given by Eqs. (1.129), (1.135) and (1.154)–(1.155). After integration over the angles in the $\boldsymbol{k}$ space in Eqs. (2.95) and (2.96), we obtain

$$\langle n'\, \boldsymbol{u} \rangle = -\frac{\langle \boldsymbol{u}^2 \rangle}{3} \left(\boldsymbol{\nabla} \overline{n} + \boldsymbol{\lambda}\, \overline{n} \right) \int_{k_0}^{k_\nu} \tau(k)\, E(k)\, dk, \tag{2.101}$$

$$\langle n'^2 \rangle = \frac{2 \langle \boldsymbol{u}^2 \rangle}{3} \left[(\boldsymbol{\nabla} \overline{n})^2 + (\boldsymbol{\lambda} \cdot \boldsymbol{\nabla}) \overline{n}^2 \right] \int_{k_0}^{k_\nu} \tau^2(k)\, E(k)\, dk. \tag{2.102}$$

After integration over k in Eqs. (2.101) and (2.102), we obtain formulae for the particle turbulent flux $\langle n' \boldsymbol{u}\rangle$ and the level of particle number density fluctuations $\langle n'^2\rangle$:

$$\boxed{\langle n' \boldsymbol{u}\rangle = \boldsymbol{V}^{\rm eff}\, \overline{n} - D_T\, \boldsymbol{\nabla}\overline{n},} \tag{2.103}$$

$$\boxed{D_T = \frac{1}{3}\, u_0\, \ell_0,} \tag{2.104}$$

$$\boxed{\boldsymbol{V}^{\rm eff} = -D_T \boldsymbol{\lambda} = D_T\, \frac{\boldsymbol{\nabla}\overline{\rho}}{\overline{\rho}},} \tag{2.105}$$

$$\boxed{\langle n'^2\rangle = \frac{8}{9}\, \ell_0^2\, \left[(\boldsymbol{\nabla}\overline{n})^2 + (\boldsymbol{\lambda}\cdot\boldsymbol{\nabla})\overline{n}^2\right].} \tag{2.106}$$

We compare these results with those obtained by dimensional analysis for ${\rm Pe}^{(n)} \gg 1$ (see Section 2.1.6). This comparison shows that Eqs. (2.18)–(2.20) and (2.24) for the particle turbulent flux and the level of particle number density fluctuations obtained by the dimensional analysis are in agreement with Eqs. (2.103)–(2.106), derived using the spectral tau approach. Physics of turbulent thermal diffusion described by the effective pumping velocity $\boldsymbol{V}^{\rm eff}$ is discussed in more detail in Sections 2.2.5 and 2.4.4.

2.4 Turbulent Transport of Inertial Particles

In this section, we discuss the turbulent transport of inertial particles with material density ρ_p being much larger than the fluid density ρ, i.e., inertial particles are solid particles moving in a turbulent fluid flow.

2.4.1 Equation of Motion for Inertial Particles

The equation of motion for inertial particles for $\rho_p \gg \rho$ reads

$$\boxed{\frac{{\rm d}\boldsymbol{U}^{\rm (p)}}{{\rm d}t} = -\frac{\boldsymbol{U}^{\rm (p)} - \boldsymbol{U}}{\tau_p} + \boldsymbol{g},} \tag{2.107}$$

where $\boldsymbol{U}^{\rm (p)}$ is the particle velocity, $\boldsymbol{U}$ is the fluid velocity, $\boldsymbol{g}$ is the acceleration caused by the gravity field, $\tau_p = m_p/(6\pi\rho\,\nu\,a_p)$ is the Stokes time of spherical particles having the radius a_p. Here ν is the kinematic viscosity of the fluid and $m_p = (4\pi/3)\,\rho_p\,a_p^3$ is the particle mass. The ratio of the Stokes time τ_p and the Kolmogorov turbulent turnover time τ_ν is the Stokes number,

$$\boxed{{\rm St} = \frac{\tau_p}{\tau_\nu} = \frac{2\rho_p\, a_p^2}{9\rho\,\ell_\nu^2},} \tag{2.108}$$

where the Kolmogorov turbulent time is $\tau_\nu = \ell_\nu/u_\nu = \tau_0/\mathrm{Re}^{1/2} = \ell_\nu^2/\nu$. Here $u_\nu = u_0/\mathrm{Re}^{1/4}$ is the characteristic fluid velocity at the Kolmogorov viscous microscale $\ell_\nu = \ell_0/\mathrm{Re}^{3/4}$, the fluid Reynolds number is $\mathrm{Re} = u_0\ell_0/\nu$ with u_0 being the characteristic turbulent velocity at the integral turbulent scale ℓ_0 and $\tau_0 = \ell_0/u_0$.

The solution of the equation of motion for small Stokes time is obtained by iterations. Indeed, we rewrite Eq. (2.107) in the following form:

$$\boldsymbol{U}^{(\mathrm{p})} = \boldsymbol{U} - \tau_p \, \frac{\mathrm{d}\boldsymbol{U}^{(\mathrm{p})}}{\mathrm{d}t} + \tau_p \boldsymbol{g}, \tag{2.109}$$

where $\tau_p \boldsymbol{g}$ is the terminal fall velocity of particles due to the gravity field. We seek a solution for Eq. (2.109) as a series

$$\boldsymbol{U}^{(\mathrm{p})} = \boldsymbol{U}^{(\mathrm{p},0)} + \boldsymbol{U}^{(\mathrm{p},1)} + \boldsymbol{U}^{(\mathrm{p},2)} + \ldots, \tag{2.110}$$

where $\boldsymbol{U}^{(\mathrm{p},0)} = \boldsymbol{U}$,

$$\begin{aligned}
\boldsymbol{U}^{(\mathrm{p},1)} &= -\tau_p \, \frac{\mathrm{d}\boldsymbol{U}^{(\mathrm{p},0)}}{\mathrm{d}t} = -\tau_p \, \frac{\mathrm{d}\boldsymbol{U}}{\mathrm{d}t}, \\
\boldsymbol{U}^{(\mathrm{p},2)} &= -\tau_p \, \frac{\mathrm{d}\boldsymbol{U}^{(\mathrm{p},1)}}{\mathrm{d}t} = \tau_p^2 \, \frac{\mathrm{d}^2\boldsymbol{U}}{\mathrm{d}t^2}, \\
\boldsymbol{U}^{(\mathrm{p,k})} &= -\tau_p \, \frac{\mathrm{d}\boldsymbol{U}^{(\mathrm{p,k-1})}}{\mathrm{d}t} = (-1)^k \, \tau_p^k \, \frac{\mathrm{d}^k\boldsymbol{U}}{\mathrm{d}t^k}.
\end{aligned} \tag{2.111}$$

Therefore,

$$\boldsymbol{U}^{(\mathrm{p})} = \boldsymbol{U} - \tau_p \, \frac{\mathrm{d}\boldsymbol{U}}{\mathrm{d}t} + \tau_p^2 \frac{\mathrm{d}^2\boldsymbol{U}}{\mathrm{d}t^2} + \tau_p \boldsymbol{g} + \mathrm{O}(\tau_p^3). \tag{2.112}$$

Using the identity for the substantial time derivative $\mathrm{d}\boldsymbol{U}/\mathrm{d}t$ for the moving fluid element,

$$\frac{\mathrm{d}\boldsymbol{U}}{\mathrm{d}t} = \frac{\partial \boldsymbol{U}}{\partial t} + (\boldsymbol{U} \cdot \boldsymbol{\nabla})\boldsymbol{U}, \tag{2.113}$$

and the Navier-Stokes equation for an incompressible fluid,

$$\frac{\partial \boldsymbol{U}}{\partial t} + (\boldsymbol{U} \cdot \boldsymbol{\nabla})\boldsymbol{U} = -\frac{\boldsymbol{\nabla} P}{\rho} + \nu \, \Delta \boldsymbol{U}, \tag{2.114}$$

we determine $\boldsymbol{\nabla} \cdot \boldsymbol{U}^{(\mathrm{p})}$ as

$$\boxed{\boldsymbol{\nabla} \cdot \boldsymbol{U}^{(\mathrm{p})} = -\tau_p \, \boldsymbol{\nabla} \cdot \, [(\boldsymbol{U} \cdot \boldsymbol{\nabla})\boldsymbol{U}] + \mathrm{O}(\tau_p^2) = \frac{\tau_p}{\rho} \, \boldsymbol{\nabla}^2 P + \mathrm{O}(\tau_p^2),} \tag{2.115}$$

where P is the fluid pressure. Equation (2.115) implies that due to particle inertia, the velocity field $\boldsymbol{U}^{(\mathrm{p})}$ of the inertial particles is compressible even in an incompressible fluid flow.

2.4.2 Physics of Turbophoresis of Inertial Particles

In this section, we consider one of the effects, namely turbophoresis, which is caused by a combined action of particle inertia and inhomogeneity of turbulence. Averaging Eq. (2.112) over an ensemble of turbulent fluid velocity, we obtain an equation for the mean particle velocity:

$$\left\langle \boldsymbol{U}^{(\mathrm{p})} \right\rangle = \overline{\boldsymbol{U}} - \tau_p \left\langle \frac{d\boldsymbol{U}}{dt} \right\rangle + \tau_p \boldsymbol{g} + \mathrm{O}(\tau_p^2). \tag{2.116}$$

We consider the case a zero mean fluid velocity, $\overline{\boldsymbol{U}} = 0$. Using the identity (2.113), we rewrite Eq. (2.116) as

$$\left\langle \boldsymbol{U}^{(\mathrm{p})} \right\rangle = -\tau_p \left\langle (\boldsymbol{u} \cdot \nabla)\boldsymbol{u} \right\rangle + \tau_p \boldsymbol{g} + \mathrm{O}(\tau_p^2). \tag{2.117}$$

For an incompressible and isotropic turbulence, $\langle (\boldsymbol{u} \cdot \nabla)\boldsymbol{u} \rangle = (1/3)\nabla \left\langle \boldsymbol{u}^2 \right\rangle$. Indeed, for an incompressible fluid $(u_j \nabla_j) u_i = \nabla_j (u_i u_j)$ and for isotropic turbulence $\left\langle u_i u_j \right\rangle = \delta_{ij} \left\langle \boldsymbol{u}^2 \right\rangle /3$. Therefore, Eq. (2.117) can be rewritten as

$$\left\langle \boldsymbol{U}^{(\mathrm{p})} \right\rangle = -\frac{\tau_p}{3} \nabla \left\langle \boldsymbol{u}^2 \right\rangle + \tau_p \boldsymbol{g} + \mathrm{O}(\tau_p^2). \tag{2.118}$$

This implies that in an inhomogeneous turbulence, particles drift out from the region of intensive turbulent fluctuations. This phenomenon is called *turbophoresis*, and it exists only for inertial particles. Even in the case of a zero mean fluid velocity, the mean particle velocity is the sum of the terminal fall velocity

$$\boxed{\boldsymbol{V}_{\mathrm{g}} = \tau_p \boldsymbol{g},} \tag{2.119}$$

caused by the gravity field and the turbophoretic velocity in an inhomogeneous turbulence (Caporaloni et al., 1975; Reeks, 1983),

$$\boxed{\overline{\boldsymbol{V}}_{\mathrm{turboph}} = -\frac{\tau_p}{3} \nabla \left\langle \boldsymbol{u}^2 \right\rangle.} \tag{2.120}$$

The physics of turbophoresis are related to effects of inertia. The inertia results in a drift of particles by centrifugal force to the boundary regions between turbulent eddies where turbulence intensity vanishes. Since there is a preferential direction due to a non-zero gradient of turbulence intensity, $\nabla \left\langle \boldsymbol{u}^2 \right\rangle$, the total effect of the particle drift to the regions with low turbulence intensity does not vanish.

2.4.3 Effective Pumping Velocity of Inertial Particles

Equation for the evolution of particle number density $n(t, \boldsymbol{x})$ reads:

$$\frac{\partial n}{\partial t} + \nabla \cdot \left(n \, \boldsymbol{U}^{(\mathrm{p})}\right) = D^{(n)} \, \Delta n. \tag{2.121}$$

Let us consider homogeneous turbulence, and average Eq. (2.121) over an ensemble of turbulent particle velocity field. This yields an equation for the mean number density $\overline{n}$ of inertial particles,

$$\boxed{\frac{\partial \overline{n}}{\partial t} + \nabla \cdot \left[\overline{n}\,(\boldsymbol{V}_{\rm g} + \boldsymbol{V}^{\rm eff}) - (D^{(n)} + D_T)\,\nabla \overline{n}\right] = 0,} \tag{2.122}$$

where $\boldsymbol{V}_{\rm g} = \tau_p \boldsymbol{g}$ is the terminal fall velocity of particles, and

$$\boxed{\boldsymbol{V}^{\rm eff} = -\left\langle \tau\, \boldsymbol{u}^{(\rm p)}\, \left(\nabla \cdot \boldsymbol{u}^{(\rm p)}\right)\right\rangle} \tag{2.123}$$

is the effective pumping velocity of particles due to turbulent thermal diffusion and $\boldsymbol{u}^{(\rm p)}$ are fluctuations of the particle velocity $\boldsymbol{U}^{(\rm p)}$. Equation (2.123) for non-inertial particles has been derived in this chapter using dimensional analysis, the quasi-linear approach and the τ approach. The derivation of Eq. (2.123) for inertial particles is similar, whereby we replace fluctuations of the fluid velocity $\boldsymbol{u}$ by fluctuations of the particle velocity $\boldsymbol{u}^{(\rm p)}$.

Let us determine the effective pumping velocity for inertial particles. We consider a low-Mach-number flow, ${\rm Ma} = u/c_{\rm s} \ll 1$, where c_s is the sound speed. For a low-Mach-number isothermal flow, the turbulent fluid mass flux $\langle \rho' \mathbf{u}\rangle$ is very small, i.e., it is of the order of ${\rm Ma}^2$, and fluctuations of the fluid density ρ' and velocity $\mathbf{u}$ are weakly correlated. This is also valid for a low-Mach-number density-stratified flow. We use the equation of state for a perfect gas, which yields

$$\frac{p}{\overline{P}} = \frac{\rho'}{\overline{\rho}} + \frac{\theta}{\overline{T}} + {\rm O}(\rho'\,\theta), \tag{2.124}$$

and $\langle p\, \mathbf{u}\rangle/\overline{P} = \langle \theta\, \mathbf{u}\rangle/\overline{T}$, where we neglect very small turbulent fluid mass flux $\langle \rho'\, \mathbf{u}\rangle$ for a low-Mach-number flow. Here $\overline{P}$, $\overline{T}$ and $\overline{\rho}$ are the mean fluid pressure, temperature and density, respectively. Therefore, the effective pumping velocity for inertial particles is

$$\begin{aligned}\boldsymbol{V}^{\rm eff} &= -\left\langle \tau\, \boldsymbol{u}^{(\rm p)}\, (\nabla \cdot \boldsymbol{u}^{(\rm p)})\right\rangle = -\langle \tau\, \boldsymbol{u}\, (\nabla \cdot \boldsymbol{u})\rangle - \frac{\tau_p}{\overline{\rho}} \left\langle \tau \mathbf{u}(\mathbf{x}) \left[\nabla^2 p(\mathbf{x})\right]\right\rangle \\ &\approx -\langle \tau\, \boldsymbol{u}\, (\nabla \cdot \boldsymbol{u})\rangle - \frac{\tau_p\, \overline{P}}{\overline{\rho}\, \overline{T}} \left\langle \tau \mathbf{u}(\mathbf{x}) \left[\nabla^2 \theta(\mathbf{x})\right]\right\rangle,\end{aligned} \tag{2.125}$$

where the second term,

$$\tilde{\boldsymbol{V}}^{\rm eff} \equiv -\frac{\tau_p\, \overline{P}}{\overline{\rho}\, \overline{T}} \left\langle \tau \mathbf{u}(\mathbf{x}) \left[\nabla^2 \theta(\mathbf{x})\right]\right\rangle = -\frac{\tau_p}{\overline{\rho}} \left\langle \tau \mathbf{u}(\mathbf{x}) \left[\nabla^2 p(\mathbf{x})\right]\right\rangle, \tag{2.126}$$

on the right-hand side of Eq. (2.125) determines the contribution to the effective pumping velocity $\boldsymbol{V}^{\rm eff}$ due to the particle inertia effect, while the first term $-\langle \tau\, \boldsymbol{u}\, (\nabla \cdot \boldsymbol{u}) \rangle$ describes the "no-inertia" contribution to $\boldsymbol{V}^{\rm eff}$ caused by the density stratification of the fluid flow. The latter contribution,

$$- \langle \tau\, \boldsymbol{u}\, (\nabla \cdot \boldsymbol{u}) \rangle = -D_T\, \frac{\nabla \overline{T}}{\overline{T}}, \tag{2.127}$$

has been derived in previous Sections by different methods. Below we derive the contribution to $\boldsymbol{V}^{\rm eff}$ due to the particle inertia effect. To this end, we determine the correlation function $\langle \tau \mathbf{u}(\mathbf{x}) \left[\nabla^2 \theta(\mathbf{x})\right] \rangle$:

$$\langle \tau \mathbf{u}(\mathbf{x}) \left[\nabla^2 \theta(\mathbf{x})\right] \rangle = \lim_{\mathbf{x}\to\mathbf{y}} \langle \tau \mathbf{u}(\mathbf{x}) \left[\nabla^2 \theta(\,\mathbf{y})\right] \rangle = - \int \tau(k)\, k^2\, \langle \mathbf{u}(\mathbf{k})\, \theta(-\mathbf{k}) \rangle\, d\mathbf{k}, \tag{2.128}$$

where the correlation function $\langle u_i(\mathbf{k})\, \theta(-\mathbf{k}) \rangle$ in the $\mathbf{k}$ space is

$$\langle u_i(\mathbf{k})\, \theta(-\mathbf{k}) \rangle = -\tau(k) \left\langle u_i(\mathbf{k})\, u_j(-\mathbf{k}) \right\rangle \nabla_j \overline{T} \tag{2.129}$$

[see Eq. (1.149)]. Therefore, Eqs. (2.125)–(2.129) yield the contribution to the effective pumping velocity due to the particle inertia effect:

$$\tilde{\boldsymbol{V}}^{\rm eff} = - \frac{\tau_p\, \overline{P}}{\overline{\rho}\, \overline{T}} \int \tau^2(k)\, k^2 \left\langle u_i(\mathbf{k})\, u_j(-\mathbf{k}) \right\rangle d\mathbf{k}\ \nabla_j \overline{T}. \tag{2.130}$$

We use the model for the second-order moment of turbulent velocity field, $\langle u_i(\boldsymbol{k})\, u_j(-\boldsymbol{k}) \rangle$, for large fluid Reynolds numbers given by Eq. (2.94), where the energy spectrum function is $E(k) = (2/3) k_0^{-1}\, (k/k_0)^{-5/3}$ in the interval of the wave numbers $k_0 \le k \le k_\nu$ and $k_\nu = k_0\, {\rm Re}^{3/4}$. The expression for the turbulent correlation time is given by $\tau(k) = 2\, \tau_0\, (k/k_0)^{-2/3}$, where $\tau_0 = \ell_0/u_0$ is the characteristic turbulent time. After the integration in the $\boldsymbol{k}$ space, we obtain:

$$\int \tau^2(k)\, k^2\, \langle u_i(\mathbf{k})\, u_j(-\mathbf{k}) \rangle\, d\mathbf{k} = \frac{\langle \boldsymbol{u}^2 \rangle}{8\pi} \int_{k_0}^{k_\nu} \tau^2(k)\, E(k)\, k^2\, dk$$
$$\times \int_0^{2\pi} d\varphi \int_0^{\pi} \sin\vartheta\, d\vartheta \left[\delta_{ij} - \frac{k_i\, k_j}{k^2}\right] = \frac{2}{3}\, \ln {\rm Re}\, \delta_{ij}, \tag{2.131}$$

where we use Eqs. (1.129) and (1.135), and the following integral

$$\int_{k_0}^{k_\nu} \tau^2(k)\, E(k)\, k^2\, dk = 2\, \ln {\rm Re}. \tag{2.132}$$

Equations (2.130) and (2.131) yield the expression for the total effective pumping velocity of inertial particles which takes into account the "no-inertia" and inertia contributions:

$$\boldsymbol{V}^{\rm eff} = -D_T\, \frac{\nabla \overline{T}}{\overline{T}} - \frac{2}{3\gamma}\, \tau_p\, V_T^2\, \ln {\rm Re}\, \frac{\nabla \overline{T}}{\overline{T}_*}, \tag{2.133}$$

where $V_T = (\gamma\, k_B \overline{T}_* / m_\mu)^{1/2}$ is the thermal velocity and $\gamma = c_p/c_v$ is the ratio of specific heats. Therefore, Eqs. (2.130)–(2.133) yield the expression for the total effective pumping velocity of the inertial particles as

$$\boxed{\mathbf{V}^{\rm eff} = -\mu_p\, D_T\, \boldsymbol{\nabla} \ln \overline{T},} \tag{2.134}$$

$$\boxed{\mu_p = 1 + 2\,\frac{m_p}{m_\mu} \left(\frac{\ln \mathrm{Re}}{\mathrm{Pe}^{(n)}}\right) \frac{\overline{T}}{\overline{T}_*} = 1 + 2\,\frac{V_g\, L_P\, \ln \mathrm{Re}}{u_0\, \ell_0}} \tag{2.135}$$

(Elperin et al., 1996c, 1998b), where we used the following identity: $\tau_p V_T^2 = \gamma\,(m_p/m_\mu) D^{(n)} = \gamma\, V_g\, L_P\, \overline{T}_*/\overline{T}$. Here the Stokes time can be written as $\tau_p = \overline{\rho}\, V_g\, L_P/\overline{P}$, with $\mathbf{V}_g = \tau_p\, \mathbf{g}$ being the terminal fall velocity of particles and $L_P = |\nabla_z \overline{P}/\overline{P}|^{-1}$ being the pressure height scale. For large Péclet numbers, $\mathrm{Pe}^{(n)} = u_0 \ell_0 / D^{(n)} \gg 1$, the turbulent thermal diffusion coefficient $\mu_p = 1$ for non-inertial particles, while for inertial particles, μ_p depends on the particle mass, the fluid Reynolds and Péclet numbers.

The non-diffusive turbulent flux of particles, $\overline{n}\, \mathbf{V}^{\rm eff}$, toward the mean temperature minimum is the main mechanism for the formation of large-scale particle clouds (inhomogeneous distributions of inertial particles) in temperature-stratified turbulence. The steady-state solution of Eq. (2.122) for the mean number density of inertial particles, satisfying the boundary condition with a zero total particle flux at the boundary, is given by

$$\boxed{\frac{\overline{n}}{\overline{n}_*} = \left(\frac{\overline{T}}{\overline{T}_*}\right)^{-\frac{\mu_p D_T}{D^{(n)} + D_T}} \exp\left[-\int_{z_0}^{z} \frac{V_g}{D^{(n)} + D_T}\, dz'\right],} \tag{2.136}$$

where the subscripts $*$ represent the values of the mean temperature and the mean particle number density at the boundary. Equation (2.136) implies that small inertial particles are accumulated in the vicinity of the mean temperature minimum. Due to the gravity field, inertial particles are accumulated below the mean temperature minimum.

2.4.4 Mechanism of Turbulent Thermal Diffusion for Inertial Particles

The mechanism for turbulent thermal diffusion for inertial particles is as follows. The inertia causes particles inside the turbulent eddies to drift out to the boundary regions between eddies due to the centrifugal inertial force. Indeed, for large Péclet numbers, when molecular diffusion of particles in Eq. (2.121) can be neglected, it follows that

Table 2.2 *Mechanism of particle transport* $\langle n' \boldsymbol{u}^{(p)} \rangle$ *for inertial particles along the direction of turbulent heat flux*

$\Longrightarrow$	$\langle \boldsymbol{u}\, \theta \rangle$
a	b
$\mathrm{u}^{(p)} > 0$	$\mathrm{u}^{(p)} < 0$
$\theta > 0$	$\theta < 0$
$p > 0$	$p < 0$
$\nabla \cdot \boldsymbol{u}^{(p)} < 0$	$\nabla \cdot \boldsymbol{u}^{(p)} > 0$
$n' > 0$	$n' < 0$
$n' \mathrm{u}^{(p)} > 0$	$n' \mathrm{u}^{(p)} > 0$
$\Longrightarrow$	$\langle n' \boldsymbol{u}^{(p)} \rangle$

$$\nabla \cdot \boldsymbol{U}^{(p)} \approx -n^{-1} \left[\frac{\partial n}{\partial t} + (\boldsymbol{U}^{(p)} \cdot \nabla) n \right] \equiv -n^{-1} \frac{\mathrm{d}n}{\mathrm{d}t}. \tag{2.137}$$

On the other hand, for inertial particles, $\nabla \cdot \boldsymbol{U}^{(p)} = (\tau_p/\rho)\, \nabla^2 P$ [see Eq. (2.115)]. Therefore, in regions with maximum fluid pressure fluctuations (where $\nabla^2 p < 0$), there is an accumulation of inertial particles, i.e., $\mathrm{d}n'/\mathrm{d}t \propto -\overline{n}\, (\tau_p/\overline{\rho})\, \nabla^2 p > 0$. These regions have low vorticity and high strain rate. Similarly, there is an outflow of inertial particles from regions with minimum fluid pressure.

In homogeneous and isotropic turbulence with a zero gradient of the mean temperature, there is no preferential direction, so that there is no large-scale effect of particle accumulation, and the pressure (temperature) of the surrounding fluid is not correlated with the turbulent velocity field. The only non-zero correlation is $\langle (\boldsymbol{u} \cdot \nabla) p \rangle$, which contributes to the flux of the turbulent kinetic energy density.

In temperature-stratified turbulence, fluctuations of fluid temperature θ and velocity $\boldsymbol{u}$ are correlated due to a non-zero turbulent heat flux, $\langle \theta\, \boldsymbol{u} \rangle \neq \boldsymbol{0}$. Fluctuations in temperature cause pressure fluctuations, which result in fluctuations in the number density of particles. Increase of the pressure of the surrounding fluid is accompanied by an accumulation of particles, and the direction of the mean flux of particles coincides with that of the turbulent heat flux. The mean flux of particles is directed toward the minimum of the mean temperature, and the particles tend to be accumulated in this region.

To demonstrate that the directions of the mean flux of particles and the turbulent heat flux coincide, we assume that the mean temperature $\overline{T}_2$ at point 2 is larger than the mean temperature $\overline{T}_1$ at point 1. We consider two small control volumes "a" and "b" located between these two points. Let the direction of the local turbulent velocity in volume "a" at some instant be the same as the direction of the turbulent heat flux $\langle \theta\, \boldsymbol{u} \rangle$ (i.e., along the x-axis toward point 1) and let the local turbulent velocity in volume "b", at the same instant, be directed opposite to the turbulent heat flux (i.e., toward point 2); see Table 2.2.

In temperature stratified turbulence with a non-zero turbulent heat flux $\langle \theta\, \boldsymbol{u} \rangle$, fluctuations of pressure p and velocity $\boldsymbol{u}$ are correlated, and regions with a higher level of pressure fluctuations have higher temperature and velocity fluctuations. Fluctuations of temperature θ and pressure p in volume "a" are positive because $\theta\, u_x > 0$, and negative in volume "b"; see Table 2.2. Fluctuations of particle number density n' are positive in volume "a" (because particles are locally accumulated in the vicinity of the maximum of pressure fluctuations, $dn'/dt \propto -\overline{n}\,(\tau_p/\overline{\rho})\,\nabla^2 p > 0$), and they are negative in volume "b" (because there is an outflow of particles from regions with low pressure). The flux of particles $n'\, \mathrm{u}^{(\mathrm{p})}$ is positive in volume "a" (i.e., it is directed toward point 1), and it is also positive in volume "b" (because both fluctuations of velocity and number density of particles are negative in volume "b"), where $\mathrm{u}^{(\mathrm{p})}$ is the particle velocity along the x-axis. Therefore the mean flux of particles $\langle n'\, \boldsymbol{u}^{(\mathrm{p})} \rangle$ is directed, as is the turbulent heat flux $\langle \theta\, \boldsymbol{u} \rangle$, toward point 1. This causes the formation of large-scale inhomogeneous structures in the spatial distribution of inertial particles in the vicinity of the mean temperature minimum.

2.5 Compressibility Effects on the Turbulent Transport of Particles

In this section, we consider non-inertial particles or gaseous admixtures in a compressible fluid flow with arbitrary Mach numbers.

2.5.1 Governing Equations

The equation for the particle number density, $n(\boldsymbol{x}, t)$, reads

$$\frac{\partial n}{\partial t} + \nabla \cdot \left(n\boldsymbol{U} - D^{(n)} \nabla n\right) = 0, \tag{2.138}$$

where $D^{(n)}$ is the microphysical diffusivity describing Brownian diffusion of particles and $\boldsymbol{U}$ is a fluid velocity. We consider a one-way coupling that corresponds to the turbulent transport of a passive scalar. We use a mean-field approach in which the number density of particles and fluid velocity are decomposed into mean and fluctuating parts, where the fluctuating parts have zero mean values. Averaging

Eq. (2.138) over an ensemble, we arrive at an equation for the mean number density, $\overline{n}(\boldsymbol{x}, t)$:

$$\frac{\partial \overline{n}}{\partial t} + \nabla \cdot \left(\langle n' \boldsymbol{u} \rangle - D^{(n)} \nabla \overline{n} \right) = 0, \tag{2.139}$$

where $\boldsymbol{F}^{(n)} = \langle n'(\boldsymbol{x}, t)\, \boldsymbol{u}(\boldsymbol{x}, t) \rangle$ is the turbulent flux of particles and we consider the case of a zero mean fluid velocity, $\overline{\boldsymbol{U}} = 0$.

To determine the turbulent flux of particles and the level of particle number density fluctuations, we use the equation for fluctuations of the particle number density, $n'(\boldsymbol{x}, t) = n - \overline{n}$, which follows from Eq. (2.138) and (2.139):

$$\frac{\partial n'}{\partial t} + \nabla \cdot \left(n' \boldsymbol{u} - \langle n' \boldsymbol{u} \rangle - D^{(n)} \nabla n' \right) = -\overline{n} \, (\nabla \cdot \boldsymbol{u}) - (\boldsymbol{u} \cdot \nabla) \overline{n}, \tag{2.140}$$

where $Q^{(n)} = \nabla \cdot \left(n'\boldsymbol{u} - \langle n'\boldsymbol{u} \rangle \right)$ are nonlinear terms and $I = -\overline{n} \, (\nabla \cdot \boldsymbol{u}) - (\boldsymbol{u} \cdot \nabla) \overline{n}$ is the source of particle number density fluctuations. The ratio of the nonlinear terms to the diffusion term is the Péclet number, which is estimated as $\mathrm{Pe}^{(n)} = u_0 \, \ell_0 / D^{(n)}$. In the following sections, we derive equations for the turbulent diffusion coefficient and effective pumping velocity for particles for small and large Péclet numbers in a compressible fluid flow with arbitrary Mach numbers.

2.5.2 Derivation of Equations for Second Moments for $\mathrm{Pe}^{(n)} \ll 1$

We derive equations for the turbulent flux of particles,

$$\langle n'(\boldsymbol{x}, t) \, u_j(\boldsymbol{x}, t) \rangle = \int F_j(\boldsymbol{k}, \omega, \boldsymbol{R}, t) \, d\omega \, d\boldsymbol{k}, \tag{2.141}$$

and the level of particle number density fluctuations,

$$\langle n'(\boldsymbol{x}, t) \, n'(\boldsymbol{x}, t) \rangle = \int E_n(\boldsymbol{k}, \omega, \boldsymbol{R}, t) \, d\omega \, d\boldsymbol{k}, \tag{2.142}$$

for small Péclet numbers using a quasi-linear approach, where

$$F_j(\boldsymbol{k}, \omega, \boldsymbol{R}, t) = \int \langle n'(\boldsymbol{k}_1, \omega_1) \, u_j(\boldsymbol{k}_2, \omega_2) \rangle \exp[\mathrm{i} \, \Omega t + \mathrm{i} \, \boldsymbol{K} \cdot \boldsymbol{R}] \, d\Omega \, d\boldsymbol{K} \tag{2.143}$$

and

$$E_n(\boldsymbol{k}, \omega, \boldsymbol{R}, t) = \int \langle n'(\boldsymbol{k}_1, \omega_1) \, n'(\boldsymbol{k}_2, \omega_2) \rangle \exp[\mathrm{i} \, \Omega t + \mathrm{i} \, \boldsymbol{K} \cdot \boldsymbol{R}] \, d\Omega \, d\boldsymbol{K}. \tag{2.144}$$

For brevity of notation, we omit below the large-scale variables t and $\boldsymbol{R}$ in $F_j(\boldsymbol{k}, \omega, \boldsymbol{R}, t)$, $E_n(\boldsymbol{k}, \omega, \boldsymbol{R}, t)$ and $\overline{n}(\boldsymbol{R}, t)$. Using Eq. (2.140), we neglect the nonlinear term $Q^{(n)}$ but keep the molecular diffusion term. We apply the multi-scale approach, assuming that there is a separation of spatial and time scales between turbulent and mean-field scales, i.e., $\ell_0 \ll L_n$ and $\tau_0 \ll t_n$, where L_n and t_n are the characteristic spatial and time scales of the mean particle number density variations.

To determine the turbulent flux of particles and the level of particle number density fluctuations for small Péclet numbers, we rewrite Eq. (2.140) in Fourier space and find the solution of Eq. (2.140) as

$$n'(\boldsymbol{k},\omega) = -\mathrm{i}\, k_p\, G_D(\boldsymbol{k},\omega) \int \overline{n}(\boldsymbol{Q})\, u_p(\boldsymbol{k}-\boldsymbol{Q},\omega)\, d\boldsymbol{Q} \tag{2.145}$$

[see Eq. (2.74)], where $G_D(\boldsymbol{k},\omega) = (D^{(n)}\boldsymbol{k}^2 + \mathrm{i}\,\omega)^{-1}$. Using Eq. (2.145), we rewrite the function $F_j(\boldsymbol{k},\boldsymbol{R})$ as

$$F_j(\boldsymbol{k},\boldsymbol{R}) = -\mathrm{i} \int \overline{n}(\boldsymbol{Q})\, k_p^{(1)} \left\langle u_p(\boldsymbol{k}_1-\boldsymbol{Q})\, u_j(\boldsymbol{k}_2)\right\rangle G_D(\boldsymbol{k}_1)\, \exp(\mathrm{i}\boldsymbol{K}\cdot\boldsymbol{R})\, d\boldsymbol{K}\, d\boldsymbol{Q}, \tag{2.146}$$

where $\boldsymbol{k}^{(1)} \equiv \boldsymbol{k}_1 = \boldsymbol{k} + \boldsymbol{K}/2$ and $\boldsymbol{k}^{(2)} \equiv \boldsymbol{k}_2 = -\boldsymbol{k} + \boldsymbol{K}/2$. Here the functions F_j, G_D and u_i depend also on ω, and the mean particle number density $\overline{n}$ depends on t as well. To simplify the notation, we do not show these dependencies. We use new variables:

$$\tilde{\boldsymbol{k}} = (\tilde{\boldsymbol{k}}_1 - \tilde{\boldsymbol{k}}_2)/2 = \boldsymbol{k} - \boldsymbol{Q}/2, \qquad \tilde{\boldsymbol{K}} = \tilde{\boldsymbol{k}}_1 + \tilde{\boldsymbol{k}}_2 = \boldsymbol{K} - \boldsymbol{Q}, \tag{2.147}$$

where

$$\tilde{\boldsymbol{k}}_1 = \boldsymbol{k} + \boldsymbol{K}/2 - \boldsymbol{Q}, \qquad \tilde{\boldsymbol{k}}_2 = -\boldsymbol{k} + \boldsymbol{K}/2. \tag{2.148}$$

Since $|\boldsymbol{Q}| \ll |\boldsymbol{k}|$ and $|\boldsymbol{K}| \ll |\boldsymbol{k}|$, we use the Taylor expansion of the functions:

$$f_{ij}(\boldsymbol{k}-\boldsymbol{Q}/2, \boldsymbol{K}-\boldsymbol{Q}) = \left[1 - \frac{Q_s}{2}\frac{\partial}{\partial k_s}\right] f_{ij}(\boldsymbol{k}, \boldsymbol{K}-\boldsymbol{Q}) + O(\boldsymbol{Q}^2), \tag{2.149}$$

$$G_D(\boldsymbol{k}+\boldsymbol{K}/2) = G_D(\boldsymbol{k})\left[1 - D^{(n)}(\boldsymbol{k}\cdot\boldsymbol{K})G_D(\boldsymbol{k})\right] + O(\boldsymbol{K}^2), \tag{2.150}$$

where $f_{ij}(\tilde{\boldsymbol{k}}_1, \tilde{\boldsymbol{k}}_2) \equiv \left\langle u_i(\tilde{\boldsymbol{k}}_1) u_j(\tilde{\boldsymbol{k}}_2)\right\rangle = f_{ij}(\boldsymbol{k}-\boldsymbol{Q}/2, \boldsymbol{K}-\boldsymbol{Q})$. Using Eqs. (2.146)–(2.150), we arrive at expression for the turbulent flux of particles in Fourier space for small Péclet numbers:

$$F_j = -\frac{1}{2}\left[\overline{n}\int G_D\left(\nabla_i - 2D^{(n)}k^2 G_D\, k_{im}\nabla_m + 2\mathrm{i}\, k_i\right) f_{ij}\, d\boldsymbol{k}\, d\omega \right.$$
$$\left. - (\nabla_i \overline{n})\int G_D\left(2D^{(n)}k^2 G_D\, k_{im} - \delta_{im} + k_m \frac{\partial}{\partial k_i}\right) f_{mj}\right] d\boldsymbol{k}\, d\omega, \tag{2.151}$$

where $G_D \equiv G_D(\boldsymbol{k}, \omega)$, $f_{ij} = f_{ij}(\boldsymbol{k}, \omega)$, δ_{ij} is the Kronecker unit tensor and $k_{ij} = k_i k_j / k^2$. Since we consider a one-way coupling, the correlation function f_{ij} in Eq. (2.151) should be replaced by $f_{ij}^{(0)}$ for the background random flow with zero turbulent heat flux.

Similarly, we rewrite the function $E_n(\boldsymbol{k}, \boldsymbol{R})$ as

$$E_n(\boldsymbol{k}, \boldsymbol{R}) = -\frac{\mathrm{i}}{2} \int \overline{n}(\boldsymbol{Q}) \Big[k_i^{(1)} G_D(\boldsymbol{k}_1) \left\langle n'(\boldsymbol{k}_2)\, u_i(\boldsymbol{k}_1 - \boldsymbol{Q}) \right\rangle + k_i^{(2)} G_D(\boldsymbol{k}_2) \left\langle n'(\boldsymbol{k}_1)\, u_i(\boldsymbol{k}_2 - \boldsymbol{Q}) \right\rangle \Big] \exp(\mathrm{i} \boldsymbol{K} \cdot \boldsymbol{R})\, d\boldsymbol{K}\, d\boldsymbol{Q}. \tag{2.152}$$

We use new variables for the first term on the right-hand side of Eq. (2.152):

$$\tilde{\boldsymbol{k}} = (\tilde{\boldsymbol{k}}_3 - \tilde{\boldsymbol{k}}_4)/2 = -\boldsymbol{k} + \boldsymbol{Q}/2, \qquad \tilde{\boldsymbol{K}} = \tilde{\boldsymbol{k}}_3 + \tilde{\boldsymbol{k}}_4 = \boldsymbol{K} - \boldsymbol{Q}, \tag{2.153}$$

where

$$\tilde{\boldsymbol{k}}_3 = -\boldsymbol{k} + \boldsymbol{K}/2, \qquad \tilde{\boldsymbol{k}}_4 = \boldsymbol{k} + \boldsymbol{K}/2 - \boldsymbol{Q}. \tag{2.154}$$

For the second term on the right-hand side of Eq. (2.152), we use another new variable:

$$\tilde{\boldsymbol{k}} = (\tilde{\boldsymbol{k}}_5 - \tilde{\boldsymbol{k}}_6)/2 = \boldsymbol{k} + \boldsymbol{Q}/2, \qquad \tilde{\boldsymbol{K}} = \tilde{\boldsymbol{k}}_5 + \tilde{\boldsymbol{k}}_6 = \boldsymbol{K} - \boldsymbol{Q}, \tag{2.155}$$

where

$$\tilde{\boldsymbol{k}}_5 = \boldsymbol{k} + \boldsymbol{K}/2, \qquad \tilde{\boldsymbol{k}}_6 = -\boldsymbol{k} + \boldsymbol{K}/2 - \boldsymbol{Q}. \tag{2.156}$$

Since $|\boldsymbol{Q}| \ll |\boldsymbol{k}|$ and $|\boldsymbol{K}| \ll |\boldsymbol{k}|$, we use the Taylor expansion of the functions:

$$\begin{aligned} F_j(-\boldsymbol{k} + \boldsymbol{Q}/2, \boldsymbol{K} - \boldsymbol{Q}) &= \left[1 + \frac{Q_s}{2} \frac{\partial}{\partial \tilde{k}_s}\right] F_j(-\boldsymbol{k}, \boldsymbol{K} - \boldsymbol{Q}) + O(\boldsymbol{Q}^2) \\ &= \left[1 - \frac{Q_s}{2} \frac{\partial}{\partial k_s}\right] F_j(-\boldsymbol{k}, \boldsymbol{K} - \boldsymbol{Q}) + O(\boldsymbol{Q}^2), \end{aligned} \tag{2.157}$$

$$F_j(\boldsymbol{k} + \boldsymbol{Q}/2, \boldsymbol{K} - \boldsymbol{Q}) = \left[1 + \frac{Q_s}{2} \frac{\partial}{\partial k_s}\right] F_j(\boldsymbol{k}, \boldsymbol{K} - \boldsymbol{Q}) + O(\boldsymbol{Q}^2), \tag{2.158}$$

$$G_D(-\boldsymbol{k} + \boldsymbol{K}/2) = G_D(\boldsymbol{k}) \left[1 + D^{(n)} (\boldsymbol{k} \cdot \boldsymbol{K}) G_D(\boldsymbol{k})\right] + O(\boldsymbol{K}^2). \tag{2.159}$$

Using Eqs. (2.150), (2.152)–(2.159), we arrive at an expression for the level of particle number density fluctuations in Fourier space for small Péclet numbers:

$$E_n = \frac{1}{4}\Big\{\overline{n}\int G_D\Big[\Big(2D^{(n)}k^2 G_D\, k_{jn}\nabla_n - \nabla_j\Big)\, F_j^{(+)} + 2\mathrm{i}\, k_j\, F_j^{(-)}\Big]\, d\boldsymbol{k}\, d\omega$$

$$+ (\nabla_n \overline{n})\int G_D\Big[2D^{(n)}k^2 G_D\, k_{jn} - \delta_{jn} + k_j \frac{\partial}{\partial k_n}\Big]\, F_j^{(+)}\Big\}\, d\boldsymbol{k}\, d\omega, \quad (2.160)$$

where $F_j^{(\pm)} = F_j(\boldsymbol{k}, \omega) \pm F_j(-\boldsymbol{k}, \omega)$.

2.5.3 Model for Compressible Random Flow

We consider a random, statistically stationary, density-stratified, inhomogeneous and compressible background flow, which is determined by the following correlation function in Fourier space:

$$f_{ij}^{(0)}(\boldsymbol{k}, \omega, \boldsymbol{R}) = \frac{\Phi(\omega)}{8\pi\, k^2\, (1+\sigma_c)}\Big[E(k)\Big(\delta_{ij} - k_{ij} + \frac{\mathrm{i}}{k^2}\,(k_j\lambda_i - k_i\lambda_j)$$

$$+\frac{\mathrm{i}}{2k^2}\,(k_i\nabla_j - k_j\nabla_i)\Big) + 2\sigma_c\, E_c(k)\,\Big(k_{ij} + \frac{\mathrm{i}}{4k^2}(k_i\nabla_j - k_j\nabla_i)$$

$$+\frac{\mathrm{i}}{2k^2}(k_j\lambda_i - k_i\lambda_j)\Big)\Big]\langle \boldsymbol{u}^2\rangle, \quad (2.161)$$

where $\boldsymbol{\lambda} = -\boldsymbol{\nabla}\ln\overline{\rho}$ characterizes the fluid density stratification and the parameter

$$\sigma_c = \frac{\langle(\boldsymbol{\nabla}\cdot\boldsymbol{u})^2\rangle}{\langle(\boldsymbol{\nabla}\times\boldsymbol{u})^2\rangle} \quad (2.162)$$

describes the degree of compressibility of the random velocity field.

In Eq. (2.161), we neglect higher-order effects $\sim\mathrm{O}(\lambda^2\,\langle \boldsymbol{u}^2\rangle,\, \nabla^2\,\langle \boldsymbol{u}^2\rangle,\, \lambda_i\nabla_i\langle \boldsymbol{u}^2\rangle)$, i.e., we assume that $\ell_0 \ll H_\rho$ and $\ell_0 \ll L_u$. Here ℓ_0 is the maximum scale of random motions, $L_u = \left|\boldsymbol{\nabla}\ln\langle \boldsymbol{u}^2\rangle\right|^{-1}$ is the characteristic scale of the inhomogeneity of random fluid flow, and $H_\rho = |\boldsymbol{\lambda}|^{-1}$ is the mean density variation scale, which is assumed to be constant. This means that in Eq. (2.161) we only take into account leading effects, which are linear in stratification ($\propto \ell_0/H_\rho$) and inhomogeneity of turbulence ($\propto \ell_0/L_u$). Generally, stratification also contributes to div $\boldsymbol{u}$, and therefore it contributes to the parameter σ_c. However, this contribution is small [$\sim\mathrm{O}(\lambda^2\,\langle \boldsymbol{u}^2\rangle)$], and it is neglected in Eq. (2.161). The degree of compressibility σ_c depends on the Mach number, but this dependence is not known for arbitrary Mach numbers and has to be determined in direct numerical simulations. This implies that we separate effects of the arbitrary Mach number, characterized by the parameter σ_c, and density stratification, described by $\boldsymbol{\lambda}$.

In Eq. (2.161), we assume that the random flow has a power-law spectrum for incompressible $E(k) = (q-1)\,(k/k_0)^{-q}\, k_0^{-1}$ and compressible

$E_c(k) = (q_c - 1)\,(k/k_0)^{-q_c}\,k_0^{-1}$ parts, where the wave number varies in the range, $k_0 \le k \le k_\nu$, $k_\nu = 1/\ell_\nu$ is the wave number based on the viscous scale ℓ_ν, and $k_0 = 1/\ell_0 \ll k_\nu$. We assume that there are no random motions for $k < k_0$. This implies that ℓ_0 is the maximum scale of random motions, and $E(k) = E_c(k) = 0$ for $k < k_0$. We consider the frequency function $\Phi(\omega)$ in the form of a Lorentz profile: $\Phi(\omega) = [\pi\,\tau_0\,(\omega^2 + \tau_0^{-2})]^{-1}$, where $\tau_0 = \ell_0/\sqrt{\langle \boldsymbol{u}^2\rangle}$ is the correlation time and $\sqrt{\langle \boldsymbol{u}^2\rangle}$ is the characteristic turbulent velocity at scale ℓ_0.

In spite of the large-scale effects caused by stratification $\boldsymbol{\lambda}$ and inhomogeneous turbulence $\boldsymbol{\nabla}\langle \boldsymbol{u}^2\rangle$, the model (2.161) describes a weakly anisotropic approximation. Indeed, the contributions of these effects to $f_{ij}^{(0)}$ are small because $\ell_0 \ll H_\rho$ and $\ell_0 \ll L_u$. Since we consider only linear effects in $\boldsymbol{\lambda}$ and $\boldsymbol{\nabla}\langle \boldsymbol{u}^2\rangle$, the second rank tensor $f_{ij}^{(0)}$ is constructed as a linear combination of symmetric tensors, δ_{ij} and k_{ij}, with respect to the indexes i and j, and non-symmetric tensors, $k_i\lambda_j$, $k_j\lambda_i$, and $k_i\nabla_j\langle \boldsymbol{u}^2\rangle$, $k_j\nabla_i\langle \boldsymbol{u}^2\rangle$. To determine unknown coefficients multiplying by these tensors, we use the following conditions in the derivation of Eq. (2.161):

$$\left\langle \boldsymbol{u}^2 \right\rangle = \int f_{ii}^{(0)}(\boldsymbol{k}, \omega, \boldsymbol{K})\, \exp(\mathrm{i}\,\boldsymbol{K}\cdot\boldsymbol{R})\, d\boldsymbol{k}\, d\omega\, d\boldsymbol{K}, \tag{2.163}$$

$$\left\langle (\mathrm{div}\,\boldsymbol{u})^2 \right\rangle = \int (k_i + K_i/2)\,(k_j - K_j/2)\, f_{ij}^{(0)}(\boldsymbol{k}, \omega, \boldsymbol{K})\, \exp(\mathrm{i}\,\boldsymbol{K}\cdot\boldsymbol{R})\, d\boldsymbol{k}\, d\omega\, d\boldsymbol{K}, \tag{2.164}$$

$$f_{ij}^{(0)}(\boldsymbol{k}, \omega, \boldsymbol{K}) = f_{ji}^{*(0)}(\boldsymbol{k}, \omega, \boldsymbol{K}) = f_{ji}^{(0)}(-\boldsymbol{k}, \omega, \boldsymbol{K}). \tag{2.165}$$

For very low Mach number and when the parameter σ_c is very small, the continuity equation can be written in the anelastic approximation, $\mathrm{div}\,(\overline{\rho}\,\boldsymbol{u}) = 0$, which implies that $(\mathrm{i}\,k_i + \mathrm{i}\,K_i/2 - \lambda_i)\, f_{ij}^{(0)}(\boldsymbol{k}, \omega, \boldsymbol{K}) = 0$ and $(-\mathrm{i}\,k_j + \mathrm{i}\,K_j/2 - \lambda_j)\, f_{ij}^{(0)}(\boldsymbol{k}, \omega, \boldsymbol{K}) = 0$. For a purely potential random flow, Eq. (2.161) is applied in the limit $\sigma_c \to \infty$, which yields

$$f_{ij}^{(0)}(\boldsymbol{k}, \omega, \boldsymbol{R}) = \frac{\Phi(\omega)\, E_c(k)}{4\pi\, k^2} \biggl[k_{ij} + \frac{\mathrm{i}}{4k^2}\,(k_i\nabla_j - k_j\nabla_i) + \frac{\mathrm{i}}{2k^2}\bigl(k_j\lambda_i - k_i\lambda_j\bigr) \biggr] \left\langle \boldsymbol{u}^2 \right\rangle. \tag{2.166}$$

2.5.4 Turbulent Particle Flux and Level of Particle Number Density Fluctuations for $Pe^{(n)} \ll 1$

Applying the model of background random flow given by Eq. (2.161), and integrating in the ω and $\boldsymbol{k}$ space by means of Eqs. (1.133)–(1.137), we arrive at an equation for the turbulent flux of particles in a compressible fluid flow with arbitrary Mach numbers:

$$\left\langle n'\,\boldsymbol{u} \right\rangle = \overline{n}\, \boldsymbol{V}^{\mathrm{eff}} - D_T\, \boldsymbol{\nabla}\overline{n}, \tag{2.167}$$

where for small Péclet numbers ($\mathrm{Pe}^{(n)} \ll 1$), the turbulent diffusion coefficient D_T and the effective pumping velocity $\boldsymbol{V}^{\rm eff}$ are given by

$$D_T = \frac{(q-1)}{3(q+1)} \, \tau_0 \left\langle \boldsymbol{u}^2 \right\rangle \mathrm{Pe}^{(n)} \left(1 - \frac{\sigma_c \, (1 + C_\sigma)}{1 + \sigma_c}\right), \tag{2.168}$$

$$\boldsymbol{V}^{\rm eff} = \frac{(q-1)}{3(q+1)} \, \frac{\tau_0 \left\langle \boldsymbol{u}^2 \right\rangle}{(1+\sigma_c)} \, \mathrm{Pe}^{(n)} \left[C_\sigma \sigma_c \boldsymbol{\lambda}_u - (1 + C_\sigma \sigma_c) \boldsymbol{\lambda}\right], \tag{2.169}$$

and

$$C_\sigma = \frac{(q_c - 1)\,(q+1)}{(q_c + 1)\,(q-1)}. \tag{2.170}$$

Here $\boldsymbol{\lambda}_u = \boldsymbol{\nabla} \ln \left\langle \boldsymbol{u}^2 \right\rangle$. Since $\tau_0 \, \mathrm{Pe}^{(n)} = \ell_0^2 / D^{(n)}$, the turbulent diffusivity and the effective pumping velocity are determined only by the microphysical diffusion time scale $\ell_0^2 / D^{(n)}$ for small Péclet numbers. Equation (2.168) implies that for small Péclet numbers, compressibility effects decrease the turbulent diffusivity. However, the total (effective) diffusivity, $D^{(n)} + D_T$, cannot be negative, because the molecular diffusivity is much larger than the turbulent one, $D^{(n)} \gg |D_T|$, for $\mathrm{Pe}^{(n)} \ll 1$.

The physics related to different terms in Eq. (2.169) will be discussed in the next section. By means of the equation of state for a perfect gas which yields $\boldsymbol{\nabla} \ln \overline{P} = \boldsymbol{\nabla} \ln \overline{\rho} + \boldsymbol{\nabla} \ln \overline{T}$, we rewrite Eq. (2.169) for the effective pumping velocity as

$$\boldsymbol{V}^{\rm eff} = \frac{(q-1)}{3(q+1)} \, \frac{\tau_0 \left\langle \boldsymbol{u}^2 \right\rangle}{(1+\sigma_c)} \, \mathrm{Pe}^{(n)} \left[(1 + C_\sigma \sigma_c) \left(\boldsymbol{\nabla} \ln \overline{P} - \boldsymbol{\nabla} \ln \overline{T}\right) + C_\sigma \, \sigma_c \, \boldsymbol{\nabla} \ln \left\langle \boldsymbol{u}^2 \right\rangle\right], \tag{2.171}$$

where $\overline{T}$ and $\overline{P}$ are the mean temperature and the mean pressure, respectively. Note that, since the density-temperature correlation $\langle \rho' \theta \rangle$ is much smaller than $\overline{\rho} \, \overline{T}$, the equation of state for perfect gas is also valid for the mean quantities, where ρ' and θ are fluctuations of the fluid density and temperature.

Using Eq. (2.160) and integrating in the ω and $\boldsymbol{k}$ space, we arrive at the expression for the level of particle number density fluctuations for small Péclet numbers as

$$\left\langle (n')^2 \right\rangle = \left(\frac{q_c - 1}{q_c + 1}\right) \left(\frac{\sigma_c}{1 + \sigma_c}\right) \mathrm{Pe}^2 \, \overline{n}^2 + \frac{q-1}{3(q+3)} \, \mathrm{Pe}^2 \, \ell_0^2 \left\{ (\boldsymbol{\nabla} \overline{n})^2 + \frac{1}{8} \left[6 (\boldsymbol{\lambda} \cdot \boldsymbol{\nabla}) + 5 \, (\boldsymbol{\lambda}_u \cdot \boldsymbol{\nabla}) \right] \overline{n}^2 \right\}. \tag{2.172}$$

The first term on the right-hand side of Eq. (2.172) determines a dominant contribution of the compressible part of velocity fluctuations to the level of particle number density fluctuations. Here we take into account only the dominant contribution and neglect much smaller contributions, $\sim \mathrm{O}[\ell_0^2/(L_n L_u)]$, $\mathrm{O}[\ell_0^2/(L_n H_\rho)]$, and $O[\ell_0^2/L_n^2]$, which are caused by the compressible part of velocity fluctuations. For small σ_c, the level of particle number density fluctuations is determined by the last three terms in Eq. (2.172) caused by the density-stratified and inhomogeneous part of velocity fluctuations.

2.5.5 Derivation of an Equation for Second Moments for $Pe^{(n)} \gg 1$

In this section, we determine the turbulent flux of particles and the level of particle number density fluctuations in a compressible fluid flow with arbitrary Mach numbers for large Péclet and Reynolds numbers. We consider fully developed turbulence, where the Strouhal number is of the order of unity and the turbulent correlation time is scale-dependent, so we apply the Fourier transform only in the $\boldsymbol{k}$ space. Using Eq. (2.140) for fluctuations n' and the Navier-Stokes equation for the velocity $\boldsymbol{u}$ written in Fourier space, we derive equations for the following correlation functions:

$$F_j(\boldsymbol{k}, \boldsymbol{R}) = \int \langle n'(\boldsymbol{k} + \boldsymbol{K}/2)\, u_j(-\boldsymbol{k} + \boldsymbol{K}/2) \rangle \exp[\mathrm{i}\boldsymbol{K} \cdot \boldsymbol{R}]\, d\boldsymbol{K}, \tag{2.173}$$

$$E_n(\boldsymbol{k}, \boldsymbol{R}) = \int \langle n'(\boldsymbol{k} + \boldsymbol{K}/2)\, n'(-\boldsymbol{k} + \boldsymbol{K}/2) \rangle \exp[\mathrm{i}\boldsymbol{K} \cdot \boldsymbol{R}]\, d\boldsymbol{K}. \tag{2.174}$$

For brevity of notation, we omit the large-scale variables t and $\boldsymbol{R}$ in the functions $F_j(\boldsymbol{k}, \boldsymbol{R}, t)$ and $E_n(\boldsymbol{k}, \boldsymbol{R}, t)$, and in the mean number density $\overline{n}(\boldsymbol{R}, t)$.

To derive evolution equations for $F_j(\boldsymbol{k}, \boldsymbol{R})$ and $E_n(\boldsymbol{k}, \boldsymbol{R})$, we rewrite equation (2.140) for fluctuations of the particle number density in the $\boldsymbol{k}$ space,

$$\frac{\partial n'(\boldsymbol{k}, t)}{\partial t} = -\mathrm{i}\, k_p \int \overline{n}(\boldsymbol{Q}, t)\, u_p(\boldsymbol{k} - \boldsymbol{Q}, t)\, d\boldsymbol{Q} + n^{(\mathrm{N})}(\boldsymbol{k}, t), \tag{2.175}$$

using (2.145), where $n^{(\mathrm{N})}(\boldsymbol{k}, t)$ includes the nonlinear terms. We use the identities:

$$\frac{\partial}{\partial t} \langle n'(\boldsymbol{k}_1, t)\, u_j(\boldsymbol{k}_2, t) \rangle = \left\langle \frac{\partial n'(\boldsymbol{k}_1, t)}{\partial t}\, u_j(\boldsymbol{k}_2, t) \right\rangle + \left\langle n'(\boldsymbol{k}_1, t)\, \frac{\partial u_j(\boldsymbol{k}_2, t)}{\partial t} \right\rangle, \tag{2.176}$$

$$\frac{\partial}{\partial t} \langle n'(\boldsymbol{k}_1, t)\, n'(\boldsymbol{k}_2, t) \rangle = \left\langle \frac{\partial n'(\boldsymbol{k}_1, t)}{\partial t}\, n'(\boldsymbol{k}_2, t) \right\rangle + \left\langle n'(\boldsymbol{k}_1, t)\, \frac{\partial n'(\boldsymbol{k}_2, t)}{\partial t} \right\rangle, \tag{2.177}$$

which yield

$$\frac{\partial F_j(\boldsymbol{k}, \boldsymbol{R})}{\partial t} = S_j(\boldsymbol{k}, \boldsymbol{R}) + \hat{\mathcal{M}} F_j^{(\mathrm{III})}(\boldsymbol{k}, \boldsymbol{R}), \tag{2.178}$$

$$\frac{\partial E_n(\boldsymbol{k},\boldsymbol{R})}{\partial t} = H(\boldsymbol{k},\boldsymbol{R}) + \hat{\mathcal{M}}E_n^{(\mathrm{III})}(\boldsymbol{k},\boldsymbol{R}), \tag{2.179}$$

where $\boldsymbol{k}_1 \equiv \boldsymbol{k}^{(1)} = \boldsymbol{k} + \boldsymbol{K}/2$ and $\boldsymbol{k}_2 \equiv \boldsymbol{k}^{(1)} = -\boldsymbol{k} + \boldsymbol{K}/2$,

$$S_j(\boldsymbol{k},\boldsymbol{R}) = -\mathrm{i} \int \overline{n}(\boldsymbol{Q})\, k_i^{(1)} \left\langle u_i(\boldsymbol{k}_1 - \boldsymbol{Q})\, u_j(\boldsymbol{k}_2)\right\rangle \exp(\mathrm{i}\,\boldsymbol{K}\cdot\boldsymbol{R})\, d\boldsymbol{K}\, d\boldsymbol{Q}, \tag{2.180}$$

$$H(\boldsymbol{k},\boldsymbol{R}) = -\mathrm{i} \int \left[k_j^{(1)} \left\langle n'(\boldsymbol{k}_2)\, u_j(\boldsymbol{k}_1 - \boldsymbol{Q})\right\rangle + k_j^{(2)} \left\langle n'(\boldsymbol{k}_1)\, u_j(\boldsymbol{k}_2 - \boldsymbol{Q})\right\rangle\right] \times \overline{n}(\boldsymbol{Q}) \exp(\mathrm{i}\,\boldsymbol{K}\cdot\boldsymbol{R})\, d\boldsymbol{K}\, d\boldsymbol{Q}, \tag{2.181}$$

and

$$\hat{\mathcal{M}}F_j^{(\mathrm{III})}(\boldsymbol{k},\boldsymbol{R}) = \int \left[\left\langle n'(\boldsymbol{k}_1)\, \frac{\partial u_j(\boldsymbol{k}_2)}{\partial t}\right\rangle + \left\langle n'^{(\mathrm{N})}(\boldsymbol{k}_1) u_j(\boldsymbol{k}_2)\right\rangle\right] \exp[\mathrm{i}\,\boldsymbol{K}\cdot\boldsymbol{R}]\, d\boldsymbol{K}, \tag{2.182}$$

$$\hat{\mathcal{M}}E_n^{(\mathrm{III})}(\boldsymbol{k},\boldsymbol{R}) = \int \left[\left\langle n'(\boldsymbol{k}_1)\, n'^{(\mathrm{N})}(\boldsymbol{k}_2)\right\rangle + \left\langle n'^{(\mathrm{N})}(\boldsymbol{k}_1) n'(\boldsymbol{k}_2)\right\rangle\right] \exp[\mathrm{i}\,\boldsymbol{K}\cdot\boldsymbol{R}]\, d\boldsymbol{K} \tag{2.183}$$

are the third-order moment terms in the $\boldsymbol{k}$ space appearing due to the non-linear terms. For brevity of notation, we omit the variable t in the functions $\overline{n}(\boldsymbol{Q}, t)$, $n'(\boldsymbol{k}, t)$, $n^{(\mathrm{N})}(\boldsymbol{k}, t)$ and $u_i(\boldsymbol{k}, t)$. We perform calculations similar to those in Eqs. (2.147)–(2.150) and in Eqs. (2.160)–(2.158), which yield the evolution equations for $F_j(\boldsymbol{k})$ and $E_n(\boldsymbol{k})$ for $\mathrm{Pe} \gg 1$ as

$$\frac{\partial F_j}{\partial t} = -\frac{1}{2}\left[\overline{n}\left(2\mathrm{i}\, k_i + \nabla_i\right) f_{ij} + \left(\nabla_i \overline{n}\right)\left(\delta_{im} - k_m \frac{\partial}{\partial k_i}\right) f_{mj}\right] + \hat{\mathcal{M}}F_j^{(\mathrm{III})}, \tag{2.184}$$

$$\frac{\partial E_n}{\partial t} = \frac{1}{2}\left[\overline{n}\left(2\mathrm{i}\, k_j F_j^{(-)} - \nabla_j F_j^{(+)}\right) + (\nabla_m \overline{n})\left(k_j \frac{\partial}{\partial k_m} - \delta_{jm}\right) F_j^{(+)}\right] + \hat{\mathcal{M}}E_n^{(\mathrm{III})}, \tag{2.185}$$

where $F_j^{(\pm)} = F_j(\boldsymbol{k},\boldsymbol{R}) \pm F_j(-\boldsymbol{k},\boldsymbol{R})$.

We use the spectral τ approximation (2.89), where functions with superscript (0) correspond to background turbulence with zero turbulent particle flux and zero level of particle number density fluctuations. Therefore, Eq. (2.89) reduces to $\hat{\mathcal{M}}F_i^{(\mathrm{III})}(\boldsymbol{k}) = -F_i(\boldsymbol{k})/\tau(k)$ and $\hat{\mathcal{M}}E_n^{(\mathrm{III})}(\boldsymbol{k}) = -E_n(\boldsymbol{k})/\tau(k)$. We also assume that the characteristic time of variation of the second moments $F_i(\boldsymbol{k})$ and $E_n(\boldsymbol{k})$ is substantially longer than the correlation time $\tau(k)$ on all turbulence scales. Therefore, we obtain

$$F_j(\boldsymbol{k}, \boldsymbol{R}) = -\frac{1}{2}\,\tau(k) \left[\overline{n}\Big(2\mathrm{i}\,k_i + \nabla_i\Big) f_{ij} + (\nabla_i \overline{n}) \left(\delta_{im} - k_m \frac{\partial}{\partial k_i}\right) f_{mj}\right], \tag{2.186}$$

$$E_n(\boldsymbol{k}, \boldsymbol{R}) = \frac{1}{2}\,\tau(k) \left[\overline{n}\left(2\mathrm{i}\,k_j F_j^{(-)} - \nabla_j F_j^{(+)}\right) + (\nabla_m \overline{n}) \left(k_j \frac{\partial}{\partial k_m} - \delta_{jm}\right) F_j^{(+)}\right], \tag{2.187}$$

In Eqs. (2.186)–(2.187), we take into account a one-way coupling, i.e., we neglect the effect of the gradient of the mean particle number density on the turbulent velocity field. This implies that we replace the correlation function f_{ij} in Eqs. (2.186)–(2.187) by $f_{ij}^{(0)}$ for the background turbulent flow with zero turbulent particle flux.

2.5.6 Model for Compressible Turbulence

We use statistically stationary, density-stratified, inhomogeneous and compressible background turbulence, which is determined by the following correlation function in the $\boldsymbol{k}$ space:

$$\begin{aligned} f_{ij}^{(0)}(\boldsymbol{k}) = \frac{1}{8\pi\,k^2\,(1+\sigma_c)} \Bigg[E(k) \bigg(\delta_{ij} - k_{ij} + \frac{\mathrm{i}}{k^2}\,(k_j \lambda_i - k_i \lambda_j) \\ + \frac{\mathrm{i}}{2k^2}\,(k_i \nabla_j - k_j \nabla_i) \bigg) + 2\sigma_c\, E_c(k) \bigg(k_{ij} + \frac{\mathrm{i}}{4k^2}\,(k_i \nabla_j - k_j \nabla_i) \\ + \frac{\mathrm{i}}{2k^2}(k_j \lambda_i - k_i \lambda_j) \bigg) \Bigg] \langle \boldsymbol{u}^2 \rangle, \end{aligned} \tag{2.188}$$

where contrary to the model adopted by Rogachevskii and Kleeorin (2021), the model (2.188) has vanishing turbulent flux $\langle \rho' \boldsymbol{u} \rangle$. Note that most statements and conditions used for the derivation of the tensorial structure of Eq. (2.161) are also valid for Eq. (2.188). We assume here that the background turbulence is of Kolmogorov type with constant energy flux over the spectrum, i.e., the kinetic energy spectrum for the incompressible part of turbulence in the range of wave numbers $k_0 < k < k_\nu$ is $E(k) = -d\bar{\tau}(k)/dk$, where the function $\bar{\tau}(k) = (k/k_0)^{1-q}$, with $1 < q < 3$ being the exponent of the kinetic energy spectrum. The condition $q > 1$ corresponds to finite kinetic energy for very large fluid Reynolds numbers, while $q < 3$ corresponds to finite dissipation of the turbulent kinetic energy at the viscous scale. Similarly, the kinetic energy spectrum for the compressible part of turbulence is $E_c(k) = -d\bar{\tau}_c(k)/dk$, where the function $\bar{\tau}_c(k) = (k/k_0)^{1-q_c}$ with $1 < q_c < 3$. For example, the exponent of the incompressible part of the energy spectrum $q{=}5/3$ (the Kolmogorov spectrum), while the exponent of the compressible part of the energy spectrum $q_c{=}2$ (like for the Burgers turbulence). For fully developed Kolmogorov-like turbulence, the degree of compressibility $\sigma_c < 1$. The turbulent correlation time in the $\boldsymbol{k}$ space is

$$\tau(k) = \frac{2\tau_0}{1+\sigma_c}\left[\bar{\tau}(k) + \sigma_c\,\bar{\tau}_c(k)\right] \tag{2.189}$$

(Rogachevskii and Kleeorin, 2021), where $\tau_0 = \ell_0/\sqrt{\langle \boldsymbol{u}^2\rangle}$ is the turbulent correlation time in physical space and $\sqrt{\langle \boldsymbol{u}^2\rangle}$ is the characteristic turbulent velocity at scale ℓ_0.

2.5.7 Turbulent Particle Flux and Level of Particle Number Density Fluctuations for $\mathrm{Pe}^{(n)} \gg 1$

Using the model of background turbulence for $f_{ij}^{(0)}$ given by Eq. (2.188), and integrating Eq. (2.186) in the $\boldsymbol{k}$ space, we obtain the turbulent flux of particles given by Eq. (2.167), where the turbulent diffusivity D_T and the effective pumping velocity $\boldsymbol{V}^{\rm eff}$ of non-inertial particles for $\mathrm{Pe}^{(n)} \gg 1$ in a compressible fluid flow with arbitrary Mach numbers are given by

$$D_T = \frac{\tau_0\,\langle \boldsymbol{u}^2\rangle}{3}\left\{1 - \frac{\sigma_c}{2(1+\sigma_c)^2}\left[\tilde{C}_\sigma\,(q-1) + \sigma_c\,(q_c-1)\right]\right\}, \tag{2.190}$$

$$\boldsymbol{V}^{\rm eff} = -\frac{\tau_0\,\langle \boldsymbol{u}^2\rangle}{3}\boldsymbol{\lambda}, \tag{2.191}$$

and

$$\tilde{C}_\sigma = \frac{2(q_c-1)}{q+q_c-2}. \tag{2.192}$$

Equation (2.190) implies that for large Péclet numbers, compressibility decreases the turbulent diffusivity of particles. Since $1 < q < 3$, the turbulent diffusivity is always positive when $\mathrm{Pe}^{(n)} \gg 1$, even for very strong compressibility, $\sigma_c \gg 1$. Here we consider a weak stratification with the density stratification height that is much larger than the turbulent integral scale. Equation (2.191) determines effective pumping velocity $\boldsymbol{V}^{\rm eff}$ of non-inertial particles. It describes pumping of particles caused by the density stratification and it is independent of the parameter σ_c.

To determine the turbulent flux of particles, we use the following integrals:

$$\int_{k_0}^{k_\nu} \tau(k)\; [E(k) + \sigma_c\, E_c(k)]\; dk = \tau_0\, (1+\sigma_c), \tag{2.193}$$

$$\int_{k_0}^{k_\nu} \tau(k)\, E(k)\, dk = \tau_0 \left[1 - \frac{\tilde{C}_\sigma\, \sigma_c}{2(1+\sigma_c)}\right], \tag{2.194}$$

$$\int_{k_0}^{k_\nu} \tau(k)\, E_c(k)\, dk = \tau_0 \left[1 + \frac{\tilde{C}_\sigma}{2(1+\sigma_c)}\right], \tag{2.195}$$

$$\int_{k_0}^{k_\nu} \tau(k)\, f_\tau(k)\, E_c(k)\, dk = -\frac{\tau_0}{1+\sigma_c} \Big[\sigma_c\, (q_c - 1) + \tilde{C}_\sigma\, (q-1)\Big], \tag{2.196}$$

where $f_\tau(k) = d \ln \tau(k)/d \ln k$.

Integrating Eq. (2.187) in the $\boldsymbol{k}$ space, we obtain the level of particle number density fluctuations for large Péclet numbers

$$\left\langle (n')^2 \right\rangle = 8\, f_c \left(\frac{\sigma_c}{1+\sigma_c}\right)^3 \overline{n}^2 + \frac{1}{9}\, \ell_0^2 \left\{8\, (\nabla \overline{n})^2 + \Big[5\, (\boldsymbol{\lambda}_u \cdot \nabla) + 6\, (\boldsymbol{\lambda} \cdot \nabla)\Big] \overline{n}^2\right\}, \tag{2.197}$$

where the function $f_c(q, q_c, \sigma_c)$ depends on the degree of compressibility and the exponents of spectra incompressible and compressible parts of velocity fluctuations:

$$f_c = \frac{q_c - 1}{3q_c - 5} + \frac{2(q_c - 1)}{\sigma_c (q + 2q_c - 5)} + \frac{q_c - 1}{\sigma_c^2 (2q + q_c - 5)}. \tag{2.198}$$

The first term on the right-hand side of Eq. (2.197) determines a dominant contribution of the compressible part of velocity fluctuations to the level of particle number density fluctuations. Here we take into account only the dominant contribution and neglect much smaller contributions $\sim$ O$[\ell_0^2/(L_n L_u)]$, O$[\ell_0^2/(L_n H_\rho)]$, caused by the compressible part of velocity fluctuations. For small σ_c, the level of particle number density fluctuations is determined by the last three terms in Eq. (2.197) caused by the density-stratified and inhomogeneous part of velocity fluctuations. To determine the level of particle number density fluctuations, we use the following integrals:

$$\int_{k_0}^{k_\nu} \tau^2(k)\, k^2\, E_c(k)\, dk = 4 f_c(q, q_c, \sigma_c) \left(\frac{\tau_0}{\ell_0}\right)^2 \left(\frac{\sigma_c}{1+\sigma_c}\right)^2, \tag{2.199}$$

$$\int_{k_0}^{k_\nu} \tau^2(k)\; [E(k) + \sigma_c\, E_c(k)]\; dk = \frac{4}{3}\, \tau_0^2\, (1+\sigma_c). \tag{2.200}$$

For large Reynolds numbers, the exponent of the incompressible part of the energy spectrum $q = 5/3$ (the Kolmogorov spectrum), while the exponent of the compressible part of the energy spectrum $q_c = 2$ (Kritsuk et al., 2007; Federrath, 2013). In this case, $\tilde{C}_\sigma = 6/5$ and the turbulent diffusivity D_T and the effective pumping velocity $\boldsymbol{V}^{\rm eff}$ of non-inertial particles for $\mathrm{Pe}^{(n)} \gg 1$ in a compressible fluid flow with arbitrary Mach numbers are given by

$$D_T = \frac{\tau_0 \left\langle \boldsymbol{u}^2 \right\rangle}{3} \left[1 - \frac{\sigma_c\,(\sigma_c + 4/5)}{2(1+\sigma_c)^2}\right], \tag{2.201}$$

$$\boldsymbol{V}^{\rm eff} = \frac{\tau_0 \left\langle \boldsymbol{u}^2 \right\rangle}{3} \boldsymbol{\nabla} \ln \overline{\rho}. \tag{2.202}$$

In this case, the level of particle number density fluctuations is given by

$$\left\langle (n')^2 \right\rangle = 8\,\left(1 + 3\sigma_c^{-1} + 3\sigma_c^{-2}\right) \left(\frac{\sigma_c}{1+\sigma_c}\right)^3 \overline{n}^2. \tag{2.203}$$

In Eq. (2.203), we neglect small contributions $\sim \mathrm{O}[\ell_0^2/(L_n L_u)]$, $\mathrm{O}[\ell_0^2/(L_n H_\rho)]$ and $\sim \mathrm{O}[\ell_0^2/L_n^2]$.

For nearly irrotational flow ($\sigma_c \gg 1$), the turbulent diffusivity for large Péclet numbers is given by

$$D_T = \frac{3 - q_c}{6} \tau_0 \left\langle \boldsymbol{u}^2 \right\rangle, \tag{2.204}$$

and the level of particle number density fluctuations is

$$\left\langle (n')^2 \right\rangle = \frac{8\,(q_c - 1)}{3q_c - 5} \overline{n}^2. \tag{2.205}$$

2.5.8 Physics of Compressibility Effects on Turbulent Particle Transport

In this section, we discuss the physics related to compressibility effects on turbulent particle transport. We start with a discussion of the physics related to the turbulent pumping of particles caused by the fluid density stratification. Substituting Eq. (2.167) into Eq. (2.139), we obtain the equation for the mean number density $\overline{n}$ as

$$\frac{\partial \overline{n}}{\partial t} + \boldsymbol{\nabla} \cdot \left[\overline{n}\, \boldsymbol{V}^{\rm eff} - (D^{(n)} + D_T)\, \boldsymbol{\nabla} \overline{n}\right] = 0, \tag{2.206}$$

where the turbulent diffusivity D_T and the effective pumping velocity $\boldsymbol{V}^{\rm eff}$ are given by Eqs. (2.168) and (2.169) for small Péclet numbers and by Eqs. (2.190) and (2.191) for large Péclet numbers.

The steady-state solution of Eq. (2.206) for the mean number density of non-inertial particles is given by $\overline{n}/\overline{n}_* = [\overline{\rho}/\overline{\rho}_*]^{D_T/(D^{(n)}+D_T)}$, where subscripts $*$ represent the values of the mean fluid density $\overline{\rho}_*$ and the mean number density of particles $\overline{n}_*$ at the boundary where the total particle flux vanishes. This solution implies that small particles are accumulated in the vicinity of the maximum of the fluid density because the non-diffusive turbulent flux of particles, $\overline{n}\,\boldsymbol{V}^{\rm eff}$, is directed toward the maximum of the fluid density.

Using the equation of state for a perfect gas, we rewrite Eq. (2.191) for the pumping effective velocity in the form

$$\boldsymbol{V}^{\rm eff} = \frac{\tau_0\,\left\langle \boldsymbol{u}^2\right\rangle}{3}\left(\boldsymbol{\nabla}\ln\overline{P} - \boldsymbol{\nabla}\ln\overline{T}\right). \tag{2.207}$$

The first term on the right-hand side of Eq. (2.207), $\propto \boldsymbol{\nabla}\ln\overline{P}$, describes turbulent barodiffusion (Elperin et al., 1997c), while the second term, $\propto -\boldsymbol{\nabla}\ln\overline{T}$, characterizes the phenomenon of turbulent thermal diffusion (Elperin et al., 1996c). The last term, $\propto \sigma_c\boldsymbol{\nabla}\ln\left\langle \boldsymbol{u}^2\right\rangle$, in Eq. (2.171) for the pumping effective velocity for small Péclet numbers describes the compressible turbophoresis for non-inertial particles. For large Péclet numbers this effect does not occur, see Eq. (2.207).

The classical turbophoresis (Caporaloni et al., 1975; Reeks, 1983) exists only for inertial particles and causes an additional mean particle velocity given by Eq. (2.120) in Section 2.4.2. This turbophoresis effect results in the formation of large-scale inhomogeneous distribution of inertial particles at local minima of the mean-squared turbulent velocity. On the other hand, the compressible turbophoresis for non-inertial particles originates from the turbulent particle flux $\left\langle n'\,\boldsymbol{u}\right\rangle$, i.e., it describes a collective statistical phenomenon, while classical turbophoresis is not related to the correlations between velocity and number density fluctuations and is obtained from the averaging of the velocity of inertial particles.

2.6 Further Reading

Turbulent transport of particles has been a subject of many studies due to numerous applications in geophysics and environmental sciences, astrophysics and various industrial flows (Csanady, 2012; Zeldovich et al., 1990; Blackadar, 2012; Seinfeld and Pandis, 2016; Zaichik et al., 2008; Crowe et al., 2011; Armitage, 2010;

Warhaft, 2000; Shaw, 2003; Khain et al., 2007; Warhaft, 2008; Toschi and Bodenschatz, 2009; Balachandar and Eaton, 2010; Khain and Pinsky, 2018). Different mechanisms of large-scale and small-scale clustering of inertial particles have been proposed. The large-scale clustering occurs in scales that are much larger than the integral scale of turbulence, while the small-scale clustering is observed in scales that are much smaller than the integral turbulence scale.

2.6.1 Large-Scale Clustering

The large-scale clustering of inertial particles in isothermal non-stratified inhomogeneous turbulence occurs, e.g., due to turbophoresis, see Section 2.4.2 (Caporaloni et al., 1975; Reeks, 1983; Guha, 1997; Elperin et al., 1998b; Guha, 2008; Mitra et al., 2018), which is a combined effect of particle inertia and inhomogeneity of turbulence. Turbophoresis results in the appearance of the additional non-diffusive turbulent flux of inertial particles $\boldsymbol{J}_{\rm turboph} = \overline{n}\, \overline{\boldsymbol{V}}_{\rm turboph}$, where the mean particle velocity caused by turbophoresis can be written in a general case as

$$\overline{\boldsymbol{V}}_{\rm turboph} = -\kappa_{\rm turboph} \frac{\nabla \left\langle \boldsymbol{u}^2 \right\rangle}{2}. \tag{2.208}$$

Here $\boldsymbol{u}$ is the turbulent fluid velocity, $\kappa_{\rm turboph}$ is the turbophoretic coefficient which generally depends on the Stokes number $\mathrm{St} = \tau_p/\tau_\nu$ and the fluid Reynolds number $\mathrm{Re} = \ell_0\, u_0/\nu$, where $\tau_\nu = \tau_0/\mathrm{Re}^{1/2}$ is the Kolmogorov time and τ_p is the Stokes time for the small spherical particles. Due to turbophoresis, inertial particles are accumulated in the vicinity of the minimum of the turbulent intensity.

Direct numerical simulations (Mitra et al., 2018) show that inertial particles in inhomogeneously forced isothermal turbulent flows are accumulated at the minima of mean-square turbulent velocity. Two turbulent transport processes, turbophoresis and turbulent diffusion together determine the spatial distribution of the particles. The non-dimensional product of the turbophoretic coefficient $\kappa_{\rm turboph}$ and the rms velocity $u_{\rm rms} \equiv \sqrt{\langle \boldsymbol{u}^2 \rangle}$ is shown to increase linearly with the parameter St_f for $\mathrm{St}_f \ll 1$, reach a maxima for $\mathrm{St}_f \sim 10$ and decrease as $\mathrm{St}_f^{-1/3}$ for large St_f (Mitra et al., 2018), where $\mathrm{St}_f = \tau_p/\tau_0$ is the Stokes number defined using the characteristic flow time scale, τ_0, based on the integral (forcing) scale of turbulence. The same clustering phenomenon has been studied numerically in turbulent Kolmogorov flows (De Lillo et al., 2016). Although the authors do not interpret their results as a balance between turbophoretic and turbulent diffusive fluxes, they do observe that clustering increases for small St_f but this trend reverses smoothly at higher values of St_f. The large-scale clustering due to turbophoresis has been also observed in DNS of turbulent channel flows (Sardina et al., 2012) and in various experimental studies (Kaftory et al., 1995a,b; Righetti and Romano, 2004).

Another example of the large-scale clustering of inertial particles is the phenomenon of turbulent thermal diffusion (see Sections 2.1.6, 2.2.4, 2.4.3, 2.4.4, 2.5.4 and 2.5.7) that is a combined effect of the temperature stratified turbulence and inertia of small particles (Elperin et al., 1996c, 1997c). Turbulent thermal diffusion is a purely collective phenomenon occurring in temperature stratified turbulence and resulting in the appearance of a non-zero mean effective velocity of particles in the direction opposite to the mean temperature gradient. This implies that this phenomenon causes a non-diffusive turbulent flux of particles in the direction of the turbulent heat flux. A competition between the turbulent thermal diffusion and turbulent diffusion determines the conditions for the formation of large-scale particle concentrations in the vicinity of the mean temperature minimum. Turbulent thermal diffusion has been intensively investigated analytically (Elperin et al., 1996c, 1997c, 1998b, 2000a,b, 2001a; Pandya and Mashayek, 2002; Reeks, 2005; Amir et al., 2017) using different theoretical approaches, in laboratory experiments in oscillating grid turbulence (Buchholz et al., 2004; Eidelman et al., 2004, 2006a; Amir et al., 2017) and in the multi-fan produced turbulence (Eidelman et al., 2006c). This effect has also been detected in direct numerical simulations (Haugen et al., 2012; Brandenburg et al., 2012a) using the pencil-code (Brandenburg and Dobler, 2002). This phenomenon has been observed in geophysical turbulence, i.e., in the atmosphere of the Earth (Sofiev et al., 2009) and the atmosphere of Titan (Elperin et al., 1997a), and it also has been discussed in astrophysical turbulence applications (Hubbard, 2016).

2.6.2 Small-Scale Clustering

Small-scale clusters (small-scale inhomogeneities) in spatial distribution of particles in turbulent flows have been observed in various laboratory experiments (Eaton and Fessler, 1994; Warhaft, 2000, 2008; Aliseda et al., 2002; Wood et al., 2005; Salazar et al., 2008; Xu and Bodenschatz, 2008), direct numerical simulations (Wang and Maxey, 1993; Boffetta et al., 2004; Collins and Keswani, 2004; Bec et al., 2007; Van Aartrijk and Clercx, 2008; Toschi and Bodenschatz, 2009; Pandey et al., 2019) and in atmospheric flows (Shaw, 2003; Britter and Hanna, 2003; Khain et al., 2007; Siebert et al., 2010; Khain and Pinsky, 2018). The small-scale clustering of inertial particles in a turbulent flow is caused by the centrifugal effect, which implies that the inertial particles are locally accumulated in regions between the turbulent eddies. These regions have a low vorticity, high strain rate and maximum fluid pressure. Turbulent vortices act as small centrifuges that push heavy particles to the boundary regions between the eddies by the inertial forces creating local concentration inhomogeneities. This effect is known as the inertia-induced particle clustering (Maxey, 1987), and it has been investigated in various analytical studies

(Elperin et al., 1996b, 2000c, 2002a, 2007a; Balkovsky et al., 2001; Mehlig et al., 2005; Olla, 2010; Gustavsson and Mehlig, 2016).

There are two types of inertia-induced particle clustering. The first type is source clustering related to the source term in the equation for the two-point correlation function of particle number density fluctuations. Most analytical studies of preferential concentration are related to the source inertial clustering (Balkovsky et al., 2001; Olla, 2010; Gustavsson and Mehlig, 2016). Another type of particle clustering is associated with a spontaneous breakdown of their homogeneous spatial distribution due to the inertial clustering instability (Elperin et al., 1996b, 2002a, 2007a). The inertial clustering instability can be of great importance in different practical applications involving particle mixing and transport.

Small-scale particle clustering in the temperature stratified turbulence can be more effective than the inertial particle clustering in isothermal turbulence (Eidelman et al., 2010). This is because the mean temperature gradient in turbulent flow can serve as a strong source of temperature fluctuations which are correlated with fluctuations of fluid velocity and pressure. The inertia effects cause particle clustering in the vicinity of maximum pressure fluctuations. Therefore, enhanced pressure fluctuations by the mean temperature gradient result in an increase in particle number density fluctuations, and thus they enhance the rate of formation of the particle clusters. Moreover, tangling of the mean gradient of particle number density formed by turbulent thermal diffusion generates additional fluctuations of particle concentrations, which in turn enhance particle clustering.

The source tangling clustering (i.e., clustering without excitation of the clustering instability) has been recently detected in laboratory experiments in temperature stratified turbulence (Eidelman et al., 2010). The source tangling clustering has been studied for inertial particles with small Stokes numbers and with the material density that is much larger than the fluid density. It has been demonstrated in these experiments that the source tangling clustering is much more effective than a pure inertial clustering observed in isothermal turbulence. In particular, in experiments with oscillating grid isothermal turbulence in air, the inertial clustering is shown to be very weak for solid particles with a diameter $\approx 10\ \mu$m and turbulence with Reynolds number based on turbulent length scale and rms velocity, $\mathrm{Re} = 250$ (Eidelman et al., 2010). As follows from measurements, the two-point correlation function for the inertial clustering in isothermal turbulence is significantly less localized than that for the tangling clustering in non-isothermal turbulence. The developed theory of the source tangling clustering is in agreement with the results of these laboratory experiments (Eidelman et al., 2010).

Another regime of the tangling clustering, namely the tangling clustering instability in a temperature stratified turbulence with small yet finite correlation time, has been also found in analytical studies (Elperin et al., 2013). The tangling mechanism in the temperature stratified turbulence strongly increases the degree of compressibility of the particle velocity field. This results in the essential decrease of the threshold for the excitation of the tangling clustering instability even for small particles.

The tangling clustering instability (Elperin et al., 2013) in the temperature stratified turbulence is essentially different from the inertial clustering instability (Elperin et al., 1996b, 2002a, 2007a) that occurs in non-stratified isotropic and homogeneous turbulence. While the inertial clustering instability is caused by the centrifugal effect of the turbulent eddies, the mechanism of the tangling clustering instability is related to temperature fluctuations generated by the tangling of the mean temperature gradient by velocity fluctuations. Temperature fluctuations produce pressure fluctuations and cause particle accumulations in regions with increased instantaneous pressure. The growth rate of the tangling clustering instability is by $\sqrt{\mathrm{Re}}\,(\ell_0/L_T)^2/(3\mathrm{Ma})^4$ times larger than that of the inertial clustering instability (Elperin et al., 2013), where $\mathrm{Ma} = u_0/c_s$ is the Mach number based on the characteristic turbulent velocity u_0 at the integral turbulence scale, ℓ_0, and L_T is the characteristic scale of the mean temperature variations. Depending on the parameters of the turbulence and the mean temperature gradient, there is a preferential particle size at which the particle clustering due to the tangling clustering instability is more effective. The particle number density inside the cluster after the saturation of this instability can be by several orders of magnitude larger than the mean particle number density. Note also that the evaporation of droplets can drastically change the tangling clustering instability, e.g., it increases the instability threshold in the droplet radius (Elperin et al., 2013).

The tangling clustering instability is of great importance in view of various applications, e.g., it can affect radiative transfer in turbulent flows where unconfined dust explosions can occur (Liberman et al., 2017, 2018). It is known that unconfined dust explosions typically start off with a relatively weak primary flame followed by a severe secondary explosion. The clustering of dust particles in a temperature stratified turbulent flow ahead of the primary flame may give rise to a significant increase in the radiation penetration length. These particle clusters, even far ahead of the flame, are sufficiently exposed and heated by the radiation from the flame to become ignition kernels capable to ignite a large volume of fuel-air mixtures. This efficiently increases the total flame surface area and the effective combustion speed, defined as the rate of reactant consumption of a given volume (Liberman et al., 2017, 2018). This mechanism explains the high rate of combustion and overpressures required to account for the observed level of damage in unconfined dust explosions, e.g., at the 2005 Buncefield vapor-cloud explosion. The effect of the strong increase of radiation transparency due to turbulent clustering of particles has also many implications in atmospheric physics and astrophysics.

Small-scale clustering is important in atmospheric turbulence. For instance, the key question in cloud physics is how turbulence affects the collisional growth of droplets. This is crucially important because the collisional growth of droplets in atmospheric clouds may explain the rapid rain formation (Khain and Pinsky, 2018; Lamb and Verlinde, 2011; Shaw, 2003; Khain et al., 2007; Siebert et al., 2010; Grabowski and Wang, 2013; Elperin et al., 2015). To form the raindrops, which can fall down, triggering rain, they must grow to about 50-micron-size droplets, which would take a very long time. Observations indicate that the average time for rainfall initiation is approximately 15–20 min, while existing theories predict that the duration of a time interval required for droplets to grow up to 50 μm in radius is of the order of hours. Small-scale clustering can cause large velocity differences between nearby droplets (Sundaram and Collins, 1997; Falkovich et al., 2002; Falkovich and Pumir, 2007; Wilkinson et al., 2006; Wilkinson, 2016; Gustavsson and Mehlig, 2014) and enhance the collision rate of small droplets (Duncan et al., 2005; Gustavsson et al., 2008; Gustavsson and Mehlig, 2016; Elperin et al., 2015; Li et al., 2018, 2020). This can increase the droplet size and trigger rain.

Jets are often used in numerical, experimental and theoretical studies for the modeling of mixing of particles in various engineering applications. Different kinds of structures are formed during mixing in turbulent jets. One of the most significant results is a detection of a sharp increase of mixing rates caused by the onset of small-scale turbulence within large-scale coherent motions (Reynolds et al., 2003; Dimotakis, 2005). Mixing of small solid particles characterized by large Schmidt number and small Stokes number at the external boundary of a submerged turbulent jet has been studied experimentally and theoretically by Eidelman et al. (2009). The spatial distributions of the jet fluid characterized by a high concentration of the particles and of the ambient fluid characterized by a low concentration of the tracer particles have been determined. In the data analysis two approaches are used: (i) the method based on the measured phase function for the study of the mixed state of two fluids; (ii) the approach based on the analysis of the two-point second-order correlation function of the particle number density fluctuations generated by tangling of the gradient of the mean particle number density by the turbulent velocity field. This gradient is formed at the external boundary of a submerged turbulent jet. Probability density function (PDF) of the phase function of a jet fluid penetrating into an external flow and the two-point second-order correlation function of the particle number density do not have universal scaling and cannot be described by a power-law function. The theoretical predictions made in this study are in a qualitative agreement with the obtained experimental results (Eidelman et al., 2009).

2.6.3 Turbulent Transport in Chemically Reacting Flows

Turbulent transport of particles and droplets in flows with chemical reactions is of great interest in various applications, ranging from combustion to physics of turbulent atmosphere of the Earth (Zeldovich et al., 1985; Fox, 2003; Peters, 2004; Liberman, 2010; Echekki and Mastorakos, 2010; Swaminathan and Bray, 2011; Lipatnikov, 2012; Seinfeld and Pandis, 2016). For instance, the significant influence of combustion on turbulent transport is well-known to cause the counter-gradient scalar transport, i.e., a flux of products from unburnt to burnt regions of a premixed flame (Lipatnikov, 2012; Libby, 1985; Bray, 1995; Swaminathan and Bray, 2011; Sabelnikov and Lipatnikov, 2017).

The counter-gradient transport can substantially reduce the flame speed and, therefore, is of great importance for calculations of burning rate and plays a key role in the premixed turbulent combustion (Sabelnikov and Lipatnikov, 2013, 2014, 2015). The counter-gradient transport appears to be an indirect manifestation of the influence of chemical reactions on turbulent fluxes, as this manifestation is controlled by density variations due to heat release in combustion reactions, rather than by the reactions themselves. As far as the straightforward influence of reactions on turbulent transport is concerned (Sreenivasan et al., 1982), such effects have yet been addressed in a few studies (Lipatnikov, 2012) of premixed flames.

There are several interesting phenomena in chemically reacting turbulent flows. The first is related to the effect of chemical reactions on turbulent diffusion which has been studied analytically by means of a path-integral approach (Elperin et al., 1998a) for the Kraichnan-Kazantsev model (Kraichnan, 1968; Kazantsev, 1968) of very short correlated in time random velocity field and using the spectral tau approach (Elperin et al., 2014). These studies have demonstrated that turbulent diffusion of the reacting species can be strongly depleted by a large factor proportional to turbulent Damköhler number (the ratio of turbulent and chemical times; Elperin et al., 1998a, 2014, 2017b).

The mean-field numerical simulations (Brandenburg et al., 2011b) have been used to validate the theoretical dependence of turbulent diffusion coefficient versus the turbulent Damköhler number (Elperin et al., 2014). In the mean-field simulations, a reactive front propagation in a turbulent flow was investigated using the Kolmogorov-Petrovskii-Piskunov (KPP) equation (Kolmogorov et al., 1937) or the Fisher equation (Fisher, 1937). This equation has also been amended by an advection term to describe the interaction with a turbulent velocity field (Benzi and Nelson, 2009; Perlekar et al., 2010). In mean-field simulations of this equation (Brandenburg et al., 2011b), memory effects of turbulent diffusion have been taken into account to determine the front speed in the case when the turbulent time is much larger than the characteristic chemical time. The memory effects saturate the front speed to values of the turbulent speed, while the nonlinearity of the reaction term increases the front speed. Comparison between the theoretical predictions (Elperin et al., 2014) and the mean-field numerical simulations (Brandenburg et al., 2011b) shows a good agreement.

The theory of turbulent diffusion of chemically reacting gaseous admixtures (Elperin et al., 2014) has been also generalized for large yet finite Reynolds numbers, where the dependence of turbulent diffusion coefficient on two parameters, the Reynolds number and Damköhler number has been obtained (Elperin et al., 2017b). Three-dimensional direct numerical simulations of a finite-thickness reaction wave for the first-order chemical reactions propagating in forced, homogeneous, isotropic and incompressible turbulence have been performed (Elperin et al., 2017b) to validate the theoretically predicted effect of chemical reactions on turbulent diffusion. It is shown that the obtained numerical results are in good quantitative agreement with the developed theory.

Other cross effects that strongly influence turbulent transport in chemically reacting turbulent flows are related to the appearance of additional turbulent fluxes – namely, (i) a non-diffusive turbulent flux of number density of gaseous admixture proportional to the mean temperature gradient multiplied by the number density of gaseous admixture (turbulent thermal diffusion) and (ii) additional turbulent heat flux proportional to the gradient of the mean number density of gaseous admixture (Elperin et al., 1998a, 2014, 2017b). These turbulent fluxes can be interpreted as "turbulent" analogs of the known molecular cross-effects in irreversible thermodynamics. In particular, the first effect, namely turbulent thermal diffusion, can be considered as a turbulent analog of the molecular thermal diffusion for gases or thermophoresis for particles, while the second effect can be interpreted as a turbulent analog of the effect of Dufour (De Groot and Mazur, 1984), i.e., it describes turbulent transport of heat due to the number density gradient of chemically reacting admixtures.

3

Turbulent Transport of Magnetic Fields

In this chapter, we consider a turbulent transport of magnetic field in isotropic inhomogeneous and helical turbulence for incompressible and compressible flows of conducting fluid as well as for a density-stratified flow in anelastic approximation. We study isotropic and anisotropic spectra of magnetic fluctuations in different subranges of turbulent scales and different magnetic Prandtl numbers applying the dimensional analysis. We derive mean-field equations for the magnetic field and obtain expressions for turbulent electromotive force, turbulent magnetic diffusion, the α effect and effective pumping velocity of mean magnetic field and level of magnetic fluctuations for small and large magnetic Reynolds numbers by means of various analytical methods, namely the dimensional analysis, the quasi-linear approach and the spectral tau approach (the relaxation approach).

3.1 Magnetohydrodynamic Equations

In this section, we start with a short introduction of magnetohydrodynamics and derive an induction equation that determines the generation and nonlinear evolution of a magnetic field in a turbulent velocity field.

3.1.1 Governing Equations

Motions of electrically conducting fluids are determined by the Navier-Stokes equation with the Lorentz[1] force:

$$\rho \left(\frac{\partial \boldsymbol{U}}{\partial t} + (\boldsymbol{U} \cdot \nabla) \boldsymbol{U} \right) = -\boldsymbol{\nabla} P + \frac{1}{c} \boldsymbol{j} \times \boldsymbol{B} + \rho\, \nu\, \Delta \boldsymbol{U} + \rho \boldsymbol{f}, \tag{3.1}$$

[1] Hendrik Antoon Lorentz (1853–1928) was a Dutch physicist who made significant contributions to electromagnetism, special and general relativity, the quantum theory and hydrodynamics. His name is associated with the Lorentz–Lorenz equation, the Lorentz force, the Lorentzian distribution and the Lorentz transformation. He was awarded the Nobel Prize in Physics in 1902 (with Pieter Zeeman) for the discovery and theoretical explanation of the Zeeman effect.

Table 3.1 *Dimensions and units*

Quantity	Notation	SI units	Conversion factor	Gaussian units
Electric field	$\boldsymbol{E}$	Volt/m	3×10^{-5}	statvolt/cm
Magnetic field	$\boldsymbol{B}$	Tesla	10^4	Gauss
Electric charge	q	Coulomb	3×10^9	statcoulomb
Charge density	ρ_q	Coulomb/m^3	3×10^3	statcoulomb/cm^3
Electric conductivity	σ	Siemens/m	9×10^9	s^{-1}
Electric current	I	Ampere	3×10^9	statampere
Current density	$\boldsymbol{j}$	Ampere/m^2	3×10^5	statampere/cm^2
Electric potential	U	Volt	3×10^{-3}	statvolt
Resistance	R	Ohm	10^{-12}	s/cm
Capacitance	C	Farad	9×10^{11}	cm
Inductance	L	Henry	10^{-12}	s^2/cm
Magnetic flux	Φ	Weber	10^8	Maxwell
Energy	W, E_U	Joule	10^7	erg
Power	P	Watt	10^7	erg/s
Force	$\boldsymbol{F}$	Newton	10^5	dyne

where $\boldsymbol{U}$ is the incompressible fluid velocity field, ν is the kinematic viscosity, $\rho \boldsymbol{f}$ is the external force, c is the speed of light, $\boldsymbol{j}$ is the density of electric current, $\boldsymbol{B}$ is the magnetic field and P and ρ are the fluid pressure and density, respectively. The continuity equation for the incompressible fluid velocity field is $\nabla \cdot \boldsymbol{U} = 0$ and the fluid density ρ is constant in time and space. All equations in this chapter are written for simplicity in the Gaussian units. In Table 3.1.1, we present quantities in SI units and Gaussian[2] units, along with their transformations from SI units to Gaussian units.

The density of electric current $\boldsymbol{j}$ is determined by the Ohm's[3] law:

$$\boxed{\boldsymbol{j} = \sigma \left(\boldsymbol{E} + \frac{1}{c} \boldsymbol{U} \times \boldsymbol{B} \right),} \tag{3.2}$$

where σ is the electric conductivity and $\boldsymbol{E}$ is the electric field. The electric and magnetic field are determined by the Maxwell[4] equations:

$$\boxed{\nabla \cdot \boldsymbol{E} = 4\pi \rho_q,} \tag{3.3}$$

[2] Johann Carl Friedrich Gauss (1777–1855) was a German mathematician and physicist who made significant contributions to many fields in mathematics, physics and astronomy.

[3] Georg Simon Ohm (1789–1854) was a German physicist and mathematician who made significant contributions to the theory of electricity.

[4] James Clerk Maxwell (1831–1879) was a Scottish scientist in mathematical physics, well-known for his classical theory of the electromagnetic field and electromagnetic radiation as well as the kinetic theory of gases.

$$\boxed{\nabla \cdot \boldsymbol{B} = 0,} \tag{3.4}$$

$$\boxed{\nabla \times \boldsymbol{E} = -\frac{1}{c}\frac{\partial \boldsymbol{B}}{\partial t},} \tag{3.5}$$

$$\boxed{\nabla \times \boldsymbol{B} = \frac{4\pi}{c}\boldsymbol{j} + \frac{1}{c}\frac{\partial \boldsymbol{E}}{\partial t},} \tag{3.6}$$

where ρ_q is the density of the electric charge and $q = \int \rho_q \, d\boldsymbol{r}$ is the electric charge.

3.1.2 Electric Field

According to the Coulomb's[5] law, the electric force describing an interaction between the two particles having the electric charges q_0 and q is given by

$$\boxed{\boldsymbol{F} = \frac{q q_0}{r^2}\boldsymbol{e}_r,} \tag{3.7}$$

where $\boldsymbol{r}$ is the radius-vector that starts at the electric charge q_0 and ends at the electric charge q and $\boldsymbol{e}_r = \boldsymbol{r}/r$ is the unit vector. The electric field produced by the electric charge q_0 and defined as the force that acts on the electric charge q is given by

$$\boxed{\boldsymbol{E} = \frac{\boldsymbol{F}}{q} = \frac{q_0}{r^2}\boldsymbol{e}_r.} \tag{3.8}$$

The electric field is produced by

- electric charge;
- variable in time magnetic field.

In particular, Eq. (3.3), $\nabla \cdot \boldsymbol{E} = 4\pi\rho_q$, implies that a source of potential electric field is electric charge. The lines of the potential electric field are always started at the positive charge and ended at the negative charge. Equation (3.5), $\nabla \times \boldsymbol{E} = -(1/c)\,\partial \boldsymbol{B}/\partial t$, implies that a source of vortical electric field is variable in time magnetic field. The lines of the vortical electric field are closed. Equation (3.5) can be considered as a differential form of the Faraday's[6] law.

[5] Charles-Augustin de Coulomb (1736–1806) was a French engineer and physicist, well-known for his theory of the electrostatic force of attraction and repulsion, along with his important work on friction.

[6] Michael Faraday (1791–1867) was an English scientist well-known for his works in electromagnetism and electrochemistry (discoveries of electromagnetic induction, diamagnetism and electrolysis), and he also discovered benzene.

3.1.3 Magnetic Field

Equation (3.4), $\nabla \cdot \boldsymbol{B} = 0$, implies that there is no magnetic charge (or a magnetic monopole). The Lorentz force that acts on the electric charge q moving with the velocity $\boldsymbol{U}$ is given by

$$\boxed{\boldsymbol{F}_L = \frac{q}{c} \boldsymbol{U} \times \boldsymbol{B}.} \tag{3.9}$$

Magnetic field $\boldsymbol{B}$ is defined as a force that acts on a particle having a unit electric charge that moves with a unit velocity. A magnetic field is produced by

- electric current or motions of conducting fluid;
- variable in time electric field.

In particular, the first term, $(4\pi/c)\boldsymbol{j}$, on the right-hand side of Eq. (3.6) implies that a source of magnetic field is electric current (the Ampère[7] law), while the second term in Eq. (3.6), namely the displacement current, $c^{-1}\, \partial \boldsymbol{E}/\partial t$, implies that a source of vortical magnetic field is variable in time electric field (the Maxwell law).

3.1.4 Induction Equation

Let us derive an equation describing the evolution of the magnetic field. We rewrite Eq. (3.5) as

$$\frac{\partial \boldsymbol{B}}{\partial t} = -c\, \nabla \times \boldsymbol{E}. \tag{3.10}$$

Using the Ohm's law given by Eq. (3.2) and considering the non-relativistic motions (with the fluid velocity that is much smaller than the speed of light, $U \ll c$), we find that the electric field is given by

$$\boldsymbol{E} = \frac{\boldsymbol{j}}{\sigma} - \frac{1}{c} \boldsymbol{U} \times \boldsymbol{B}, \tag{3.11}$$

where the condition $U \ll c$ allows us to neglect the displacement current, $c^{-1}\, (\partial \boldsymbol{E}/\partial t)$, given by the last term on the right-hand side of Eq. (3.6). In this case, Eq. (3.6) yields the density of the electric current:

$$\boldsymbol{j} = \frac{c}{4\pi} \nabla \times \boldsymbol{B}. \tag{3.12}$$

Substituting Eq. (3.12) into Eq. (3.11), we find the electric field as

$$\boldsymbol{E} = \frac{c}{4\pi\sigma} \nabla \times \boldsymbol{B} - \frac{1}{c} \boldsymbol{U} \times \boldsymbol{B}. \tag{3.13}$$

[7] André-Marie Ampère (1775–1836) was a French physicist and mathematician well-known for his works in classical electrodynamics. He also invented the solenoid and the electrical telegraph.

Substituting the electric field (3.13) into Eq. (3.10), we arrive at the induction equation,

$$\boxed{\frac{\partial \boldsymbol{B}}{\partial t} = \nabla \times (\boldsymbol{U} \times \boldsymbol{B} - \eta \nabla \times \boldsymbol{B}),} \tag{3.14}$$

where $\eta = c^2/4\pi\sigma$ is the coefficient of magnetic diffusion due to the electrical conductivity of the fluid. Using the identity

$$\nabla \times (\nabla \times \boldsymbol{B}) = \nabla(\nabla \cdot \boldsymbol{B}) - \Delta \boldsymbol{B} = -\Delta \boldsymbol{B}, \tag{3.15}$$

we rewrite the induction equation (3.14) as

$$\frac{\partial \boldsymbol{B}}{\partial t} = \nabla \times (\boldsymbol{U} \times \boldsymbol{B}) + \eta \Delta \boldsymbol{B}. \tag{3.16}$$

This equation was derived by H. Hertz.[8] If a given velocity field is independent of the magnetic field, the induction equation (3.14) is linear in the magnetic field. Using the identity

$$\nabla \times (\boldsymbol{U} \times \boldsymbol{B}) = -(\boldsymbol{U} \cdot \nabla)\boldsymbol{B} + (\boldsymbol{B} \cdot \nabla)\boldsymbol{U} + \boldsymbol{U}(\nabla \cdot \boldsymbol{B}) - \boldsymbol{B}(\nabla \cdot \boldsymbol{U}), \tag{3.17}$$

we rewrite the induction equation as

$$\frac{\partial \boldsymbol{B}}{\partial t} + (\boldsymbol{U} \cdot \nabla)\boldsymbol{B} = (\boldsymbol{B} \cdot \nabla)\boldsymbol{U} - \boldsymbol{B}(\nabla \cdot \boldsymbol{U}) + \eta\, \Delta \boldsymbol{B}. \tag{3.18}$$

For incompressible fluid flow, $\nabla \cdot \boldsymbol{U} = 0$, the induction equation reads:

$$\boxed{\frac{\partial \boldsymbol{B}}{\partial t} + (\boldsymbol{U} \cdot \nabla)\boldsymbol{B} = (\boldsymbol{B} \cdot \nabla)\boldsymbol{U} + \eta\, \Delta \boldsymbol{B}.} \tag{3.19}$$

The form of Eq. (3.19) is similar to that of the equation for the particle number density,

$$\frac{\partial n}{\partial t} + (\boldsymbol{U} \cdot \nabla)\, n = -n\, (\nabla \cdot \boldsymbol{U}) + D^{(n)}\, \Delta n, \tag{3.20}$$

except the magnetic field $\boldsymbol{B}$ is the vector field, while the particle number density n is the scalar field. The term $(\boldsymbol{B} \cdot \nabla)\boldsymbol{U} \equiv B_j\, (\nabla_j \mathrm{U}_i)$, which is responsible for the generation of the magnetic field, is different from the term $-n(\nabla \cdot \boldsymbol{U}) \equiv -n\, \nabla_j U_j$, for the particle number density. The latter term causes the formation of inhomogeneous particle spatial distributions when $\nabla \cdot \boldsymbol{U} \neq 0$. Other terms in Eqs. (3.19) and (3.20) have a similar form and describe similar physics:

- diffusion of the fields described by the last terms on the right-hand side in both equations; and

[8] Heinrich Rudolf Hertz (1857–1894) was a German physicist who is well-known for his works on electromagnetic waves, electromagnetic induction, photoelectric effect and meteorology.

- advection of the fields determined by the second terms on the left-hand side in both equations.

In this chapter, we will use the induction equation (3.19) to study the turbulent transport of a magnetic field.

3.2 Spectra of Magnetic Fluctuations: Dimensional Analysis

In this section, we consider different spectra of magnetic fluctuations derived by means of the dimensional analysis.

3.2.1 Governing Equations and Basic Parameters

The equation for the evolution of magnetic field $\boldsymbol{B}(t, \boldsymbol{x})$ in an incompressible fluid velocity field $\boldsymbol{U}(t, \boldsymbol{x})$ reads:

$$\frac{\partial \boldsymbol{B}}{\partial t} = \nabla \times (\boldsymbol{U} \times \boldsymbol{B}) + \eta \, \Delta \boldsymbol{B}. \tag{3.21}$$

Averaging Eq. (3.21) over an ensemble of turbulent velocity field, we obtain the equation for the mean magnetic field $\overline{\boldsymbol{B}}(t, \boldsymbol{x})$:

$$\boxed{\frac{\partial \overline{\boldsymbol{B}}}{\partial t} = \nabla \times \left(\overline{\boldsymbol{U}} \times \overline{\boldsymbol{B}} + \langle \boldsymbol{u} \times \boldsymbol{b} \rangle\right) + \eta \, \Delta \overline{\boldsymbol{B}},} \tag{3.22}$$

where $\overline{\boldsymbol{\mathcal{E}}} \equiv \langle \boldsymbol{u} \times \boldsymbol{b} \rangle$ is the turbulent electromotive force that determines generation and diffusion of the mean magnetic field caused by turbulent motions. The equation for magnetic fluctuations $\boldsymbol{b} = \boldsymbol{B} - \overline{\boldsymbol{B}}$ is obtained by subtracting Eq. (3.22) from Eq. (3.21):

$$\boxed{\frac{\partial \boldsymbol{b}}{\partial t} - \nabla \times (\boldsymbol{u} \times \boldsymbol{b} - \langle \boldsymbol{u} \times \boldsymbol{b} \rangle) - \eta \, \Delta \boldsymbol{b} = (\overline{\boldsymbol{B}} \cdot \nabla) \boldsymbol{u} - (\boldsymbol{u} \cdot \nabla) \overline{\boldsymbol{B}},} \tag{3.23}$$

where for simplicity we consider the case of a vanishing mean velocity ($\overline{\boldsymbol{U}} = 0$). A dimensionless parameter, namely the magnetic Reynolds number, is defined as the ratio of the absolute values of the nonlinear term $|\nabla \times (\boldsymbol{u} \times \boldsymbol{b} - \langle \boldsymbol{u} \times \boldsymbol{b} \rangle)\,|$ to the magnetic diffusion term $|\eta \, \Delta \boldsymbol{b}|$:

$$\mathrm{Rm} = \frac{|\nabla \times (\boldsymbol{u} \times \boldsymbol{b} - \langle \boldsymbol{u} \times \boldsymbol{b} \rangle)\,|}{|\eta \, \Delta \boldsymbol{b}|}. \tag{3.24}$$

The magnetic Reynolds number (3.24) can be estimated using the dimensional analysis as

$$\boxed{\mathrm{Rm} = \frac{\ell_0 u_0}{\eta},} \tag{3.25}$$

where ℓ_0 is the integral scale of turbulence and u_0 is the characteristic turbulent velocity in the integral scale of turbulence ℓ_0. In a fully developed turbulence, the rate of the dissipation ε of the turbulent kinetic energy density can be estimated as

$$\varepsilon = \frac{u_\ell^2}{\tau_\ell} = \frac{u_\ell^3}{\ell}, \tag{3.26}$$

where u_ℓ is the characteristic turbulent velocity in the scale ℓ inside the inertial range of turbulent scales, $\ell_\nu \le \ell \le \ell_0$, and $\tau_\ell = \ell/u_\ell$ is the characteristic turbulent time in the scale ℓ and ℓ_ν is the viscous Kolmogorov scale. Here, for simplicity, we choose the constant fluid density as unity. For Kolmogorov turbulence, the rate of dissipation of turbulent kinetic energy density is constant in the inertial range of scales,

$$\varepsilon = \frac{u_0^3}{\ell_0} = \frac{u_\ell^3}{\ell} = \ldots = \frac{u_\nu^3}{\ell_\nu} = \text{const}, \tag{3.27}$$

where u_ν is the characteristic velocity in the viscous scale ℓ_ν. Equation (3.27) yields

$$u_0 = (\varepsilon \ell_0)^{1/3}, \quad u_\ell = (\varepsilon \ell)^{1/3}, \quad u_\nu = (\varepsilon \ell_\nu)^{1/3}. \tag{3.28}$$

Using Eq. (3.28), we rewrite the magnetic Reynolds number as

$$\text{Rm} \equiv \frac{\ell_0 u_0}{\eta} = \frac{\varepsilon^{1/3} \ell_0^{4/3}}{\eta}. \tag{3.29}$$

Similarly, we introduce the local magnetic Reynolds number Rm_ℓ as

$$\text{Rm}_\ell \equiv \frac{\ell\, u_\ell}{\eta} = \frac{\varepsilon^{1/3}\, \ell^{4/3}}{\eta}. \tag{3.30}$$

We determine the ratio Rm_ℓ/Rm using Eqs. (3.29)–(3.30) as

$$\frac{\text{Rm}_\ell}{\text{Rm}} = \left(\frac{\ell}{\ell_0}\right)^{4/3}. \tag{3.31}$$

We define the magnetic diffusion scale ℓ_η (or the resistive scale) as the scale in which the local magnetic Reynolds number is 1, i.e., the nonlinear terms in the equation for magnetic fluctuations in the resistive scale are of the order of the magnetic diffusion term. Therefore, this condition and Eq. (3.31) yield the magnetic diffusion scale ℓ_η as

$$\boxed{\ell_\eta = \frac{\ell_0}{\text{Rm}^{3/4}}.} \tag{3.32}$$

Substituting the magnetic diffusion scale ℓ_η given by Eq. (3.32) into the expression for the velocity at the resistive scale, $u_\eta = (\varepsilon \ell_\eta)^{1/3}$ [see Eq. (3.28)], we obtain the characteristic velocity u_η as

$$\boxed{u_\eta = \frac{(\varepsilon \ell_0)^{1/3}}{\mathrm{Rm}^{1/4}} = \frac{u_0}{\mathrm{Rm}^{1/4}}.} \tag{3.33}$$

Using Eqs. (3.32) and (3.33), we determine the characteristic magnetic diffusion time $\tau_\eta = \ell_\eta / u_\eta$ as

$$\boxed{\tau_\eta = \frac{\tau_0}{\mathrm{Rm}^{1/2}}.} \tag{3.34}$$

Since the local magnetic Reynolds number at the magnetic diffusion scale equals 1 (which implies that $u_\eta \ell_\eta = \eta$), we can rewrite the resistive time τ_η as

$$\boxed{\tau_\eta = \frac{\ell_\eta}{u_\eta} = \frac{\ell_\eta^2}{\eta}.} \tag{3.35}$$

Now we determine the ratio ℓ_η / ℓ_ν of the resistive and viscous scales as

$$\frac{\ell_\eta}{\ell_\nu} = \left(\frac{\mathrm{Re}}{\mathrm{Rm}}\right)^{3/4} = \left(\frac{\eta}{\nu}\right)^{3/4} = \mathrm{Pr}_m^{-3/4}, \tag{3.36}$$

where

$$\boxed{\mathrm{Pr}_m = \frac{\nu}{\eta}} \tag{3.37}$$

is the magnetic Prandtl number. Small magnetic Prandtl numbers, $\mathrm{Pr}_m \ll 1$, imply that the viscous scale ℓ_ν is much smaller than the resistive scale $\ell_\nu \ll \ell_\eta$. In the opposite case of large magnetic Prandtl numbers, $\mathrm{Pr}_m \gg 1$, the resistive scale ℓ_η is the smallest scale.

3.2.2 Spectra of Isotropic Magnetic Fluctuations

Let us find spectra of magnetic fluctuations for various cases. First, we consider isotropic magnetic fluctuations inside the inertial range of turbulence, $\ell_\nu \ll \ell_\eta < \ell < \ell_0$, with small magnetic Prandtl numbers, $\mathrm{Pr}_m \ll 1$. The energy spectrum function E_u of velocity fluctuations is defined as $u_\ell^2 = \int_{k_0}^{k} E_\mathrm{u}(k')\, dk'$, where $k_0 = \ell_0^{-1}$ and $k = \ell^{-1}$. The Kolmogorov-Obukhov spectrum of velocity fluctuations in the inertial range of turbulence reads

$$\boxed{E_\mathrm{u}(k) = \frac{u_\ell^2}{k} = \varepsilon^{2/3} k^{-5/3},} \tag{3.38}$$

which corresponds to the condition

$$\varepsilon = \frac{u_\ell^2}{\tau_\ell} = \frac{E_\mathrm{u}(k)\, k}{\tau(k)} = \mathrm{const}, \tag{3.39}$$

where $u_\ell = (\varepsilon/k)^{1/3}$ and $\tau_\ell \equiv \tau(k)$. In inertial range of turbulence scales, the turbulent time $\tau(k) = \ell/u_\ell$ has the following scaling:

$$\tau(k) = \frac{\ell}{(\varepsilon\ell)^{1/3}} = \varepsilon^{-1/3}\, k^{-2/3}. \tag{3.40}$$

The spectrum function $M_\mathrm{b}(k)$ for magnetic fluctuations is defined as

$$\left\langle \boldsymbol{b}^2 \right\rangle_\ell = \int_{k_0}^{k} M_\mathrm{b}(k')\, dk'. \tag{3.41}$$

Using dimensional analysis, we rewrite this expression as $\left\langle \boldsymbol{b}^2 \right\rangle_\ell = M_\mathrm{b}(k)\, k$ and assume that the dissipation rate of the turbulent magnetic energy density $\left\langle \boldsymbol{b}^2 \right\rangle_\ell / \tau_\ell$ is constant inside the subrange of scales $\ell_\eta < \ell < \ell_0$,

$$\frac{\left\langle \boldsymbol{b}^2 \right\rangle_\ell}{\tau_\ell} = \frac{M_\mathrm{b}(k) k}{\tau(k)} = \mathrm{const}. \tag{3.42}$$

This condition is analogous to condition (3.39) for velocity fluctuations and corresponds to the scale-dependent turbulent magnetic diffusion subrange of scales. Equations (3.38) and (3.42) yield the following spectrum of isotropic magnetic fluctuations in the inertial–turbulent diffusion range of scales:

$$\boxed{M_\mathrm{b}(k) \sim E_\mathrm{u}(k) \sim \varepsilon^{2/3} k^{-5/3}.} \tag{3.43}$$

This spectrum is analogous to the Kolmogorov-Obukhov spectrum for velocity fluctuations.

Next, we consider isotropic magnetic fluctuations in the viscous range of scales, $\ell_\eta < \ell < \ell_\nu$, where the magnetic Prandtl number $\mathrm{Pr}_m > 1$. In this range of scales, condition (3.42) is valid, except the turbulent time $\tau(k)$ and the turbulent kinetic energy density do not have universal scalings. Assuming that the turbulent time is independent of scales in $\ell_\eta < \ell < \ell_\nu$, i.e., $\tau(k) = \mathrm{const}$, the spectrum of magnetic fluctuations in the viscous but scale-dependent magnetic diffusion subrange of scales is

$$\boxed{M_\mathrm{b}(k) \sim k^{-1}.} \tag{3.44}$$

This spectrum has been obtained by Kraichnan (1968).

3.2.3 Anisotropic Magnetic Fluctuations in the Inertial–Magnetic Diffusion Range

In this section, we consider anisotropic magnetic fluctuations caused by the tangling of the mean magnetic field by velocity fluctuations in the inertial–magnetic diffusion range of scales $\ell_\nu < \ell < \ell_\eta$, where we consider the case $\mathrm{Pr}_m < 1$. We use dimensional analysis and take into account the fact that the magnetic diffusion is a key effect in this range of scales. This implies that the magnetic diffusion term $|\eta \, \Delta \boldsymbol{b}|$ in Eq. (3.23) should be balanced by the source term $|(\overline{\boldsymbol{B}} \cdot \nabla)\boldsymbol{u}|$ for magnetic fluctuations, i.e.,

$$|\eta \, \Delta \boldsymbol{b}| \sim \left|(\overline{\boldsymbol{B}} \cdot \nabla)\boldsymbol{u}\right| , \tag{3.45}$$

which yields

$$\left\langle \boldsymbol{b}^2 \right\rangle_\ell \sim u_\ell^2 \, \ell^2 \left(\frac{\overline{\boldsymbol{B}}}{\eta}\right)^2 . \tag{3.46}$$

In the $\boldsymbol{k}$ space, Eq. (3.46) implies that

$$M_{\mathrm{b}}(k) \sim E_{\mathrm{u}}(k) \, k^{-2} \left(\frac{\overline{\boldsymbol{B}}}{\eta}\right)^2 , \tag{3.47}$$

where $E_{\mathrm{u}}(k)$ is the spectrum function for velocity fluctuations. In the scales $\ell_\nu \ll \ell \ll \ell_\eta$, velocity fluctuations obey the Kolmogorov-Obukhov spectrum (3.38), i.e., $E_{\mathrm{u}}(k) = \varepsilon^{2/3} \, k^{-5/3}$. Therefore, Eq. (3.47) implies that the spectrum of anisotropic magnetic fluctuations in the inertial-diffusion subrange of scales is given by

$$\boxed{M_{\mathrm{b}}(k) \sim \varepsilon^{2/3} \, k^{-11/3} \left(\frac{\overline{\boldsymbol{B}}}{\eta}\right)^2 .} \tag{3.48}$$

This spectrum has been obtained by Golitsyn (1960) and Moffatt (1961).

3.2.4 Anisotropic Fluctuations in the Inertial–Turbulent Magnetic Diffusion Range

We consider anisotropic magnetic fluctuations caused by the tangling of the mean magnetic field by velocity fluctuations in the inertial–turbulent magnetic diffusion subrange of scales, $\ell_\eta < \ell < \ell_0$, with small magnetic Prandtl numbers, $\mathrm{Pr}_m \ll 1$. We take into account that the main effect of turbulent incompressible flow is the scale-dependent turbulent magnetic diffusion that is much larger than the microscopic magnetic diffusion η for large magnetic Reynolds numbers. Let us perform a spatial averaging in Eq. (3.23) for magnetic fluctuations up to the scale ℓ_* that is inside the turbulent magnetic diffusion interval: $\ell_\eta \ll \ell_* \ll \ell_0$. This yields an effective equation for magnetic fluctuations as

$$\frac{\partial \boldsymbol{b}}{\partial t} - \eta_{_T}(\ell)\Delta \boldsymbol{b} = (\overline{\boldsymbol{B}} \cdot \nabla)\boldsymbol{u} - (\boldsymbol{u} \cdot \nabla)\overline{\boldsymbol{B}}, \tag{3.49}$$

where $\eta_{_T}(\ell)$ is the scale-dependent turbulent magnetic diffusion coefficient that can be estimated as

$$\eta_{_T}(\ell) = \ell\, u_\ell. \tag{3.50}$$

In this range of scales, we neglect the small microscopic magnetic diffusion η. We also take into account that the turbulent diffusion term $|\eta_{_T}(\ell)\Delta \boldsymbol{b}|$ in Eq. (3.49) is balanced by the source term $|(\overline{\boldsymbol{B}} \cdot \nabla)\boldsymbol{u}|$ for magnetic fluctuations, i.e.,

$$\left|\eta_{_T}(\ell)\Delta \boldsymbol{b}\right| \sim \left|(\overline{\boldsymbol{B}} \cdot \nabla)\boldsymbol{u}\right|. \tag{3.51}$$

Equation (3.51) implies that

$$\left\langle \boldsymbol{b}^2 \right\rangle_\ell \sim u_\ell^2\, \ell^2 \left(\frac{\overline{\boldsymbol{B}}}{\eta_{_T}(\ell)}\right)^2. \tag{3.52}$$

Using Eqs. (3.50) and (3.52) written in the $\boldsymbol{k}$ space, we obtain the spectrum of anisotropic magnetic fluctuations in the inertial–turbulent magnetic diffusion range of scales $\ell_\eta \ll \ell \ll \ell_0$:

$$\boxed{M_{\rm b}(k) \sim k^{-1}\,\overline{\boldsymbol{B}}^2.} \tag{3.53}$$

This spectrum is universal because it is independent of the spectrum of the turbulent velocity field. This spectrum was obtained by Ruzmaikin and Shukurov (1982) using the dimensional analysis. This spectrum has been also derived using different analytical approaches, e.g., by Kleeorin and Rogachevskii (1994a) applying the renormalization approach, and by Kleeorin et al. (1996) using the spectral tau approach.

3.3 Turbulent Transport of Magnetic Fields: Dimensional Analysis

3.3.1 Mean-Field Approach

In this section, we study the generation and dynamics of magnetic fields in a turbulent fluid flow. The equation for the evolution of a magnetic field $\boldsymbol{B}(t, \boldsymbol{x})$ in an incompressible fluid velocity field $\boldsymbol{U}(t, \boldsymbol{x})$ reads:

$$\frac{\partial \boldsymbol{B}}{\partial t} = \nabla \times (\boldsymbol{U} \times \boldsymbol{B}) + \eta\, \Delta \boldsymbol{B}, \tag{3.54}$$

where η is the coefficient of magnetic diffusion due to the electrical conductivity of the fluid that is assumed to be independent of coordinates. An incompressible velocity ($\nabla \cdot \boldsymbol{U} = 0$) is a random field that is produced by some forcing in the scales from small scales to the integral scale ℓ_0.

To study the long-term evolution of the magnetic field, we use a mean-field approach in which all quantities are decomposed into the mean and fluctuating parts, where the fluctuating parts have zero mean values. In particular, the magnetic field $\boldsymbol{B} = \overline{\boldsymbol{B}} + \boldsymbol{b}$, where $\overline{\boldsymbol{B}} = \langle \boldsymbol{B} \rangle$ is the mean magnetic field, $\boldsymbol{b}$ are magnetic fluctuations and $\langle \boldsymbol{b} \rangle = 0$. The angular brackets $\langle \ldots \rangle$ denote ensemble averaging. Similarly, we decompose the velocity field $\boldsymbol{U} = \overline{\boldsymbol{U}} + \boldsymbol{u}$, where $\overline{\boldsymbol{U}} = \langle \boldsymbol{U} \rangle$ is the mean fluid velocity, $\boldsymbol{u}$ are velocity fluctuations and $\langle \boldsymbol{u} \rangle = 0$, i.e., we apply the Reynolds averaging. We assume that there is a separation of spatial ($L_B \gg \ell_0$) and time ($t_B \gg \tau_0$) scales, where L_B is the characteristic scale of mean magnetic field variations and t_B is the characteristic time of mean magnetic field variations.

Averaging Eq. (3.54) over an ensemble of turbulent velocity field, we obtain the mean-field equation for evolution of the mean magnetic field:

$$\boxed{\frac{\partial \overline{\boldsymbol{B}}}{\partial t} = \nabla \times \left(\overline{\boldsymbol{U}} \times \overline{\boldsymbol{B}} + \langle \boldsymbol{u} \times \boldsymbol{b} \rangle \right) + \eta \, \Delta \overline{\boldsymbol{B}},} \tag{3.55}$$

where $\boldsymbol{\mathcal{E}} \equiv \langle \boldsymbol{u} \times \boldsymbol{b} \rangle$ is the turbulent electromotive force or turbulent EMF. For simplicity, we consider the case of a vanishing mean velocity ($\overline{\boldsymbol{U}} = 0$).

3.3.2 Equation for Magnetic Fluctuations

Equation (3.55) is not closed because we do not know the turbulent electromotive force $\boldsymbol{\mathcal{E}}$. To determine the turbulent electromotive force, we use the equation for magnetic fluctuations obtained by subtracting Eq. (3.55) from Eq. (3.54):

$$\boxed{\frac{\partial \boldsymbol{b}}{\partial t} - \nabla \times (\boldsymbol{u} \times \boldsymbol{b} - \langle \boldsymbol{u} \times \boldsymbol{b} \rangle) - \eta \, \Delta \boldsymbol{b} = (\overline{\boldsymbol{B}} \cdot \nabla)\boldsymbol{u} - (\boldsymbol{u} \cdot \nabla)\overline{\boldsymbol{B}}.} \tag{3.56}$$

The second term, $-\nabla \times (\boldsymbol{u} \times \boldsymbol{b} - \langle \boldsymbol{u} \times \boldsymbol{b} \rangle)$, on the left-hand side of Eq. (3.56), is the nonlinear term, while the terms $(\overline{\boldsymbol{B}} \cdot \nabla)\boldsymbol{u} - (\boldsymbol{u} \cdot \nabla)\overline{\boldsymbol{B}}$, on the right-hand side of Eq. (3.56), determine the source of magnetic fluctuations produced by the tangling of the mean magnetic field $\overline{\boldsymbol{B}}$ or by the tangling of the gradient of the mean magnetic field $\nabla_j \overline{B}_i$ by random velocity fluctuations $\boldsymbol{u}$. The ratio of the nonlinear terms to the diffusion terms in Eq. (3.56) is magnetic Reynolds number, $\mathrm{Rm} = u_0 \, \ell_0 / \eta$.

Since Eq. (3.56) is a nonlinear equation, it cannot be solved exactly for an arbitrary range of parameters and arbitrary velocity field. Therefore, we have to use different approximate methods for the solution of Eq. (3.56). We consider a one-way coupling, i.e., we neglect the effect of the magnetic field on the turbulent fluid flow, but we take into account the effect of the turbulent velocity on the magnetic field. In this sense, the magnetic field is a passive vector. This assumption is valid when the energy of the mean magnetic field is much smaller than the turbulent kinetic energy.

3.3.3 Dimensional Analysis

The first method that we use here is the dimensional analysis. The dimension of the left-hand side of Eq. (3.56) is the rate of change of magnetic field $\boldsymbol{b}$, i.e., $\boldsymbol{b}/\tau_b$, where τ_b is the characteristic time of magnetic fluctuations. We consider two cases of large and small magnetic Reynolds numbers:

- *Large magnetic Reynolds numbers*, Rm $\gg 1$, and *large fluid Reynolds numbers*, Re $\gg 1$. This implies that we consider fully developed magnetohydrodynamic (MHD) turbulence. In this case, the characteristic time of magnetic fluctuations, τ_b, can be identified with the correlation time τ_0 of the turbulent velocity field. Therefore, in the framework of the dimensional analysis, we replace the left-hand side of Eq. (3.56) with $\boldsymbol{b}/\tau_0$. This implies that magnetic fluctuations $\boldsymbol{b}$ for large Reynolds numbers are given by

$$\boldsymbol{b} = \tau_0 \left[(\overline{\boldsymbol{B}} \cdot \nabla)\boldsymbol{u} - (\boldsymbol{u} \cdot \nabla)\overline{\boldsymbol{B}} \right]. \tag{3.57}$$

- *Small magnetic Reynolds numbers*, Rm $\ll 1$. In this case, the nonlinear terms are much smaller than the magnetic diffusion term. This implies that the microscopic magnetic diffusion η for Rm $\ll 1$ is the main effect, which determines the evolution of the magnetic fluctuations. Therefore, we assume that the time τ_b can be identified with the magnetic diffusion time $t_\eta = \ell_0^2/\eta$, so that the magnetic fluctuations $\boldsymbol{b}$ for small magnetic Reynolds numbers are given by

$$\boldsymbol{b} = \frac{\ell_0^2}{\eta} \left[(\overline{\boldsymbol{B}} \cdot \nabla)\boldsymbol{u} - (\boldsymbol{u} \cdot \nabla)\overline{\boldsymbol{B}} \right]. \tag{3.58}$$

3.3.4 Turbulent Electromotive Force and Level of Magnetic Fluctuations

The turbulent electromotive force, $\boldsymbol{\mathcal{E}} \equiv \langle \boldsymbol{u} \times \boldsymbol{b} \rangle$, can be rewritten as

$$\boxed{\mathcal{E}_m = \varepsilon_{mij} \left\langle u_i b_j \right\rangle,} \tag{3.59}$$

where ε_{mij} is the Levi-Civita tensor, which is the fully antisymmetric unit tensor defined by the following properties:

$$\varepsilon_{mij} = \varepsilon_{ijm} = \varepsilon_{jmi}, \quad \varepsilon_{mij} = -\varepsilon_{imj}, \quad \varepsilon_{imj} = -\varepsilon_{ijm}, \tag{3.60}$$

$$\varepsilon_{xyz} = \varepsilon_{yzx} = \varepsilon_{zxy} = 1, \quad \varepsilon_{xxz} = \varepsilon_{xyy} = \varepsilon_{zyz} = 0, \tag{3.61}$$

$$\varepsilon_{ijp}\varepsilon_{mnp} = \delta_{im}\delta_{jn} - \delta_{in}\delta_{jm}. \tag{3.62}$$

We consider two cases of large and small magnetic Reynolds numbers.

Large Magnetic Reynolds Numbers

Using Eqs. (3.57), we determine the turbulent electromotive force:

$$\mathcal{E}_m = \tau_0\, \varepsilon_{mij} \left(\left\langle u_i \nabla_n u_j \right\rangle \overline{B}_n - \langle u_i u_n \rangle\, \nabla_n \overline{B}_j\right). \tag{3.63}$$

In isotropic turbulence, $\langle u_i u_j \rangle = \delta_{ij} \left\langle \boldsymbol{u}^2 \right\rangle /3$, where δ_{ij} is the Kronecker unit tensor. Substituting this expression in Eq. (3.63), we obtain

$$\mathcal{E}_m = \tau_0\, \varepsilon_{mij} \left\langle u_i \nabla_n u_j \right\rangle \overline{B}_n - \eta_{_T} \left(\boldsymbol{\nabla} \times \overline{\boldsymbol{B}}\right)_m, \tag{3.64}$$

where we use the following identity for isotropic turbulence:

$$\tau_0\, \varepsilon_{mij}\, \langle u_i u_n \rangle\, \nabla_n \overline{B}_j = \frac{1}{3} \tau_0 \left\langle \boldsymbol{u}^2 \right\rangle \varepsilon_{mij} \nabla_i \overline{B}_j = \eta_{_T} \left(\boldsymbol{\nabla} \times \overline{\boldsymbol{B}}\right)_m. \tag{3.65}$$

Therefore, the coefficient of turbulent magnetic diffusion for large magnetic Reynolds numbers is given by

$$\boxed{\eta_{_T} = \frac{\tau_0}{3} \left\langle \boldsymbol{u}^2 \right\rangle.} \tag{3.66}$$

The first term on the right-hand side of Eq. (3.64) can be rewritten as

$$a_{mn} \overline{B}_n = \alpha_{mn}\, \overline{B}_n + \left(\boldsymbol{V}^{\rm eff} \times \overline{\boldsymbol{B}}\right)_m, \tag{3.67}$$

where the tensor $a_{mn} = \tau_0\, \varepsilon_{mij}\, \langle u_i \nabla_n u_j \rangle$. To derive Eq. (3.67), the tensor a_{mn} is presented as a sum of symmetric part, $(a_{mn} + a_{nm})/2 \equiv \alpha_{mn}$, and anti symmetric part, $(a_{mn} - a_{nm})/2$, where the anti symmetric part of the tensor a_{mn} is related to the effective pumping velocity as $V_p^{\rm eff} = -\varepsilon_{mnp}\, a_{mn}/2$. Equations (3.64)–(3.67) yield the turbulent electromotive force for large magnetic Reynolds numbers as

$$\boxed{\boldsymbol{\mathcal{E}} = \boldsymbol{\alpha}\, \overline{\boldsymbol{B}} + \boldsymbol{V}^{\rm eff} \times \overline{\boldsymbol{B}} - \eta_{_T} (\boldsymbol{\nabla} \times \overline{\boldsymbol{B}}),} \tag{3.68}$$

where the tensor $\boldsymbol{\alpha} \equiv \alpha_{mn}$ is given by

$$\boxed{\alpha_{mn} = \frac{\tau_0}{2} \left(\varepsilon_{mij} \left\langle u_i \nabla_n u_j \right\rangle + \varepsilon_{nij} \left\langle u_i \nabla_m u_j \right\rangle\right)} \tag{3.69}$$

and the effective pumping velocity $\boldsymbol{V}^{\rm eff}$ is

$$\boxed{\boldsymbol{V}^{\rm eff} = -\frac{1}{2} \boldsymbol{\nabla} \eta_{_T}.} \tag{3.70}$$

To derive Eq. (3.70), we perform the following calculations:

$$\begin{aligned} \varepsilon_{mnp}\, a_{mn} &= \tau_0\, \varepsilon_{mnp}\, \varepsilon_{mij}\, \left\langle u_i \nabla_n u_j \right\rangle = \tau_0\, \varepsilon_{npm}\, \varepsilon_{ijm}\, \left\langle u_i \nabla_n u_j \right\rangle \\ &= \tau_0 \left(\delta_{ni}\delta_{pj} - \delta_{nj}\delta_{pi}\right) \left\langle u_i \nabla_n u_j \right\rangle = \tau_0 \left(\left\langle u_i \nabla_i u_p \right\rangle - \left\langle u_p \nabla_j u_j \right\rangle \right) \\ &= \tau_0\, \nabla_i \langle u_i u_p \rangle = \frac{\tau_0}{3}\, \nabla_p \left\langle \boldsymbol{u}^2 \right\rangle = \nabla_p \eta_T , \end{aligned} \tag{3.71}$$

where we take into account that τ_0 is independent of coordinates and consider an isotropic turbulence, $\langle u_i\, u_j \rangle = \delta_{ij} \left\langle \boldsymbol{u}^2 \right\rangle /3$. The velocity $\boldsymbol{V}^{\rm eff}$ describes the turbulent diamagnetic effect predicted by Zeldovich (1957).[9] This effect causes the mean magnetic field is pushed out from the region of intense velocity fluctuations (i.e., it is pushed out from regions occupied by turbulence).

Equation (3.69) for the tensor α_{mn} is for an arbitrary anisotropic turbulence. To derive an equation for the tensor α_{mn} for isotropic turbulence, we rewrite the tensor $\varepsilon_{mij} \left\langle u_i \nabla_n u_j \right\rangle$ using the identity: $\nabla_n u_j = (\partial u)_{nj} + \varepsilon_{njp} \boldsymbol{\omega}/2$, so that

$$\varepsilon_{mij} \left\langle u_i \nabla_n u_j \right\rangle = \frac{1}{2} \Big(\langle u_n \omega_m \rangle - \langle \boldsymbol{u} \cdot \boldsymbol{\omega} \rangle\, \delta_{mn} \Big) + \varepsilon_{mij} \left\langle u_i\, (\partial u)_{nj} \right\rangle , \tag{3.72}$$

where $\boldsymbol{\omega} = \boldsymbol{\nabla} \times \boldsymbol{u}$ and $(\partial u)_{ij} = (\nabla_i u_j + \nabla_j u_i)/2$. Substituting Eq. (3.72) into Eq. (3.69), we obtain the tensor α_{mn} as

$$\begin{aligned} \alpha_{mn} = \frac{\tau_0}{2} \Bigg[\frac{1}{2} \Big(\langle u_n\, \omega_m \rangle + \langle u_m\, \omega_n \rangle \Big) - \langle \boldsymbol{u} \cdot \boldsymbol{\omega} \rangle\, \delta_{mn} \\ + \varepsilon_{mij} \left\langle u_i\, (\partial u)_{nj} \right\rangle + \varepsilon_{nij} \left\langle u_i\, (\partial u)_{mj} \right\rangle \Bigg] . \end{aligned} \tag{3.73}$$

Assuming that for isotropic turbulence $\langle u_i\, \omega_j \rangle = \delta_{ij}\, \langle \boldsymbol{u} \cdot \boldsymbol{\omega} \rangle /3$ and $\langle u_i\, (\partial u)_{mj} \rangle = \delta_{ij}\, \langle u_n\, (\partial u)_{mn} \rangle /3$, we obtain the tensor $\alpha_{ij} = \alpha\, \delta_{ij}$ for an isotropic turbulence, where α for large magnetic Reynolds numbers is given by

$$\boxed{\alpha = -\frac{\tau_0}{3}\, \langle \boldsymbol{u} \cdot (\boldsymbol{\nabla} \times \boldsymbol{u}) \rangle\, .} \tag{3.74}$$

Here $\langle \boldsymbol{u} \cdot (\boldsymbol{\nabla} \times \boldsymbol{u}) \rangle$ is the kinetic helicity. The symmetric tensor α_{ij} describes the α effect. We will show in Section 3.7 that this effect is responsible for the generation of a large-scale magnetic field in a small-scale helical turbulence.

Next, we determine the level of magnetic fluctuations $\left\langle \boldsymbol{b}^2 \right\rangle$. We consider magnetic fluctuations caused by the tangling of the mean magnetic field by velocity fluctuations in the inertial–turbulent magnetic diffusion range of scales, $\ell_\nu \ll \ell_\eta \ll \ell \ll \ell_0$. The spectrum of anisotropic magnetic fluctuations in this range of scales obtained using the dimensional analysis is given by

$$M_{\rm b}(k) \sim k^{-1}\, \overline{\boldsymbol{B}}^2 . \tag{3.75}$$

[9] Yakov Zeldovich (1914–1987) was a Russian physicist well-known for his works in cosmology, fluid dynamics, magnetohydrodynamics, nuclear and thermonuclear physics.

[See Eq. (3.53).] Thus the level of magnetic fluctuations $\langle \boldsymbol{b}^2 \rangle$ is given by

$$\langle \boldsymbol{b}^2 \rangle = \int_{k_0}^{k_\eta} M_{\rm b}(k)\, dk = \int_{k_0}^{k_0\,{\rm Rm}^{3/4}} k^{-1}\, dk\, \overline{\boldsymbol{B}}^2, \tag{3.76}$$

where $k_\eta = k_0 {\rm Rm}^{3/4}$. After integration over k in Eq. (3.76), we obtain the level of magnetic fluctuations $\langle \boldsymbol{b}^2 \rangle$ for large magnetic and fluid Reynolds numbers as

$$\boxed{\langle \boldsymbol{b}^2 \rangle \approx (\ln {\rm Rm})\, \overline{\boldsymbol{B}}^2.} \tag{3.77}$$

Small Magnetic Reynolds Numbers

Now we consider the case of small magnetic Reynolds numbers, using Eq. (3.58) and performing calculations that are similar to those given by Eqs. (3.63)–(3.71), except we replace in these equations the turbulent time τ_0 with the resistive time ℓ_0^2/η. This yields Eq. (3.68) for the turbulent electromotive force $\boldsymbol{\mathcal{E}}$ and Eq. (3.70) for the effective pumping velocity $\boldsymbol{V}^{\rm eff}$ with the following expressions for the coefficient of turbulent magnetic diffusion $\eta_{_T}$ and the tensor α_{mn} for small magnetic Reynolds numbers:

$$\boxed{\eta_{_T} = \frac{\tau_0}{3} \left\langle \boldsymbol{u}^2 \right\rangle {\rm Rm}\,,} \tag{3.78}$$

$$\boxed{\alpha_{mn} = \frac{\tau_0}{2} \left(\varepsilon_{mij} \left\langle u_i \nabla_n u_j \right\rangle + \varepsilon_{nij} \left\langle u_i \nabla_m u_j \right\rangle\right) {\rm Rm}\,.} \tag{3.79}$$

Calculations, similar to Eqs. (3.72)–(3.73), yield the α effect for small magnetic Reynolds numbers:

$$\boxed{\alpha = -\frac{\tau_0}{3} \left\langle \boldsymbol{u} \cdot (\nabla \times \boldsymbol{u}) \right\rangle {\rm Rm}.} \tag{3.80}$$

Using Eq. (3.58), we determine the level of magnetic fluctuations $\langle \boldsymbol{b}^2 \rangle$ for small magnetic Reynolds numbers:

$$\begin{aligned}\langle \boldsymbol{b}^2 \rangle &= \left(\frac{\ell_0^2}{\eta}\right)^2 \left\langle \left[(\overline{\boldsymbol{B}} \cdot \nabla)\boldsymbol{u}\right]^2 \right\rangle \left[1 + O\left(\frac{\ell_0}{L_B}\right)\right] \\ &\approx \left(\frac{\ell_0^2}{\eta}\right)^2 \overline{B}_m \overline{B}_n \left\langle (\nabla_m u_i)(\nabla_n u_i) \right\rangle \approx \frac{\ell_0^2 u_0^2}{\eta^2}\, \overline{\boldsymbol{B}}^2,\end{aligned} \tag{3.81}$$

where we take into account that for isotropic turbulence and in the framework of dimensional analysis $\langle (\nabla_m u_i)(\nabla_n u_i) \rangle = \left(u_0^2/\ell_0^2\right) \delta_{mn}$. Here L_B is the characteristic scale of the mean magnetic field variations. Therefore, the level of magnetic fluctuations for small magnetic Reynolds numbers is given by

$$\boxed{\langle b^2 \rangle = \mathrm{Rm}^2\, \overline{B}^2.} \tag{3.82}$$

Substituting the turbulent electromotive force (3.68) into Eq. (3.55), we arrive at the following mean-field equation:

$$\boxed{\frac{\partial \overline{\boldsymbol{B}}}{\partial t} = \nabla \times \left(\overline{\boldsymbol{U}} \times \overline{\boldsymbol{B}} + \boldsymbol{\alpha} \overline{\boldsymbol{B}} + \boldsymbol{V}^{\rm eff} \times \overline{\boldsymbol{B}} - \eta_{_T} \nabla \times \overline{\boldsymbol{B}}\right) + \eta\, \Delta \overline{\boldsymbol{B}},} \tag{3.83}$$

where $\boldsymbol{\alpha} \equiv \alpha_{mn}$. Equation (3.83) describes the generation of the mean magnetic field by small-scale helical turbulence. In particular, this equation is used to study the generation of solar, stellar and galactic large-scale magnetic fields.

3.4 Turbulent Transport of Magnetic Fields: Quasi-linear Approach

In this and the following sections, we consider the turbulent transport of magnetic fields using more rigorous approaches. We will start with the quasi-linear approach, which is valid for small magnetic and fluid Reynolds numbers.

3.4.1 Governing Equations

We consider the evolution of magnetic fields in isotropic inhomogeneous incompressible and helical turbulence. The equation for the magnetic field $\boldsymbol{B}(t, \boldsymbol{x})$ is determined by the induction equation (3.54). To study the large-scale evolution of the magnetic field, we apply a mean-field approach in which all quantities are decomposed into the mean and fluctuating parts. Averaging the induction equation over an ensemble of turbulent velocity field, we obtain the mean-field equation (3.55) for the mean magnetic field. To determine the turbulent electromotive force $\boldsymbol{\mathcal{E}} \equiv \langle \boldsymbol{u} \times \boldsymbol{b} \rangle$, we use the equation for magnetic fluctuations obtained by subtracting Eq. (3.55) from Eq. (3.54):

$$\frac{\partial \boldsymbol{b}}{\partial t} - \nabla \times (\boldsymbol{u} \times \boldsymbol{b} - \langle \boldsymbol{u} \times \boldsymbol{b} \rangle) - \eta\, \Delta \boldsymbol{b} = (\overline{\boldsymbol{B}} \cdot \nabla)\boldsymbol{u} - (\boldsymbol{u} \cdot \nabla)\overline{\boldsymbol{B}}. \tag{3.84}$$

We consider a one-way coupling, i.e., we take into account the effect of the turbulent velocity on the magnetic field, but we neglect the effect of the magnetic field on the turbulent fluid flow. The ratio of the nonlinear terms to the diffusion terms in Eq. (3.84) is the magnetic Reynolds number, $\mathrm{Rm} = u_0\, \ell_0/\eta$. First, we consider the case when the nonlinear terms in Eq. (3.84) are much smaller than the magnetic diffusion terms, i.e., $\mathrm{Rm} \ll 1$.

3.4.2 Quasi-linear Approach

We use a quasi-linear approach that is valid for small hydrodynamic and magnetic Reynolds numbers, $\mathrm{Re} = \ell_0 \sqrt{\langle \boldsymbol{u}^2 \rangle}/\nu \ll 1$ and $\mathrm{Rm} = \ell_0 \sqrt{\langle \boldsymbol{u}^2 \rangle}/\eta \ll 1$, respectively. Here $\sqrt{\langle \boldsymbol{u}^2 \rangle}$ is the characteristic turbulent velocity in the integral turbulent scale ℓ_0. In the high conductivity limit (very small microscopic magnetic diffusivity η), the quasi-linear approach is only valid for small Strouhal numbers, which is defined as the ratio of turbulent correlation time τ_0 to turnover time $\ell_0/\sqrt{\langle \boldsymbol{u}^2 \rangle}$. This implies that the quasi-linear approach in this limit is only valid for very short turbulent correlation times. Often the quasi-linear approach is applied only to the induction equation and the origin of the velocity field is not discussed. In this case, the evolution of the magnetic field is considered for a prescribed velocity field, and formally the validity of the quasi-linear approach only requires small magnetic Reynolds numbers.

To determine the turbulent electromotive force, which is a one-point correlation function, and to take into account small-scale properties of the turbulence, e.g., the turbulent spectrum, one must use two-point correlation functions. For fully developed turbulence, scalings for the turbulent correlation time and energy spectrum are related via the Kolmogorov scalings. This is the reason why for calculations of the turbulent electromotive force in fully developed turbulence, we use instantaneous two-point correlation functions. On the other hand, for a random flow with small Re and Rm, there are no universal scalings for the correlation time and energy spectrum. This is the reason why we have to use non-instantaneous two-point correlation functions in this case.

In the framework of the mean-field approach, we assume that there is a separation of spatial and temporal scales, i.e., $\ell_0 \ll L_B$ and $\tau_0 \ll t_B$, where L_B and t_B are the characteristic spatial and temporal scales characterizing the variations of the mean magnetic field. The mean fields depend on "slow" variables, while fluctuations depend on "fast" variables. Separation into slow and fast variables is widely used in theoretical physics, and all calculations are reduced to Taylor expansions of all functions using small parameters ℓ_0/L_B and τ_0/t_B. The findings are further truncated to leading order terms.

Separation to slow and fast variables is performed by means of the standard multi-scale approach. In the framework of this approach, the non-instantaneous two-point second-order correlation function of magnetic $\boldsymbol{b}$ and velocity $\boldsymbol{u}$ turbulent fields is written as follows:

$$\begin{aligned}\big\langle b_i(\boldsymbol{x}, t_1)\, u_j(\boldsymbol{y}, t_2)\big\rangle &= \int d\omega_1\, d\omega_2\, d\boldsymbol{k}_1\, d\boldsymbol{k}_2\, \big\langle b_i(\boldsymbol{k}_1, \omega_1) u_j(\boldsymbol{k}_2, \omega_2)\big\rangle \\ &\quad \times \exp\big[\mathrm{i}\,(\boldsymbol{k}_1 \cdot \boldsymbol{x} + \boldsymbol{k}_2 \cdot \boldsymbol{y}) + \mathrm{i}\,(\omega_1 t_1 + \omega_2 t_2)\big] \\ &= \int g_{ij}(\boldsymbol{k}, \omega, t, \boldsymbol{R}) \exp[\mathrm{i}\,\boldsymbol{k} \cdot \boldsymbol{r} + \mathrm{i}\,\omega\,\tau]\, d\omega\, d\boldsymbol{k}, \qquad (3.85)\end{aligned}$$

where

$$g_{ij}(\boldsymbol{k}, \omega, \boldsymbol{R}, t) = \int \langle b_i(\boldsymbol{k}_1, \omega_1)\, u_j(\boldsymbol{k}_2, \omega_2)\rangle \exp[\mathrm{i}\,\Omega t + \mathrm{i}\,\boldsymbol{K}\cdot\boldsymbol{R}]\, d\Omega\, d\boldsymbol{K}. \tag{3.86}$$

Here we introduced large-scale variables, $\boldsymbol{R} = (\boldsymbol{x}+\boldsymbol{y})/2$, $\boldsymbol{K} = \boldsymbol{k}_1 + \boldsymbol{k}_2$, $t = (t_1 + t_2)/2$, $\Omega = \omega_1 + \omega_2$, and small-scale variables, $\boldsymbol{r} = \boldsymbol{x} - \boldsymbol{y}$, $\boldsymbol{k} = (\boldsymbol{k}_1 - \boldsymbol{k}_2)/2$, $\tau = t_1 - t_2$, $\omega = (\omega_1 - \omega_2)/2$. This implies that $\omega_1 = \omega + \Omega/2$, $\omega_2 = -\omega + \Omega/2$, $\boldsymbol{k}_1 = \boldsymbol{k} + \boldsymbol{K}/2$ and $\boldsymbol{k}_2 = -\boldsymbol{k} + \boldsymbol{K}/2$. Mean fields depend on the large-scale variables, while fluctuations depend on the small-scale variables. Similarly to Eqs. (3.85)–(3.86), we obtain the second-order correlation function of velocity fluctuations in the Fourier space in small-scale variables as

$$f_{ij}(\boldsymbol{k}, \omega, \boldsymbol{R}, t) = \int \langle u_i(\boldsymbol{k}_1, \omega_1)\, u_j(\boldsymbol{k}_2, \omega_2)\rangle \exp[\mathrm{i}\,\Omega t + \mathrm{i}\,\boldsymbol{K}\cdot\boldsymbol{R}]\, d\Omega\, d\boldsymbol{K}. \tag{3.87}$$

After separation into slow and fast variables and calculating the function g_{ij}, Eq. (3.85) allows us to determine the cross-helicity tensor in the physical space and to calculate the limit of $\boldsymbol{r} \to \boldsymbol{0}$ and $\tau \to 0$, which yields

$$\langle b_i(\boldsymbol{x}, t)\, u_j(\boldsymbol{x}, t)\rangle = \int g_{ij}(\boldsymbol{k}, \omega, \boldsymbol{R}, t)\, d\omega\, d\boldsymbol{k}. \tag{3.88}$$

To consider magnetic fluctuations for small magnetic Reynolds numbers, we use the quasi-linear approach. This implies that we neglect the nonlinear terms but keep the magnetic diffusion term in Eq. (3.84). We apply the Fourier transform of the magnetic and velocity fields,

$$b_j(t, \boldsymbol{x}) = \int\!\!\int b_j(\omega, \boldsymbol{k}) \exp(\mathrm{i}\boldsymbol{k}\cdot\boldsymbol{x} + \mathrm{i}\,\omega t)\, d\boldsymbol{k}\, d\omega, \tag{3.89}$$

$$u_i(t, \boldsymbol{x}) = \int\!\!\int u_i(\omega, \boldsymbol{k}) \exp(\mathrm{i}\boldsymbol{k}\cdot\boldsymbol{x} + \mathrm{i}\,\omega t)\, d\boldsymbol{k}\, d\omega, \tag{3.90}$$

and use the following identities:

$$\frac{\partial b_j(t, \boldsymbol{x})}{\partial t} = \int\!\!\int (\mathrm{i}\,\omega)\, b_j(\omega, \boldsymbol{k}) \exp(\mathrm{i}\boldsymbol{k}\cdot\boldsymbol{x} + \mathrm{i}\,\omega t)\, d\boldsymbol{k}\, d\omega, \tag{3.91}$$

$$\Delta b_j(t, \boldsymbol{x}) = \int\!\!\int (-\boldsymbol{k}^2)\, b_j(\omega, \boldsymbol{k}) \exp(\mathrm{i}\boldsymbol{k}\cdot\boldsymbol{x} + \mathrm{i}\,\omega t)\, d\boldsymbol{k}\, d\omega, \tag{3.92}$$

$$\nabla_n u_j(t, \boldsymbol{x}) = \int\!\!\int (\mathrm{i}\,k_n)\, u_j(\omega, \boldsymbol{k}) \exp(\mathrm{i}\boldsymbol{k}\cdot\boldsymbol{x} + i\mathrm{i}\,\omega t)\, d\boldsymbol{k}\, d\omega. \tag{3.93}$$

Substituting Eqs. (3.91)–(3.93) into Eq. (3.84) and neglecting the nonlinear terms, we obtain the following equation:

$$\int\int \left[\left(\eta k^2 + \mathrm{i}\,\omega\right)\, b_j(\omega, \boldsymbol{k}) + u_n(\omega, \boldsymbol{k})(\nabla_n \overline{B}_j) - \mathrm{i}\, k_n\, u_j(\omega, \boldsymbol{k})\, \overline{B}_n\right] \times \exp(\mathrm{i}\boldsymbol{k}\cdot\boldsymbol{x} + \mathrm{i}\,\omega\, t)\, d\boldsymbol{k}\, d\omega = 0. \tag{3.94}$$

Equation (3.94) implies that

$$\left(\eta k^2 + \mathrm{i}\,\omega\right)\, b_j(\omega, \boldsymbol{k}) + u_n(\omega, \boldsymbol{k})\,(\nabla_n \overline{B}_j) - \mathrm{i}\, k_n\, u_j(\omega, \boldsymbol{k})\, \overline{B}_n = 0, \tag{3.95}$$

which yields

$$b_j(\omega, \boldsymbol{k}) = -G_\eta(\omega, \boldsymbol{k})\left[u_n(\omega, \boldsymbol{k})\,(\nabla_n \overline{B}_j) - \mathrm{i}\, k_n\, u_j(\omega, \boldsymbol{k})\, \overline{B}_n\right], \tag{3.96}$$

where the function $G_\eta(\omega, \boldsymbol{k}) = \left(\eta k^2 + \mathrm{i}\,\omega\right)^{-1}$. Equation (3.96) represents a solution of Eq. (3.84) in the Fourier space.

Note that Eqs. (3.94)–(3.96) are not valid in a general case. However, in this section, we take into account only the effects that are linear in $\nabla_j \overline{\boldsymbol{B}}$ in the turbulent electromotive force, and we neglect higher-order spatial derivatives of the mean magnetic field. In addition, we do not study here the compressibility effects of a turbulent velocity field in the evolution of the mean magnetic field (i.e., we consider an incompressible fluid flow). We will show in Section 3.4.5 that in this case, we can use Eq. (3.96) to determine the turbulent electromotive force for small magnetic Reynolds numbers. The expression for magnetic fluctuations $\boldsymbol{b}(\omega, \boldsymbol{k})$ in a more general case in the Fourier space for small magnetic Reynolds numbers is given by Eq. (3.117) in Section 3.4.5.

We determine the cross-helicity tensor, $\langle u_i(t, \boldsymbol{x})\, b_j(t, \boldsymbol{x})\rangle$, using the multi-scale approach,

$$\begin{aligned}\left\langle u_i(t, \boldsymbol{x})\, b_j(t, \boldsymbol{x})\right\rangle &= \lim_{t_1 \to t_2, \boldsymbol{x} \to \boldsymbol{y}} \left\langle u_i(t_1, \boldsymbol{x})\, b_j(t_2, \boldsymbol{y})\right\rangle \\ &= \lim_{\tau \to 0, \boldsymbol{r} \to 0} \int\int \left\langle u_i(\omega, \boldsymbol{k})\, b_j(-\omega, -\boldsymbol{k})\right\rangle \exp[\mathrm{i}\boldsymbol{k}\cdot\boldsymbol{r} + \mathrm{i}\,\omega\,\tau]\, d\omega\, d\boldsymbol{k} \\ &= \int\int \left\langle u_i(\omega, \boldsymbol{k})\, b_j(-\omega, -\boldsymbol{k})\right\rangle\, d\omega\, d\boldsymbol{k},\end{aligned} \tag{3.97}$$

where $\boldsymbol{r} = \boldsymbol{x} - \boldsymbol{y}$ and $\tau = t_1 - t_2$, and we take into account that turbulence is inhomogeneous and stationary in the statistical sense. The latter implies that the correlation function $\langle u_i\, u_j\rangle$ depends on the large-scale variable $\boldsymbol{R}$, but it is independent of the "slow" time t. Similarly, we determine the level of fluctuations $\left\langle \boldsymbol{b}^2\right\rangle$:

$$\begin{aligned}\left\langle \boldsymbol{b}^2\right\rangle \equiv \left\langle b_j(t, \boldsymbol{x})\, b_j(t, \boldsymbol{x})\right\rangle &= \lim_{t_1 \to t_2, \boldsymbol{x} \to \boldsymbol{y}} \left\langle b_j(t_1, \boldsymbol{x})\, b_j(t_2, \boldsymbol{y})\right\rangle \\ &= \lim_{\tau \to 0, \boldsymbol{r} \to 0} \int\int \left\langle b_j(\omega, \boldsymbol{k})\, b_j(-\omega, -\boldsymbol{k})\right\rangle \exp[\mathrm{i}\boldsymbol{k}\cdot\boldsymbol{r} + \mathrm{i}\,\omega\,\tau]\, d\omega\, d\boldsymbol{k}\end{aligned}$$

$$= \int\int \langle b_j(\omega, \boldsymbol{k})\, b_j(-\omega, -\boldsymbol{k})\rangle \, d\omega \, d\boldsymbol{k}. \tag{3.98}$$

Using Eqs. (3.96)–(3.98), we determine the cross-helicity tensor $\langle u_i\, b_j\rangle$ and the level of magnetic fluctuations $\langle \boldsymbol{b}^2\rangle$:

$$\langle u_i\, b_j\rangle = -\int\int G^*_\eta \left[\nabla_m \overline{B}_j + \mathrm{i}\, k_n\, \delta_{mj} \overline{B}_n\right] f_{im}(\omega, \boldsymbol{k})\, d\omega\, d\boldsymbol{k}, \tag{3.99}$$

$$\begin{aligned}\langle \boldsymbol{b}^2\rangle = \int\int G_\eta G^*_\eta \Big[& \left(\nabla_n \overline{B}_l\right) \left(\nabla_m \overline{B}_l\right) + k_l k_q\, \delta_{nm}\, \overline{B}_l \overline{B}_q \\ & + \mathrm{i}\, k_l\, \overline{B}_l \left(\nabla_n \overline{B}_m - \nabla_m \overline{B}_n\right)\Big] f_{nm}(\omega, \boldsymbol{k})\, d\omega\, d\boldsymbol{k},\end{aligned} \tag{3.100}$$

where $f_{ij}(\omega, \boldsymbol{k}) = \langle u_i(\omega, \boldsymbol{k})\, u_j(-\omega, -\boldsymbol{k})\rangle$. For brevity of notation, we omit that below the large-scale variable $\boldsymbol{R}$ in the correlation functions $f_{ij}(\boldsymbol{k}, \omega, \boldsymbol{R})$ and $g_{ij}(\boldsymbol{k}, \omega, \boldsymbol{R})$. Since we consider a one-way coupling when there is no effect of magnetic field on the fluid flow, the correlation function f_{ij} in Eqs. (3.99) and (3.100) should be replaced by $f^{(0)}_{ij}$ for the background random flow.

3.4.3 Model of Isotropic Inhomogeneous Incompressible and Helical Turbulence

We use the model for the second moments, $f^{(0)}_{ij}(\boldsymbol{k}, \boldsymbol{R})$, of isotropic inhomogeneous incompressible and helical turbulence in a Fourier space in the following form:

$$\begin{aligned} f^{(0)}_{ij}(\omega, \boldsymbol{k}, \boldsymbol{R}) = \frac{E(k)\, \Phi(\omega)}{8\pi k^2} \Bigg\{ & \left[\delta_{ij} - k_{ij} + \frac{\mathrm{i}}{2k^2} \left(k_i \nabla_j - k_j \nabla_i\right)\right] \langle \mathbf{u}^2\rangle \\ & - \frac{\mathrm{i}}{k^2}\, \varepsilon_{ijp}\, k_p\, \langle \mathbf{u} \cdot (\nabla \times \mathbf{u})\rangle \Bigg\}. \end{aligned} \tag{3.101}$$

Here δ_{ij} is the Kronecker tensor, $k_{ij} = k_i\, k_j / k^2$, $\langle \mathbf{u} \cdot (\nabla \times \mathbf{u})\rangle$ is the kinetic helicity, the energy spectrum function is $E(k) = (q-1)\, k_0^{-1}\, (k/k_0)^{-q}$ for the range of wave numbers $k_0 < k < k_d$, the wave number $k_0 = 1/\ell_0$, the length ℓ_0 is the maximum scale of random motions, the exponent q varies in the interval $1 < q < 3$ and the wave number $k_d = 1/\ell_d$ based on the dissipative scale ℓ_d satisfies to $k_d \gg k_0$. We use the frequency function $\Phi(\omega)$ in the form of the Lorentz profile: $\Phi(\omega) = [\pi \tau_0\, (\omega^2 + \tau_0^{-2})]^{-1}$, where τ_0 is the correlation time of a random velocity field. This model for the frequency function corresponds to the following non-instantaneous correlation function: $\langle u_i(t) u_j(t+\tau)\rangle \propto \exp(-\tau/\tau_0)$.

The part of the correlation function (3.101) of the turbulent velocity field that is proportional to $(i/2k^2)\left(k_i \nabla_j - k_j \nabla_i\right)$ describes inhomogeneous turbulence. To derive Eq. (3.101), we use the multi-scale approach, in which we make a change of variables, $\boldsymbol{k}_1 = \boldsymbol{k} + \boldsymbol{K}/2$ and $\boldsymbol{k}_2 = -\boldsymbol{k} + \boldsymbol{K}/2$, where $\boldsymbol{k}$ corresponds to small-scale variables and $\boldsymbol{K}$ corresponds to large-scale variables. We consider the model of the isotropic inhomogeneous and incompressible turbulence in the following form:

$$\langle u_i(\boldsymbol{k}_1)\, u_j(\boldsymbol{k}_2)\rangle = \frac{E(k)}{8\pi\, k^2}\left[\delta_{ij} - k_{ij} + A_1 k_i K_j + A_2 k_j K_i + O(\boldsymbol{K}^2)\right]\langle \boldsymbol{u}^2\rangle, \tag{3.102}$$

taking into account that $|\boldsymbol{K}| \ll |\boldsymbol{k}|$. The incompressibility condition, $\nabla \cdot \boldsymbol{u} = 0$, in the Fourier space implies the following equation: $\boldsymbol{k}_1 \cdot \boldsymbol{u}(\boldsymbol{k}_1) \equiv k_i^{(1)} u_i(\boldsymbol{k}_1) = 0$, where $\boldsymbol{k}_1 \equiv k_i^{(1)}$. Multiplying this equation by $u_j(\boldsymbol{k}_2)$ and averaging it over an ensemble of turbulent velocity field, we arrive at the following condition:

$$k_i^{(1)}\,\langle u_i(\mathbf{k}_1)\, u_j(\mathbf{k}_2)\rangle = 0. \tag{3.103}$$

Similarly, $k_j^{(2)}\,\langle u_i(\mathbf{k}_1)\, u_j(\mathbf{k}_2)\rangle = 0$, where $\boldsymbol{k}_2 \equiv k_j^{(2)}$. Condition (3.103) yields:

$$\left(k_i + \frac{1}{2}K_i\right)\langle u_i(\mathbf{k}_1)\, u_j(\mathbf{k}_2)\rangle = \left[\frac{1}{2}K_i(\delta_{ij} - k_{ij}) + A_1 k_i^2 K_j + A_2 k_i k_j K_i + O(\mathbf{K}^2)\right]\frac{E(k)}{8\pi\, k^2}\langle \mathbf{u}^2\rangle = \frac{E(k)}{8\pi\, k^2}\left[\left(\frac{1}{2} + A_1 k^2\right)K_j + \left(A_2 - \frac{1}{2k^2}\right)k_i k_j K_i + O(\mathbf{K}^2)\right]\langle \mathbf{u}^2\rangle = 0. \tag{3.104}$$

Equation (3.104) is valid for arbitrary $\mathbf{k}$ and $\mathbf{K}$. This implies that $A_1 = -1/2k^2$ and $A_2 = 1/2k^2$. Since $\nabla = \mathrm{i}\,\boldsymbol{K}$, we obtain $K_j = -\mathrm{i}\,\nabla_j$. This yields Eq. (3.101) for the model of the isotropic inhomogeneous and incompressible turbulence. The term proportional to $\varepsilon_{ijp}\, k_p\, \langle \mathbf{u} \cdot (\nabla \times \mathbf{u})\rangle$ describes the helical part of the turbulence [see Eq. (3.101)], which is proportional to the kinetic helicity. A non-zero kinetic helicity $\langle \mathbf{u} \cdot (\nabla \times \mathbf{u})\rangle$ is responsible for the generation of the mean magnetic field in a helical turbulence.

3.4.4 Turbulent Electromotive Force and Level of Magnetic Fluctuations

Substituting Eq. (3.101) into Eq. (3.99) and (3.100), we determine the cross-helicity tensor, $\langle u_i\, b_j\rangle$, and the level of magnetic fluctuations $\langle \boldsymbol{b}^2\rangle$:

$$\langle u_i\, b_j\rangle = -\frac{1}{8\pi^2\, \tau_0}\int_{k_0}^{k_\nu} E(k)\, dk \int_0^{2\pi} d\varphi \int_0^{\pi} \sin\vartheta\, d\vartheta \left[\nabla_m \overline{B}_j + \mathrm{i}\, k_n\, \delta_{mj} \overline{B}_n\right] \times \left\{\left[\delta_{im} - \frac{k_i\, k_m}{k^2} + \frac{\mathrm{i}}{2k^2}\left(k_i \nabla_m - k_m \nabla_i\right)\right]\langle \mathbf{u}^2\rangle - \frac{\mathrm{i}}{k^2}\,\varepsilon_{imp}\, k_p\, \langle \mathbf{u} \cdot (\nabla \times \mathbf{u})\rangle\right\} \times \int \frac{G_\eta^*}{(\omega^2 + \tau_0^{-2})}\, d\omega = -\frac{1}{6\eta}\left\{\left[2\nabla_i \overline{B}_j + \frac{1}{2}\left(\overline{B}_j \nabla_i - \overline{B}_i \nabla_j\right)\right]\langle \mathbf{u}^2\rangle\right.$$

$$+\varepsilon_{ijn}\overline{B}_n \, \langle \mathbf{u} \cdot (\nabla \times \mathbf{u})\rangle\Big\} \int_{k_0}^{k_\nu} \frac{E(k)}{k^2}\, dk, \tag{3.105}$$

$$\begin{aligned}\langle b^2\rangle &= \frac{1}{8\pi^2\tau_0}\int_{k_0}^{k_\nu} E(k)\,dk \int_0^{2\pi} d\varphi \int_0^{\pi} \sin\vartheta\, d\vartheta \, \Big[(\nabla_n \overline{B}_l)\,(\nabla_m \overline{B}_l) \\ &+ k_l k_q\, \delta_{nm}\, \overline{B}_l \overline{B}_q + \mathrm{i}\, k_l\, \overline{B}_l \left(\nabla_n \overline{B}_m - \nabla_m \overline{B}_n\right)\Big] \Big\{ \Big[\delta_{nm} - \frac{k_n k_m}{k^2} \\ &+ \frac{\mathrm{i}}{2k^2}\left(k_n \nabla_m - k_m \nabla_n\right)\Big] \langle \mathbf{u}^2\rangle - \frac{\mathrm{i}}{k^2}\varepsilon_{nmp}\, k_p\, \langle \mathbf{u} \cdot (\nabla \times \mathbf{u})\rangle \Big\} \int \frac{G_\eta\, G_\eta^*}{(\omega^2 + \tau_0^{-2})}\, d\omega \\ &= \frac{1}{3\eta^2}\Big\{ \langle \mathbf{u}^2\rangle\, \overline{\boldsymbol{B}}^2 \int_{k_0}^{k_\nu} \frac{E(k)}{k^2}\, dk + \Big[\frac{1}{4}\Big(\nabla \overline{\boldsymbol{B}}^2 - 2(\overline{\boldsymbol{B}} \cdot \nabla)\overline{\boldsymbol{B}}\Big) \cdot \nabla \langle \mathbf{u}^2\rangle \\ &+ 2\overline{\boldsymbol{B}} \cdot (\nabla \times \overline{\boldsymbol{B}})\, \langle \mathbf{u} \cdot (\nabla \times \mathbf{u})\rangle\Big] \int_{k_0}^{k_\nu} \frac{E(k)}{k^4}\, dk \Big\}.\end{aligned} \tag{3.106}$$

For the integration over ω and $\boldsymbol{k}$ in Eqs. (3.105) and (3.106), we use identities (1.133)–(1.137). After integration in the ω space and $\boldsymbol{k}$ space, we obtain formulae for the turbulent electromotive force, $\boldsymbol{\mathcal{E}} \equiv \langle \boldsymbol{u} \times \boldsymbol{b}\rangle$, and the level of magnetic fluctuations $\langle \boldsymbol{b}^2\rangle$:

$$\boxed{\boldsymbol{\mathcal{E}} = \alpha \overline{\boldsymbol{B}} + \boldsymbol{V}^{\rm eff} \times \overline{\boldsymbol{B}} - \eta_{_T} \nabla \times \overline{\boldsymbol{B}},} \tag{3.107}$$

$$\boxed{\alpha = -\frac{q-1}{3(q+1)}\, \tau_0\, \langle \mathbf{u} \cdot (\nabla \times \mathbf{u})\rangle\, \mathrm{Rm},} \tag{3.108}$$

$$\boxed{\eta_{_T} = \frac{q-1}{3(q+1)}\, u_0\, \ell_0\, \mathrm{Rm},} \tag{3.109}$$

$$\boxed{\boldsymbol{V}^{\rm eff} = -\frac{1}{2} \nabla \eta_{_T},} \tag{3.110}$$

$$\boxed{\langle \boldsymbol{b}^2\rangle = \frac{q-1}{3(q+1)}\, \mathrm{Rm}^2\, \overline{\boldsymbol{B}}^2 \left[1 + O\left(\frac{\ell_0^2}{L_B L_u}\right)\right],} \tag{3.111}$$

where L_B and L_u are characteristic scales of spatial variations of the mean magnetic field and the inhomogeneity of the turbulence, respectively, and $L_B, L_u \gg \ell_0$. Here $\boldsymbol{V}^{\rm eff}$ is the effective pumping velocity of the mean magnetic field that describes the turbulent diamagnetic effect (the mean magnetic field is pushed out from the regions with higher turbulence intensity), the parameter α in Eqs. (3.107) and (3.108) describes the α-effect caused by the kinetic helicity $\langle \mathbf{u} \cdot (\nabla \times \mathbf{u}) \rangle$ and η_T is the coefficient of turbulent magnetic diffusion. Equations (3.107)–(3.110) have been derived by M. Steenbeck, F. Krause and K.-H. Rädler using the quasi-linear approach (Steenbeck et al., 1966). Note also that the results given by Eqs. (3.107)–(3.111) derived using the quasi-linear approach are in agreement with results (3.78)–(3.79) and (3.82), obtained by the dimensional analysis.

Substituting the turbulent electromotive force (3.107) into Eq. (3.55), we obtain the mean-field induction equation:

$$\boxed{\frac{\partial \overline{\boldsymbol{B}}}{\partial t} = \nabla \times \left(\overline{\boldsymbol{U}} \times \overline{\boldsymbol{B}} + \alpha \overline{\boldsymbol{B}} + \boldsymbol{V}^{\rm eff} \times \overline{\boldsymbol{B}} \right) + (\eta + \eta_T)\, \Delta \overline{\boldsymbol{B}}.} \tag{3.112}$$

Using Eq. (3.112), we can study the large-scale dynamics of the magnetic field (i.e., we can investigate the large-scale dynamo instability) that results in the generation of the large-scale magnetic field by random flow of the electrically conducting fluid.

3.4.5 Rigorous Derivation of Equation for Turbulent Electromotive Force for Small Magnetic Reynolds Numbers

In Section 3.4.2, we used approximate solution (3.96) for magnetic fluctuations $\boldsymbol{b}(\omega, \boldsymbol{k})$ in the Fourier space to derive equation for the turbulent electromotive force for small magnetic Reynolds numbers. In this section, we give a rigorous derivation of the turbulent electromotive force. To this end, we use the following identities in the $\boldsymbol{k}$ space:

$$\left[(\overline{\boldsymbol{B}} \cdot \nabla) u_i \right]_{\boldsymbol{k}} = {\rm i}\, k_p \int \overline{B}_p(\boldsymbol{Q})\, u_i(\boldsymbol{k} - \boldsymbol{Q})\, d\boldsymbol{Q}, \tag{3.113}$$

$$\left[(\boldsymbol{u} \cdot \nabla) \overline{B}_i \right]_{\boldsymbol{k}} = {\rm i} \int Q_p\, \overline{B}_i(\boldsymbol{Q})\, u_p(\boldsymbol{k} - \boldsymbol{Q})\, d\boldsymbol{Q}, \tag{3.114}$$

where $[\ldots]_{\boldsymbol{k}}$ denotes the Fourier transform. To derive Eq. (3.113), we rewrite the term $(\overline{\boldsymbol{B}} \cdot \nabla)\boldsymbol{u}$ in the $\boldsymbol{k}$ space using the Fourier transform for $\overline{\boldsymbol{B}}(\boldsymbol{x}, t)$ and $\nabla_p u_i(\boldsymbol{x}, t)$, i.e.,

$$(\overline{\boldsymbol{B}} \cdot \nabla) u_i = \int \overline{B}_p(\boldsymbol{Q}) \exp({\rm i}\boldsymbol{Q} \cdot \boldsymbol{x})\, d\boldsymbol{Q} \int {\rm i}\, k'_p\, u_i(\boldsymbol{k}') \exp(i\boldsymbol{k}' \cdot \boldsymbol{x})\, d\boldsymbol{k}'. \tag{3.115}$$

For brevity of notation, we omit hereafter the arguments t and $\boldsymbol{x}$ in the functions $\overline{\boldsymbol{B}}(\boldsymbol{x}, t)$ and $\boldsymbol{u}(\boldsymbol{x}, t)$, and the variable t in the functions $\overline{B}_p(\boldsymbol{Q}, t)$ and $u_i(\boldsymbol{k}, t)$. We use a new variable $\boldsymbol{k} = \boldsymbol{k}' + \boldsymbol{Q}$ and take into account that the condition, div $\overline{\boldsymbol{B}} = 0$, in the $\boldsymbol{k}$ space reads i $Q_p\, \overline{B}_p(\boldsymbol{Q}) = 0$. Therefore, Eq. (3.115) can be rewritten as

$$(\overline{\boldsymbol{B}} \cdot \nabla) u_i = \int \left(\mathrm{i}\, k_p \int \overline{B}_p(\boldsymbol{Q})\, u_i(\boldsymbol{k} - \boldsymbol{Q})\, d\boldsymbol{Q} \right) \exp(\mathrm{i}\, \boldsymbol{k} \cdot \boldsymbol{x})\, d\boldsymbol{k}$$
$$\equiv \int \left[(\overline{\boldsymbol{B}} \cdot \nabla) u_i \right]_{\boldsymbol{k}} \exp(\mathrm{i}\, \boldsymbol{k} \cdot \boldsymbol{x})\, d\boldsymbol{k}, \tag{3.116}$$

so that $\left[(\overline{\boldsymbol{B}} \cdot \nabla) u_i \right]_{\boldsymbol{k}} = \mathrm{i}\, k_p \int \overline{B}_p(\boldsymbol{Q})\, u_i(\boldsymbol{k} - \boldsymbol{Q})\, d\boldsymbol{Q}$, as given by Eq. (3.113). Similar derivation is also performed to arrive at Eq. (3.114). Using Eqs. (3.113)–(3.114), we obtain a solution of Eq. (3.84) in the $\boldsymbol{k}$ space as

$$b_i(\boldsymbol{k}, \omega) = \mathrm{i}\, G_\eta(\boldsymbol{k}, \omega) \biggl[k_p \int \overline{B}_p(\boldsymbol{Q}) u_i(\boldsymbol{k} - \boldsymbol{Q}, \omega)\, d\boldsymbol{Q}$$
$$- \int Q_p\, \overline{B}_i(\boldsymbol{Q})\, u_p(\boldsymbol{k} - \boldsymbol{Q}, \omega)\, d\boldsymbol{Q} \biggr], \tag{3.117}$$

where $G_\eta(\boldsymbol{k}, \omega) = (\eta \boldsymbol{k}^2 + \mathrm{i}\, \omega)^{-1}$.

In the framework of the multi-scale approach, the non-instantaneous two-point second-order correlation function of $\boldsymbol{b}$ and $\boldsymbol{u}$ is written as follows:

$$\langle b_i(\boldsymbol{x}, t_1)\, u_j(\boldsymbol{y}, t_2) \rangle = \int d\omega_1\, d\omega_2\, d\boldsymbol{k}_1\, d\boldsymbol{k}_2\, \langle b_i(\boldsymbol{k}_1, \omega_1) u_j(\boldsymbol{k}_2, \omega_2) \rangle$$
$$\times \exp\left[\mathrm{i}\, (\boldsymbol{k}_1 \cdot \boldsymbol{x} + \boldsymbol{k}_2 \cdot \boldsymbol{y}) + \mathrm{i}\, (\omega_1 t_1 + \omega_2 t_2) \right]$$
$$= \int \tilde{g}_{ij}(\boldsymbol{k}, \omega, t, \boldsymbol{R}) \exp[\mathrm{i}\, \boldsymbol{k} \cdot \boldsymbol{r} + \mathrm{i}\, \omega\, \tau]\, d\omega\, d\boldsymbol{k}, \tag{3.118}$$

where

$$\tilde{g}_{ij}(\boldsymbol{k}, \omega, \boldsymbol{R}, t) = \int \langle b_i(\boldsymbol{k}_1, \omega_1)\, u_j(\boldsymbol{k}_2, \omega_2) \rangle \exp[\mathrm{i}\, \Omega t + \mathrm{i}\, \boldsymbol{K} \cdot \boldsymbol{R}]\, d\Omega\, d\boldsymbol{K}, \tag{3.119}$$

and for brevity of notation, we omit hereafter the arguments t and $\boldsymbol{x}$ in the function $\tilde{g}_{ij}(\boldsymbol{k}, \omega, \boldsymbol{R}, t)$. Using Eq. (3.117), we determine the tensor $\tilde{g}_{ij}(\boldsymbol{k}, \boldsymbol{R})$ in the $\boldsymbol{k}$ space as

$$\tilde{g}_{ij}(\boldsymbol{k}, \boldsymbol{R}) = J_{ij}(\boldsymbol{k}, \boldsymbol{R}) - (\nabla_p \overline{B}_i)\, G_\eta(\boldsymbol{k})\, f_{pj}(\boldsymbol{k}, \boldsymbol{R}), \tag{3.120}$$

where

$$J_{ij}(\boldsymbol{k}, \boldsymbol{R}) = \mathrm{i} \int (k_p + K_p/2)\, \langle u_i(\boldsymbol{k} + \boldsymbol{K}/2 - \boldsymbol{Q}) u_j(-\boldsymbol{k} + \boldsymbol{K}/2) \rangle$$
$$\times G_\eta(\boldsymbol{k} + \boldsymbol{K}/2)\, \overline{B}_p(\boldsymbol{Q}) \exp(\mathrm{i}\, \boldsymbol{K} \cdot \boldsymbol{R})\, d\boldsymbol{K}\, d\boldsymbol{Q}, \tag{3.121}$$

and the functions J_{ij}, G_η and u_i depend also on ω. To simplify the notation, we do not show this dependence in the following calculations. We study an evolution of a magnetic field in an incompressible random flow, and we take into account in the turbulent electromotive force only the effects that are linear in the gradient of the mean magnetic field and neglect higher-order spatial derivatives of the mean magnetic field. The second term on the right-hand side of Eqs. (3.117) and (3.120) has a simple form. This term is caused by $(\boldsymbol{u} \cdot \nabla)\overline{\boldsymbol{B}}$ term in the induction equation, i.e., it has already the spatial derivative of the mean magnetic field. Now let us determine the function J_{ij} caused by $(\overline{\boldsymbol{B}} \cdot \nabla)\boldsymbol{u}$ term in the induction equation. To this end, we use new variables:

$$\tilde{\boldsymbol{k}} = (\tilde{\boldsymbol{k}}_1 - \tilde{\boldsymbol{k}}_2)/2 = \boldsymbol{k} - \boldsymbol{Q}/2 \,, \;\; \tilde{\boldsymbol{K}} = \tilde{\boldsymbol{k}}_1 + \tilde{\boldsymbol{k}}_2 = \boldsymbol{K} - \boldsymbol{Q},$$

where

$$\tilde{\boldsymbol{k}}_1 = \boldsymbol{k} + \boldsymbol{K}/2 - \boldsymbol{Q} \,, \quad \tilde{\boldsymbol{k}}_2 = -\boldsymbol{k} + \boldsymbol{K}/2. \tag{3.122}$$

Therefore,

$$J_{ij}(\boldsymbol{k}, \boldsymbol{R}) = \mathrm{i} \int (k_p + K_p/2)\, f_{ij}(\boldsymbol{k} - \boldsymbol{Q}/2, \boldsymbol{K} - \boldsymbol{Q})\, G_\eta(\boldsymbol{k} + \boldsymbol{K}/2)\, \overline{B}_i(\boldsymbol{Q}) \times \exp(\mathrm{i}\boldsymbol{K} \cdot \boldsymbol{R})\, d\boldsymbol{K}\, d\boldsymbol{Q}. \tag{3.123}$$

Since $|\boldsymbol{Q}| \ll |\boldsymbol{k}|$ and $|\boldsymbol{K}| \ll |\boldsymbol{k}|$, we use the Taylor expansion

$$f_{ij}(\boldsymbol{k} - \boldsymbol{Q}/2, \boldsymbol{K} - \boldsymbol{Q}) = f_{ij}(\boldsymbol{k}, \boldsymbol{K} - \boldsymbol{Q}) - \frac{1}{2}\frac{\partial f_{ij}}{\partial k_s} Q_s + O(\boldsymbol{Q}^2), \tag{3.124}$$

$$G_\eta(\boldsymbol{k} + \boldsymbol{K}/2) = G_\eta(\boldsymbol{k}) \left[1 - \eta(\boldsymbol{k} \cdot \boldsymbol{K}) G_\eta(\boldsymbol{k})\right] + O(\boldsymbol{K}^2). \tag{3.125}$$

Substituting Eqs. (3.124)–(3.125) into Eq. (3.123), we get

$$J_{ij}(\boldsymbol{k}, \boldsymbol{R}) = \mathrm{i}\, G_\eta(\boldsymbol{k}) \int \Bigg[\left(k_p - \eta G_\eta(\boldsymbol{k}) k_p k_s K_s + \frac{1}{2} K_p\right) \times \int f_{ij}(\boldsymbol{k}, \boldsymbol{K} - \boldsymbol{Q}) \overline{B}_p(\boldsymbol{Q})\, d\boldsymbol{Q} - \frac{1}{2} k_p \int Q_s \overline{B}_p(\boldsymbol{Q}) \frac{\partial}{\partial k_s} f_{ij}(\boldsymbol{k}, \boldsymbol{K} - \boldsymbol{Q})\, d\boldsymbol{Q}\Bigg] \times \exp(\mathrm{i}\boldsymbol{K} \cdot \boldsymbol{R})\, d\boldsymbol{K} + O(\boldsymbol{K}^2, \boldsymbol{Q}^2, K_p Q_s). \tag{3.126}$$

We use the following identities:

$$\nabla_p [f_{ij}(\boldsymbol{k}, \boldsymbol{R}) \overline{B}_p(\boldsymbol{R})] = \int \mathrm{i}\, K_p\, [f_{ij}(\boldsymbol{k}, \boldsymbol{K}) \overline{B}_p(\boldsymbol{K})]_{\boldsymbol{K}} \exp(\mathrm{i}\boldsymbol{K} \cdot \boldsymbol{R})\, d\boldsymbol{K}, \tag{3.127}$$

where

$$[f_{ij}(\boldsymbol{k}, \boldsymbol{K}) \overline{B}_p(\boldsymbol{K})]_{\boldsymbol{K}} = \int f_{ij}(\boldsymbol{k}, \boldsymbol{K} - \boldsymbol{Q}) \overline{B}_p(\boldsymbol{Q})\, d\boldsymbol{Q}, \tag{3.128}$$

and

$$f_{ij}(\boldsymbol{k}, \boldsymbol{R}) \, \nabla_s \overline{B}_p(\boldsymbol{R}) = \int \left[\int f_{ij}(\boldsymbol{k}, \boldsymbol{K} - \boldsymbol{Q}) \, \mathrm{i} \, Q_s \overline{B}_p(\boldsymbol{Q}) \, d\boldsymbol{Q} \right]_{\boldsymbol{K}} \exp(\mathrm{i} \boldsymbol{K} \cdot \boldsymbol{R}) \, d\boldsymbol{K}. \tag{3.129}$$

Therefore, Eqs. (3.126)–(3.129) yield

$$J_{ij}(\boldsymbol{k}, \boldsymbol{R}) = G_\eta \left\{ \overline{B}_p \left[\mathrm{i} \, k_p f_{ij} + \frac{1}{2} \nabla_p f_{ij} - \eta k^2 G_\eta \, k_{sp} \nabla_s f_{ij} \right] - (\nabla_s \overline{B}_p) \left[\frac{1}{2} k_p \frac{\partial f_{ij}}{\partial k_s} + \eta k^2 G_\eta \, k_{sp} f_{ij} \right] \right\} + O(\nabla^2). \tag{3.130}$$

Substituting Eq. (3.130) in Eq. (3.120), we obtain an expression for $\tilde{g}_{ij}(\boldsymbol{k}, \omega)$:

$$\tilde{g}_{ij}(\boldsymbol{k}, \omega) = G_\eta \left\{ \overline{B}_p \left[\mathrm{i} \, k_p f_{ij} + \frac{1}{2} \nabla_p f_{ij} - \eta k^2 G_\eta \, k_{sp} \nabla_s f_{ij} \right] - (\nabla_s \overline{B}_p) \left[\delta_{ip} \, f_{sj} + \frac{1}{2} k_p \frac{\partial f_{ij}}{\partial k_s} + \eta k^2 G_\eta \, k_{sp} f_{ij} \right] \right\}, \tag{3.131}$$

where $k_{ij} = k_i \, k_j / k^2$ and $G_\eta \equiv G_\eta(\boldsymbol{k}, \omega) = (\eta \boldsymbol{k}^2 + \mathrm{i} \, \omega)^{-1}$. When $\overline{\boldsymbol{B}}^2/4\pi \ll \overline{\rho} \langle \boldsymbol{u}^2 \rangle$, the correlation function $f_{ij}(\boldsymbol{k}, \omega)$ in Eq. (3.131) should be replaced by the correlation function $f_{ij}^{(0)}(\boldsymbol{k}, \omega)$ given below by Eq. (3.101) for a random flow with a zero mean magnetic field, called a background flow. Since turbulent electromotive force $\mathcal{E}_m = \varepsilon_{mji} \int \tilde{g}_{ij}(\boldsymbol{k}, \omega) \, d\boldsymbol{k} \, d\omega$, the contributions from the symmetric part of the tensor $\tilde{g}_{ij}$ with respect to the indexes "i" and "j" to $\boldsymbol{\mathcal{E}}$ vanish. We remind that we take into account in the turbulent electromotive force only the effects which are linear in $\nabla_j \overline{\boldsymbol{B}}$ and neglect higher-order spatial derivatives of the mean magnetic field. We also consider an incompressible fluid flow. This implies that

$$\tilde{g}_{ij}(\boldsymbol{k}, \omega) = G_\eta \left[\mathrm{i} \, k_p f_{ij}^{(0)} \, \overline{B}_p - (\nabla_s \overline{B}_i) \, f_{sj}^{(0)} \right], \tag{3.132}$$

where $\tilde{g}_{ij}(\boldsymbol{k}, \omega) = \langle b_i(\boldsymbol{k}, \omega) \, u_j(-\boldsymbol{k}, -\omega) \rangle$ [see Eqs. (3.119)]. To obtain the correlation function $\langle u_i(\boldsymbol{k}, \omega) \, b_j(-\boldsymbol{k}, -\omega) \rangle$ from Eq. (3.132), we have to replace the vector $\boldsymbol{k}$ by $-\boldsymbol{k}$, indexes i by j and j by i. This yields an equation that coincides with Eq. (3.99) derived using an approximate solution (3.96) for magnetic fluctuations $\boldsymbol{b}(\omega, \boldsymbol{k})$. Therefore, the turbulent electromotive force derived using Eq. (3.132) coincides with that for $\boldsymbol{\mathcal{E}} \equiv \langle \boldsymbol{u} \times \boldsymbol{b} \rangle$ given by Eqs. (3.107)–(3.109), which are derived using approximate solution (3.96) for magnetic fluctuations for small magnetic Reynolds numbers.

3.5 Turbulent Transport of Magnetic Fields: Spectral Tau Approach

In this section, we consider the case of large fluid and magnetic Reynolds numbers, so that turbulence is fully developed, the Strouhal number is of the order of unity, and the turbulent correlation time is scale-dependent, like in Kolmogorov type turbulence. In this case, we perform the Fourier transform only in the $\boldsymbol{k}$ space but not in the ω space, as is usually done in studies of turbulent transport in a fully developed Kolmogorov-type turbulence. We take into account the nonlinear terms in equations for velocity and magnetic fluctuations and apply the spectral τ approach.

The τ approach is a universal tool in turbulent transport for strongly nonlinear systems that allow us to obtain closed results and compare them with the results of laboratory experiments, observations and numerical simulations. The τ approximation reproduces many well-known phenomena found by other methods in turbulent transport of particles and magnetic fields, in turbulent convection and stably stratified turbulent flows for large fluid and magnetic Reynolds and Péclet numbers. This approach is different from the quasi-linear approach applied in the high-conductivity limit. The latter approach is only valid for small Strouhal numbers, i.e., for very short correlation time. Therefore, the final results found with this approach are different from those obtained using the τ approach.

3.5.1 Governing Equations

Averaging induction equation (3.54) for the evolution of magnetic field $\boldsymbol{B}(t, \boldsymbol{x})$ over an ensemble of turbulent incompressible velocity field, we obtain the mean-field equation for $\overline{\boldsymbol{B}}$:

$$\frac{\partial \overline{\boldsymbol{B}}}{\partial t} = \nabla \times \left(\overline{\boldsymbol{U}} \times \overline{\boldsymbol{B}} + \langle \boldsymbol{u} \times \boldsymbol{b} \rangle\right) + \eta \, \Delta \overline{\boldsymbol{B}}, \tag{3.133}$$

where $\boldsymbol{\mathcal{E}} \equiv \langle \boldsymbol{u} \times \boldsymbol{b} \rangle$ is the turbulent electromotive force or turbulent EMF. The equation for magnetic fluctuations is obtained by subtracting Eq. (3.133) from Eq. (3.54) for the field $\boldsymbol{B}(t, \boldsymbol{x})$:

$$\frac{\partial \boldsymbol{b}}{\partial t} - \nabla \times (\boldsymbol{u} \times \boldsymbol{b} - \langle \boldsymbol{u} \times \boldsymbol{b} \rangle) - \eta \, \Delta \boldsymbol{b} = (\overline{\boldsymbol{B}} \cdot \nabla)\boldsymbol{u} - (\boldsymbol{u} \cdot \nabla)\overline{\boldsymbol{B}}, \tag{3.134}$$

where for simplicity we assume that $\overline{\boldsymbol{U}} = 0$. We consider the case when nonlinear terms are much larger than the magnetic diffusion term, i.e., $\mathrm{Rm} \gg 1$.

3.5.2 Equations for Second Moments

Using Eq. (3.134) written in a Fourier space, we derive equations for the cross-helicity tensor $g_{ij}(\boldsymbol{k}) = \langle u_i(t, \boldsymbol{k}) \, b_j(t, -\boldsymbol{k}) \rangle$ and the tensor $h_{ij}(\boldsymbol{k}) = \langle b_i(t, \boldsymbol{k}) \, b_j(t, -\boldsymbol{k}) \rangle$ for magnetic fluctuations:

$$\frac{\partial g_{ij}(\boldsymbol{k})}{\partial t} = -\left[\mathrm{i}\,\left(\boldsymbol{k}\cdot\overline{\boldsymbol{B}}\right) - \frac{1}{2}(\overline{\boldsymbol{B}}\cdot\nabla) + \frac{1}{2}(\nabla_s\overline{B}_n)\,k_n\,\frac{\partial}{\partial k_s}\right] f_{ij}(\boldsymbol{k}) - f_{in}(\boldsymbol{k})\,\nabla_n\overline{B}_j,$$
$$+\hat{\mathcal{M}}^{(b)} g_{ij}^{(III)}(\boldsymbol{k}), \tag{3.135}$$

$$\frac{\partial h_{ij}(\boldsymbol{k})}{\partial t} = \mathrm{i}\,\left(\boldsymbol{k}\cdot\overline{\boldsymbol{B}}\right)\,\left[g_{ij}(\boldsymbol{k}) - g_{ji}(-\boldsymbol{k})\right] - \left[g_{nj}(\boldsymbol{k})\,\delta_{mi} + g_{ni}(-\boldsymbol{k})\,\delta_{mj}\right](\nabla_n\overline{B}_m),$$
$$+\frac{1}{2}\left[\left(\overline{\boldsymbol{B}}\cdot\nabla\right) - \left(\nabla_s\overline{B}_n\right)\,k_n\,\frac{\partial}{\partial k_s}\right] f_{ij}(\boldsymbol{k}) + \hat{\mathcal{M}}^{(b)} h_{ij}^{(III)}(\boldsymbol{k}), \tag{3.136}$$

where $f_{ij}(\boldsymbol{k}) = \langle u_i(t,\boldsymbol{k})\,u_j(t,-\boldsymbol{k})\rangle$, and

$$\hat{\mathcal{M}}^{(b)} g_{ij}^{(III)}(\boldsymbol{k}) = -\left\langle u_i(t,\boldsymbol{k})\,N_j^{(b)}(t,-\boldsymbol{k})\right\rangle + \left\langle\frac{\partial u_i(t,\boldsymbol{k})}{\partial t}\,b_j(t,-\boldsymbol{k})\right\rangle, \tag{3.137}$$

$$\hat{\mathcal{M}}^{(b)} h_{ij}^{(III)}(\boldsymbol{k}) = -\left\langle b_i(t,\boldsymbol{k})\,N_j^{(b)}(t,-\boldsymbol{k})\right\rangle - \left\langle N_i^{(b)}(t,\boldsymbol{k})\,b_j(t,-\boldsymbol{k})\right\rangle \tag{3.138}$$

are the third-order moment terms appearing due to the nonlinear terms

$$N_j^{(b)} = [\nabla\times(\boldsymbol{u}\times\boldsymbol{b} - \langle\boldsymbol{u}\times\boldsymbol{b}\rangle)]_j - \eta\Delta b_j, \tag{3.139}$$

which include also magnetic diffusion term.

For the derivation of Eqs. (3.135)–(3.136), we perform several calculations that are similar to the following. The first source term, $(\overline{\boldsymbol{B}}\cdot\nabla)\boldsymbol{u}$, on the right-hand side of Eq. (3.134) in a Fourier space, reads:

$$\left[\left(\overline{\boldsymbol{B}}\cdot\nabla\right)u_j\right]_{\boldsymbol{k}} = \mathrm{i}\,k_p\int\overline{B}_p(\boldsymbol{Q})\,u_j(\boldsymbol{k}-\boldsymbol{Q})\,d\boldsymbol{Q} \tag{3.140}$$

[see Eq. (3.116)], so that the induction equation for $b_j(\boldsymbol{k}_2)$ in the $\boldsymbol{k}$ space is given by

$$\frac{\partial b_j(\boldsymbol{k}_2)}{\partial t} = \mathrm{i}\,k_p^{(2)}\int\overline{B}_p(\boldsymbol{Q})\,u_j(\boldsymbol{k}_2-\boldsymbol{Q})\,d\boldsymbol{Q} - u_n(\boldsymbol{k}_2)\,\nabla_n\overline{B}_j + N_j^{(b)}(\boldsymbol{k}_2), \tag{3.141}$$

where $\boldsymbol{k}^{(2)} \equiv \boldsymbol{k}_2 = -\boldsymbol{k} + \boldsymbol{K}/2$. We use the identity

$$\frac{\partial}{\partial t}\langle u_i(\boldsymbol{k}_1,t)\,b_j(\boldsymbol{k}_2,t)\rangle = \left\langle\frac{\partial u_i(\boldsymbol{k}_1,t)}{\partial t}\,b_j(\boldsymbol{k}_2,t)\right\rangle + \left\langle u_i(\boldsymbol{k}_1,t)\,\frac{\partial b_j(\boldsymbol{k}_2,t)}{\partial t}\right\rangle. \tag{3.142}$$

First we derive an equation for the second term on the right-hand side of Eq. (3.142). To this end, we multiply Eq. (3.141) by $u_i(\boldsymbol{k}_1)$ and average over an ensemble of turbulent velocity field, where $\boldsymbol{k}_1 = \boldsymbol{k} + \boldsymbol{K}/2$. This yields:

$$\left\langle u_i(\boldsymbol{k}_1)\,\frac{\partial b_j(\boldsymbol{k}_2)}{\partial t}\right\rangle = \mathrm{i}\,\left(-k_p + K_p/2\right)\,\int \overline{B}_p(\boldsymbol{Q})\,\langle u_i(\boldsymbol{k}_1)\,u_j(\boldsymbol{k}_2 - \boldsymbol{Q})\,d\boldsymbol{Q}\rangle$$
$$- \langle u_i(\boldsymbol{k}_1)\,u_n(\boldsymbol{k}_2)\rangle\,\nabla_n \overline{B}_j + \left\langle u_i(\boldsymbol{k}_1)\,N_j^{(b)}(\boldsymbol{k}_2)\right\rangle, \qquad (3.143)$$

where for brevity of notation we omit the argument t in the velocity and magnetic fields. Next, we perform in Eq. (3.143) the Fourier transform in the large-scale variable $\boldsymbol{K}$, i.e., we use the transformation

$$F(\boldsymbol{R}) = \int F(\boldsymbol{K})\exp(\mathrm{i}\boldsymbol{K}\cdot\boldsymbol{R})\,d\boldsymbol{K}.$$

The first term $Y_{ij}(\boldsymbol{k},\boldsymbol{R})$ on the right-hand side of the obtained equation [which originates from the first term on the right-hand side of Eq. (3.141)] is given by

$$Y_{ij}(\boldsymbol{k},\boldsymbol{R}) = \mathrm{i}\int\int \overline{B}_p(\boldsymbol{Q})\langle u_i(\boldsymbol{k}+\boldsymbol{K}/2)u_j(-\boldsymbol{k}+\boldsymbol{K}/2-\boldsymbol{Q})\rangle$$
$$\times\left(-k_p + K_p/2\right)\,\exp(\mathrm{i}\boldsymbol{K}\cdot\boldsymbol{R})\,d\boldsymbol{K}\,d\boldsymbol{Q}. \qquad (3.144)$$

Next, we introduce new variables:

$$\tilde{\boldsymbol{k}} = (\tilde{\boldsymbol{k}}_1 - \tilde{\boldsymbol{k}}_2)/2 = \boldsymbol{k} + \boldsymbol{Q}/2\,,\ \ \tilde{\boldsymbol{K}} = \tilde{\boldsymbol{k}}_1 + \tilde{\boldsymbol{k}}_2 = \boldsymbol{K} - \boldsymbol{Q}, \qquad (3.145)$$

where

$$\tilde{\boldsymbol{k}}_1 = \boldsymbol{k} + \boldsymbol{K}/2, \quad \tilde{\boldsymbol{k}}_2 = -\boldsymbol{k} + \boldsymbol{K}/2 - \boldsymbol{Q}. \qquad (3.146)$$

Therefore, Eq. (3.144) in the new variables reads

$$Y_{ij}(\boldsymbol{k},\boldsymbol{R}) = \mathrm{i}\int\int f_{ij}\,(\boldsymbol{k}+\boldsymbol{Q}/2,\boldsymbol{K}-\boldsymbol{Q})\ \overline{B}_p(\boldsymbol{Q})$$
$$\times\left(-k_p + K_p/2\right)\,\exp\left(\mathrm{i}\boldsymbol{K}\cdot\boldsymbol{R}\right)d\boldsymbol{K}\,d\boldsymbol{Q}. \qquad (3.147)$$

Since $|\boldsymbol{Q}| \ll |\boldsymbol{k}|$, we use the Taylor expansion

$$f_{ij}(\boldsymbol{k}+\boldsymbol{Q}/2,\boldsymbol{K}-\boldsymbol{Q}) \simeq f_{ij}(\boldsymbol{k},\boldsymbol{K}-\boldsymbol{Q}) + \frac{1}{2}\frac{\partial f_{ij}(\boldsymbol{k},\boldsymbol{K}-\boldsymbol{Q})}{\partial k_s}Q_s + O(Q^2) \qquad (3.148)$$

and the following identity

$$\nabla_p[f_{ij}(\boldsymbol{k},\boldsymbol{R})\overline{B}_p(\boldsymbol{R})] = \int \mathrm{i}\,K_p[f_{ij}(\boldsymbol{k},\boldsymbol{R})\overline{B}_p(\boldsymbol{R})]_{\boldsymbol{K}}\exp\left(\mathrm{i}\boldsymbol{K}\cdot\boldsymbol{R}\right)d\boldsymbol{K}, \qquad (3.149)$$

where

$$[f_{ij}(\boldsymbol{k},\boldsymbol{R})\overline{B}_p(\boldsymbol{R})]_{\boldsymbol{K}} = \int f_{ij}(\boldsymbol{k},\boldsymbol{K}-\boldsymbol{Q})\overline{B}_p(\boldsymbol{Q})\,d\boldsymbol{Q}. \qquad (3.150)$$

Therefore, Eqs. (3.147)–(3.149) yield

$$Y_{ij}(\boldsymbol{k},\boldsymbol{R}) \simeq \left[-\mathrm{i}\,(\boldsymbol{k}\cdot\overline{\boldsymbol{B}}) + \frac{1}{2}(\overline{\boldsymbol{B}}\cdot\nabla)\right] f_{ij}(\boldsymbol{k},\boldsymbol{R}) - \frac{1}{2}k_p \frac{\partial f_{ij}(\boldsymbol{k})}{\partial k_s}\nabla_s \overline{B}_p. \tag{3.151}$$

We take into account that the terms in $g_{ij}(\boldsymbol{k},\boldsymbol{R})$ with symmetric tensors with respect to the indexes "i" and "j" do not contribute to the turbulent electromotive force because $\mathcal{E}_m = \varepsilon_{mij}\int g_{ij}(\boldsymbol{k},\boldsymbol{R})\,d\boldsymbol{k}$. In $g_{ij}(\boldsymbol{k},\boldsymbol{R})$, we also neglect the second and higher derivatives over $\boldsymbol{R}$. This procedure yields Eq. (3.135). Similar calculations are performed to derive Eq. (3.136).

3.5.3 Spectral Tau Approach

Equations (3.135)–(3.136) for the second moment include the first-order spatial differential operators applied to the third-order moments $\hat{\mathcal{M}}^{(b)} g_{ij}^{(III)}(\boldsymbol{k})$ and $\hat{\mathcal{M}}^{(b)} h_{ij}^{(III)}(\boldsymbol{k})$. The problem arises of how to close the system, i.e., how to express the third-order terms through the lower moments $F^{(II)}$. We use the spectral τ approximation, which postulates that the deviations of the third-moment terms, $\hat{\mathcal{M}}F^{(III)}(\boldsymbol{k})$, from the contributions to these terms afforded by the background turbulence, $\hat{\mathcal{M}}F^{(III,0)}(\boldsymbol{k})$, can be expressed through the similar deviations of the second moments, $F^{(II)}(\boldsymbol{k}) - F^{(II,0)}(\boldsymbol{k})$:

$$\hat{\mathcal{M}}F^{(III)}(\boldsymbol{k}) - \hat{\mathcal{M}}F^{(III,0)}(\boldsymbol{k}) = -\frac{1}{\tau_r(k)}\left[F^{(II)}(\boldsymbol{k}) - F^{(II,0)}(\boldsymbol{k})\right], \tag{3.152}$$

where $\tau_r(k)$ is the scale-dependent relaxation time, which can be identified with the correlation time $\tau(k)$ of the turbulent velocity field for large fluid and magnetic Reynolds numbers. The functions with the superscript (0) correspond to the background turbulence with a zero mean magnetic field. Validation of the τ approximation for different situations has been performed in numerous numerical simulations and analytical studies. The contributions of the terms with the superscript (0) vanish because when the mean magnetic field is zero, the turbulent electromotive force vanishes. Here we assume that magnetic fluctuations caused by a small-scale dynamo (the dynamo with a zero mean magnetic field) are not generated. Consequently, Eq. (3.152) reduces to $\hat{\mathcal{M}}^{(b)} g_{ij}^{(III)}(\boldsymbol{k}) = -g_{ij}(\boldsymbol{k})/\tau(k)$ and $\hat{\mathcal{M}}^{(b)} h_{ij}^{(III)}(\boldsymbol{k}) = -h_{ij}(\boldsymbol{k})/\tau(k)$.

3.5.4 Cross-helicity Tensor and Magnetic Fluctuations

We assume that the characteristic time of variation of the second moments $g_{ij}(\boldsymbol{k})$ and $h_{ij}(\boldsymbol{k})$ is substantially larger than the correlation time $\tau(k)$ for all turbulence scales. Therefore, in a steady state Eqs. (3.135) and (3.136) yield the following formulae for the cross-helicity tensor $\langle u_i\, b_j\rangle$ and the function $\langle \boldsymbol{b}^2\rangle$:

$$\langle u_i\, b_j\rangle = -\int \tau(k)\left[\nabla_m \overline{B}_j + \mathrm{i}\, k_n\, \delta_{mj}\, \overline{B}_n\right] f_{im}(\boldsymbol{k})\, d\boldsymbol{k}, \tag{3.153}$$

$$\langle \boldsymbol{b}^2\rangle = \int \tau^2(k)\left(\boldsymbol{k}\cdot\overline{\boldsymbol{B}}\right)^2 \left[f_{ii}(\boldsymbol{k}) + f_{ii}(-\boldsymbol{k})\right] d\boldsymbol{k}. \tag{3.154}$$

In Eq. (3.154), we neglect small contributions proportional to spatial derivatives of the mean magnetic field and spatial derivatives of $\langle u^2\rangle$. Since we consider a one-way coupling (i.e., there is no effect of magnetic field on the fluid flow), the correlation function f_{ij} in Eqs. (3.153)–(3.154) should be replaced by $f_{ij}^{(0)}$ for the background random flow.

3.5.5 Model of Isotropic Inhomogeneous Incompressible and Helical Turbulence

We use the model for the second moments, $f_{ij}^{(0)}(\boldsymbol{k}, \boldsymbol{R})$, of isotropic inhomogeneous incompressible and helical turbulence in a Fourier space in the following form:

$$f_{ij}^{(0)}(\boldsymbol{k}, \boldsymbol{R}) = \frac{E(k)}{8\pi k^2}\left\{\left[\delta_{ij} - \frac{k_i\, k_j}{k^2} + \frac{\mathrm{i}}{2k^2}\left(k_i \nabla_j - k_j \nabla_i\right)\right]\langle \boldsymbol{u}^2\rangle \right.$$
$$\left. -\frac{\mathrm{i}}{k^2}\, \varepsilon_{ijp}\, k_p\, \langle \boldsymbol{u}\cdot(\nabla\times\boldsymbol{u})\rangle\right\}, \tag{3.155}$$

where δ_{ij} is the Kronecker tensor, $\langle \mathbf{u}\cdot(\nabla\times\mathbf{u})\rangle$ is the kinetic helicity, the energy spectrum function is $E(k) = (2/3)\, k_0^{-1}\, (k/k_0)^{-5/3}$ in the inertial range of turbulence $k_0 \le k \le k_\nu$, the wave number $k_0 = 1/\ell_0$, the length ℓ_0 is the maximum scale of random motions, the wave number $k_\nu = \ell_\nu^{-1}$, the length $\ell_\nu = \ell_0\, \mathrm{Re}^{-3/4}$ is the Kolmogorov (viscous) scale, and the expression for the turbulent correlation time is given by $\tau(k) = 2\,\tau_0\,(k/k_0)^{-2/3}$, where $\tau_0 = \ell_0/u_0$ is the characteristic turbulent time.

3.5.6 Turbulent Electromotive Force and Level of Magnetic Fluctuations

Substituting Eq. (3.155) into Eqs. (3.153)–(3.154), we determine the cross-helicity tensor $\langle u_i\, b_j\rangle$, and the level of magnetic fluctuations $\langle \boldsymbol{b}^2\rangle$:

$$\langle u_i\, b_j\rangle = -\frac{1}{8\pi}\int_{k_0}^{k_\nu} \tau(k)\, E(k)\, dk \int_0^{2\pi} d\varphi \int_0^{\pi} \sin\vartheta\, d\vartheta \left\{\left[\left(\delta_{im} - \frac{k_i\, k_m}{k^2}\right)\nabla_m \overline{B}_j \right.\right.$$
$$\left.\left. +\frac{1}{2}\,\overline{B}_n\left(\frac{k_n\, k_j}{k^2}\nabla_i - \frac{k_i\, k_n}{k^2}\nabla_j\right)\right]\langle \boldsymbol{u}^2\rangle + \overline{B}_n\, \frac{k_n\, k_p}{k^2}\, \varepsilon_{ijp}\, \langle \boldsymbol{u}\cdot(\nabla\times\boldsymbol{u})\rangle\right\}, \tag{3.156}$$

$$\left\langle b^2\right\rangle = \frac{\left\langle u^2\right\rangle}{2\pi}\, \overline{B}_m\, \overline{B}_n \int_{k_0}^{k_\nu} \tau^2(k)\, E(k)\, k^2\, dk \int_0^{2\pi} d\varphi \int_0^{\pi} \sin\vartheta\, d\vartheta \left(\frac{k_m\, k_n}{k^2}\right). \tag{3.157}$$

For the integration over $\boldsymbol{k}$, we use the integrals given by Eqs. (1.129), (1.135) and (1.154)–(1.155). After integration in the $\boldsymbol{k}$ space in Eqs. (3.156) and (3.157), we obtain formulae for the turbulent electromotive force $\boldsymbol{\mathcal{E}} \equiv \langle \boldsymbol{u} \times \boldsymbol{b}\rangle$ and the level of magnetic fluctuations $\langle \boldsymbol{b}^2\rangle$:

$$\boxed{\boldsymbol{\mathcal{E}} = \alpha\, \overline{\boldsymbol{B}} + \boldsymbol{V}^{\rm eff} \times \overline{\boldsymbol{B}} - \eta_{_T} \boldsymbol{\nabla} \times \overline{\boldsymbol{B}},} \tag{3.158}$$

$$\boxed{\alpha = -\frac{\tau_0}{3}\, \langle \boldsymbol{u} \cdot (\boldsymbol{\nabla} \times \boldsymbol{u})\rangle\,,} \tag{3.159}$$

$$\boxed{\eta_{_T} = \frac{1}{3}\, u_0\, \ell_0,} \tag{3.160}$$

$$\boxed{\boldsymbol{V}^{\rm eff} = -\frac{\tau_0}{6}\, \boldsymbol{\nabla} \left\langle \boldsymbol{u}^2\right\rangle,} \tag{3.161}$$

$$\boxed{\left\langle \boldsymbol{b}^2\right\rangle = \frac{4}{3}\, \ln \mathrm{Re}\, \overline{\boldsymbol{B}}^2,} \tag{3.162}$$

where $\boldsymbol{V}^{\rm eff}$ is the effective pumping velocity that describes the turbulent diamagnetic effect, the coefficient α determines the α effect caused by the kinetic helicity in a helical turbulence and $\eta_{_T}$ is the coefficient of turbulent magnetic diffusion. Equations (3.158)–(3.162), derived using the spectral tau approach, are in agreement with Eqs. (3.66)–(3.70) and (3.77) obtained using the dimensional analysis for large fluid and magnetic Reynolds numbers. Equation (3.162) is valid for large magnetic Prandtl numbers. For small magnetic Prandtl numbers, the level of magnetic fluctuations reads

$$\boxed{\left\langle \boldsymbol{b}^2\right\rangle = \frac{4}{3}\, \ln \mathrm{Rm}\, \overline{\boldsymbol{B}}^2,} \tag{3.163}$$

where the integration over k in Eqs. (3.156) and (3.157) for small magnetic Prandtl numbers should be performed up to the wave number $k_\eta = \ell_\eta^{-1}$, and $\ell_\eta = \ell_0\, \mathrm{Rm}^{-3/4}$ is the resistive scale.

Substituting the turbulent electromotive force (3.158) into the mean-field equation (3.133), we obtain the final form of the mean-field equation:

$$\frac{\partial \overline{\boldsymbol{B}}}{\partial t} = \boldsymbol{\nabla} \times \left(\overline{\boldsymbol{U}} \times \overline{\boldsymbol{B}} + \alpha\, \overline{\boldsymbol{B}} + \boldsymbol{V}^{\rm eff} \times \overline{\boldsymbol{B}} - \eta_{_T} \boldsymbol{\nabla} \times \overline{\boldsymbol{B}}\right) + \eta\, \Delta \overline{\boldsymbol{B}}. \tag{3.164}$$

This equation allows us to study the large-scale evolution of the magnetic field. In particular, this equation allows us to study the dynamo instability, i.e., to describe the generation of a magnetic field via turbulent helical motions of an electrically conducting fluid.

3.6 Compressibility Effects

In this section, we study turbulent transport of the magnetic field in a compressible helical inhomogeneous turbulence for small and large magnetic Reynolds numbers.

3.6.1 Governing Equations

The magnetic field $\boldsymbol{B}(\boldsymbol{x}, t)$ is governed by the induction equation

$$\frac{\partial \boldsymbol{B}}{\partial t} = \nabla \times (\boldsymbol{U} \times \boldsymbol{B} - \eta \nabla \times \boldsymbol{B}), \tag{3.165}$$

where η is the magnetic diffusivity due to the electrical conductivity of the fluid, and fluid velocity $\boldsymbol{U}(\boldsymbol{x}, t)$ is determined by the Navier-Stokes equation,

$$\rho \left(\frac{\partial}{\partial t} + \boldsymbol{U} \cdot \nabla\right) \boldsymbol{U} = -\nabla P + \frac{1}{4\pi} (\nabla \times \boldsymbol{B}) \times \boldsymbol{B} + \nabla \cdot \left(2\nu\, \rho\, \mathbf{S}^{(\mathrm{U})}\right), \tag{3.166}$$

where $P(\boldsymbol{x}, t)$ is the fluid pressure, $\mathsf{S}_{ij}^{(\mathrm{U})} = \frac{1}{2}(\nabla_j U_i + \nabla_i U_j) - \frac{1}{3} \delta_{ij}\, \nabla \cdot \boldsymbol{U}$ are the components of the traceless rate-of-strain-tensor $\mathbf{S}^{(\mathrm{U})}$, and ν is the kinematic viscosity. The fluid density is determined by the continuity equation:

$$\frac{\partial \rho}{\partial t} + \nabla \cdot (\rho\, \boldsymbol{U}) = 0. \tag{3.167}$$

We apply a mean-field approach and use Reynolds averaging. In the framework of this approach, velocity and magnetic fields, fluid density, and pressure are decomposed into mean (denoted by overbars) and fluctuating parts (lowercase symbols). Ensemble averaging of Eq. (3.165) yields an equation for the mean magnetic field $\overline{\boldsymbol{B}}(\boldsymbol{x}, t)$:

$$\frac{\partial \overline{\boldsymbol{B}}}{\partial t} = \nabla \times \left(\overline{\boldsymbol{U}} \times \overline{\boldsymbol{B}} + \boldsymbol{\mathcal{E}} - \eta \nabla \times \overline{\boldsymbol{B}}\right), \tag{3.168}$$

where $\overline{\boldsymbol{U}}(\boldsymbol{x}, t)$ is the mean velocity and $\boldsymbol{\mathcal{E}} = \langle \boldsymbol{u} \times \boldsymbol{b} \rangle$ is the turbulent electromotive force. For simplicity, we consider in this study the case $\overline{\boldsymbol{U}} = \boldsymbol{0}$. We determine $\boldsymbol{\mathcal{E}}$ for compressible inhomogeneous and helical turbulence. The procedure of the derivation of the equation for the turbulent electromotive force is as follows. The momentum and induction equations for fluctuations are given by

$$\overline{\rho} \frac{\partial \boldsymbol{u}}{\partial t} - \nabla \cdot \left(2\nu \overline{\rho} \mathbf{S}^{(\mathrm{u})}\right) = \frac{1}{4\pi} \left[(\boldsymbol{b} \cdot \nabla) \overline{\boldsymbol{B}} + (\overline{\boldsymbol{B}} \cdot \nabla) \boldsymbol{b}\right] - \nabla p_{\mathrm{tot}} + \boldsymbol{u}^{(\mathrm{N})}, \tag{3.169}$$

$$\frac{\partial \boldsymbol{b}}{\partial t} - \eta \Delta \boldsymbol{b} = (\overline{\boldsymbol{B}} \cdot \nabla)\boldsymbol{u} - (\boldsymbol{u} \cdot \nabla)\overline{\boldsymbol{B}} - \overline{\boldsymbol{B}}\,(\nabla \cdot \boldsymbol{u}) + \boldsymbol{b}^{(\mathrm{N})}, \tag{3.170}$$

where $\boldsymbol{u}(\boldsymbol{x}, t)$ and $\boldsymbol{b}(\boldsymbol{x}, t)$ are fluctuations of velocity and magnetic fields, $\boldsymbol{u}^{(\mathrm{N})}$ and $\boldsymbol{b}^{(\mathrm{N})}$ are terms that are nonlinear in fluctuations, $p_{\mathrm{tot}} = p + (4\pi)^{-1}\,(\overline{\boldsymbol{B}} \cdot \boldsymbol{b})$ are fluctuations of the total pressure, $\mathsf{S}^{(\mathrm{u})} \equiv \mathsf{S}^{(\mathrm{u})}_{ij} = \frac{1}{2}(\nabla_j u_i + \nabla_i u_j) - \frac{1}{3}\,\delta_{ij}\,\nabla \cdot \boldsymbol{u}$, and $p(\boldsymbol{x}, t)$ are fluctuations of the fluid pressure. We rewrite Eqs. (3.169) and (3.170) in the $\boldsymbol{k}$ space. In the following, we derive equations for the turbulent electromotive force for small and large magnetic Reynolds numbers.

3.6.2 Derivation of Equations for Second Moments for *Rm* $\ll 1$

To derive the equation for the turbulent electromotive force for small magnetic Reynolds numbers, we apply the quasi-linear approach. In the framework of the quasi-linear approach, we neglect nonlinear terms, but keep molecular dissipative terms in Eq. (3.170) for magnetic fluctuations. This allows us to obtain a solution for Eq. (3.170) in the limit of the small magnetic Reynolds number as

$$\begin{aligned} b_i(\boldsymbol{k}, \omega) = \mathrm{i}\, k_p G_\eta(\boldsymbol{k}, \omega) \bigg[&\int \overline{B}_p(\boldsymbol{Q}) u_i(\boldsymbol{k} - \boldsymbol{Q}, \omega)\, d\boldsymbol{Q} \\ &- \int \overline{B}_i(\boldsymbol{Q})\, u_p(\boldsymbol{k} - \boldsymbol{Q}, \omega)\, d\boldsymbol{Q} \bigg], \end{aligned} \tag{3.171}$$

where $G_\eta(\boldsymbol{k}, \omega) = (\eta k^2 + \mathrm{i}\,\omega)^{-1}$. To derive Eq. (3.171), we consider the term, $\overline{B}_i\,(\nabla \cdot \boldsymbol{u})$, in the induction equation (3.170), which can be rewritten using the Fourier transform as

$$\overline{B}_i\,(\nabla \cdot \boldsymbol{u}) = \int \overline{B}_i(\boldsymbol{Q})\, \exp(\mathrm{i}\,\boldsymbol{Q} \cdot \boldsymbol{x})\, d\boldsymbol{Q} \int \mathrm{i}\, k'_p\, u_p(\boldsymbol{k}')\, \exp(\mathrm{i}\,\boldsymbol{k}' \cdot \boldsymbol{x})\, d\boldsymbol{k}'. \tag{3.172}$$

For brevity of notation, we omit hereafter the arguments t and $\boldsymbol{x}$ in the functions $\overline{\boldsymbol{B}}(\boldsymbol{x}, t)$ and $\boldsymbol{u}(\boldsymbol{x}, t)$, and also the variable t in the functions $\overline{B}_i(\boldsymbol{Q}, t)$ and $u_p(\boldsymbol{k}, t)$. We introduce a new variable $\boldsymbol{k} = \boldsymbol{k}' + \boldsymbol{Q}$, which yields

$$\begin{aligned} \overline{B}_i(\nabla \cdot \boldsymbol{u}) &= \int \left(\int \mathrm{i}\, \left(k_p - Q_p\right) \overline{B}_i(\boldsymbol{Q})\, u_p(\boldsymbol{k} - \boldsymbol{Q})\, d\boldsymbol{Q} \right) \exp(\mathrm{i}\,\boldsymbol{k} \cdot \boldsymbol{x})\, d\boldsymbol{k} \\ &\equiv \int \left[\overline{B}_i(\nabla \cdot \boldsymbol{u})\right]_{\boldsymbol{k}} \exp(\mathrm{i}\,\boldsymbol{k} \cdot \boldsymbol{x})\, d\boldsymbol{k}, \end{aligned} \tag{3.173}$$

where $[\ldots]_{\boldsymbol{k}}$ denotes the Fourier transform. Therefore,

$$\left[\overline{B}_i\,(\nabla \cdot \boldsymbol{u})\right]_{\boldsymbol{k}} = \mathrm{i} \int \left(k_p - Q_p\right) \overline{B}_i(\boldsymbol{Q})\, u_p(\boldsymbol{k} - \boldsymbol{Q})\, d\boldsymbol{Q}. \tag{3.174}$$

Similar derivations are used for the other terms in the induction equation:

$$\left[(\boldsymbol{u}\cdot\nabla)\overline{B}_i\right]_{\boldsymbol{k}} = \mathrm{i}\int Q_p\,\overline{B}_i(\boldsymbol{Q})\,u_p(\boldsymbol{k}-\boldsymbol{Q})\,d\boldsymbol{Q}, \tag{3.175}$$

$$\left[(\overline{\boldsymbol{B}}\cdot\nabla)u_i\right]_{\boldsymbol{k}} = \mathrm{i}\,k_p\int \overline{B}_p(\boldsymbol{Q})\,u_i(\boldsymbol{k}-\boldsymbol{Q})\,d\boldsymbol{Q}, \tag{3.176}$$

which yield Eq. (3.171). We determine the turbulent electromotive force

$$\overline{\mathcal{E}}_m = \varepsilon_{mji}\int g_{ij}(\boldsymbol{k},\omega,\boldsymbol{R},t)\,d\boldsymbol{k}\,d\omega,$$

where the cross-helicity tensor $g_{ij}(\boldsymbol{k},\omega,\boldsymbol{R},t)$ is

$$g_{ij}(\boldsymbol{k},\omega,\boldsymbol{R},t) = \int \langle b_i(\boldsymbol{k}_1,\omega_1)\,u_j(\boldsymbol{k}_2,\omega_2)\rangle\,\exp[\mathrm{i}\boldsymbol{K}\cdot\boldsymbol{R}+\mathrm{i}\,\Omega t]\,d\boldsymbol{K}\,d\Omega.$$

Using Eq. (3.171), we derive an equation for the cross-helicity tensor g_{ij} as

$$g_{ij}(\boldsymbol{k},\boldsymbol{R}) = J_{ij}^{(2)}(\boldsymbol{k},\boldsymbol{R}) - J_{ij}^{(1)}(\boldsymbol{k},\boldsymbol{R}), \tag{3.177}$$

where

$$J_{ij}^{(1)}(\boldsymbol{k},\boldsymbol{R}) = \mathrm{i}\int (k_p+K_p/2)\,\langle u_p(\boldsymbol{k}+\boldsymbol{K}/2-\boldsymbol{Q})u_j(-\boldsymbol{k}+\boldsymbol{K}/2)\rangle \times G_\eta(\boldsymbol{k}+\boldsymbol{K}/2)\,\overline{B}_i(\boldsymbol{Q})\exp(\mathrm{i}\boldsymbol{K}\cdot\boldsymbol{R})\,d\boldsymbol{K}\,d\boldsymbol{Q}, \tag{3.178}$$

$$J_{ij}^{(2)}(\boldsymbol{k},\boldsymbol{R}) = \mathrm{i}\int (k_p+K_p/2)\,\langle u_i(\boldsymbol{k}+\boldsymbol{K}/2-\boldsymbol{Q})u_j(-\boldsymbol{k}+\boldsymbol{K}/2)\rangle \times G_\eta(\boldsymbol{k}+\boldsymbol{K}/2)\,\overline{B}_p(\boldsymbol{Q})\exp(\mathrm{i}\boldsymbol{K}\cdot\boldsymbol{R})\,d\boldsymbol{K}\,d\boldsymbol{Q}, \tag{3.179}$$

and the functions $J_{ij}^{(1)}$, $J_{ij}^{(2)}$, G_η and u_i depend also on ω. To simplify the notation, we do not show this dependence in the calculations. First, we determine the function $J_{ij}^{(1)}$ using new variables:

$$\tilde{\boldsymbol{k}} = (\tilde{\boldsymbol{k}}_1-\tilde{\boldsymbol{k}}_2)/2 = \boldsymbol{k}-\boldsymbol{Q}/2\ ,\ \tilde{\boldsymbol{K}} = \tilde{\boldsymbol{k}}_1+\tilde{\boldsymbol{k}}_2 = \boldsymbol{K}-\boldsymbol{Q}, \tag{3.180}$$

where

$$\tilde{\boldsymbol{k}}_1 = \boldsymbol{k}+\boldsymbol{K}/2-\boldsymbol{Q}\ ,\quad \tilde{\boldsymbol{k}}_2 = -\boldsymbol{k}+\boldsymbol{K}/2. \tag{3.181}$$

Therefore,

$$J_{ij}^{(1)}(\boldsymbol{k},\boldsymbol{R}) = \mathrm{i}\int (k_p+K_p/2)\,f_{pj}(\boldsymbol{k}-\boldsymbol{Q}/2,\boldsymbol{K}-\boldsymbol{Q})\,G_\eta(\boldsymbol{k}+\boldsymbol{K}/2)\,\overline{B}_i(\boldsymbol{Q}) \times \exp(\mathrm{i}\boldsymbol{K}\cdot\boldsymbol{R})\,d\boldsymbol{K}\,d\boldsymbol{Q}. \tag{3.182}$$

Since $|\boldsymbol{Q}| \ll |\boldsymbol{k}|$ and $|\boldsymbol{K}| \ll |\boldsymbol{k}|$, we use the Taylor expansion

$$f_{pj}(\boldsymbol{k}-\boldsymbol{Q}/2, \boldsymbol{K}-\boldsymbol{Q}) = f_{pj}(\boldsymbol{k}, \boldsymbol{K}-\boldsymbol{Q}) - \frac{1}{2}\frac{\partial f_{pj}}{\partial k_s} Q_s + O(\boldsymbol{Q}^2), \tag{3.183}$$

$$G_\eta(\boldsymbol{k}+\boldsymbol{K}/2) = G_\eta(\boldsymbol{k})\left[1-\eta(\boldsymbol{k}\cdot\boldsymbol{K})G_\eta(\boldsymbol{k})\right] + O(\boldsymbol{K}^2). \tag{3.184}$$

Substituting Eqs. (3.183) and (3.184) into Eq. (3.182), we get

$$\begin{aligned} J_{ij}^{(1)}(\boldsymbol{k},\boldsymbol{R}) &= \mathrm{i}\, G_\eta(\boldsymbol{k}) \int \Bigg[\left(k_p - \eta G_\eta(\boldsymbol{k}) k_p k_s K_s + \frac{1}{2}K_p\right) \\ &\times \left(\int f_{pj}(\boldsymbol{k}, \boldsymbol{K}-\boldsymbol{Q})\overline{B}_i(\boldsymbol{Q})\, d\boldsymbol{Q}\right) - \frac{1}{2}k_p \left(\int Q_s \overline{B}_i(\boldsymbol{Q}) \frac{\partial}{\partial k_s} f_{pj}(\boldsymbol{k}, \boldsymbol{K}-\boldsymbol{Q})\, d\boldsymbol{Q}\right)\Bigg] \\ &\times \exp(\mathrm{i}\boldsymbol{K}\cdot\boldsymbol{R})\, d\boldsymbol{K} + O(\boldsymbol{K}^2, \boldsymbol{Q}^2, K_p Q_s). \end{aligned} \tag{3.185}$$

We use the following identity:

$$\nabla_p[f_{pj}(\boldsymbol{k},\boldsymbol{R})\overline{B}_i(\boldsymbol{R})] = \int \mathrm{i}\, K_p\, [f_{pj}(\boldsymbol{k},\boldsymbol{K})\overline{B}_i(\boldsymbol{K})]_{\boldsymbol{K}} \exp(\mathrm{i}\boldsymbol{K}\cdot\boldsymbol{R})\, d\boldsymbol{K}, \tag{3.186}$$

where

$$[f_{pj}(\boldsymbol{k},\boldsymbol{K})\overline{B}_i(\boldsymbol{K})]_{\boldsymbol{K}} = \int f_{pj}(\boldsymbol{k}, \boldsymbol{K}-\boldsymbol{Q})\overline{B}_i(\boldsymbol{Q})\, d\boldsymbol{Q}. \tag{3.187}$$

Similarly,

$$f_{pj}(\boldsymbol{k},\boldsymbol{R})\,\nabla_p \overline{B}_i(\boldsymbol{R}) = \int \left[\int f_{pj}(\boldsymbol{k},\boldsymbol{K}-\boldsymbol{Q})\,\mathrm{i}\, Q_p \overline{B}_i(\boldsymbol{Q})\, d\boldsymbol{Q}\right]_{\boldsymbol{K}} \exp(\mathrm{i}\boldsymbol{K}\cdot\boldsymbol{R})\, d\boldsymbol{K}. \tag{3.188}$$

Therefore, Eqs. (3.185)–(3.188) yield

$$\begin{aligned} J_{ij}^{(1)}(\boldsymbol{k},\boldsymbol{R}) = G_\eta \Bigg\{ &\left[\mathrm{i}\, k_p \left(1 + \mathrm{i}\,\eta\, G_\eta k_s \nabla_s\right)\left(\overline{B}_i\, f_{pj}\right) + \overline{B}_i \frac{1}{2}\nabla_p f_{pj}\right] \\ &+ \frac{1}{2}\left(f_{pj} - k_m \frac{\partial f_{mj}}{\partial k_p}\right)(\nabla_p \overline{B}_i)\Bigg\} + O(\nabla^2). \end{aligned} \tag{3.189}$$

A similar derivation is also performed for $J_{ij}^{(2)}$, which yields

$$\begin{aligned} J_{ij}^{(2)}(\boldsymbol{k},\boldsymbol{R}) = G_\eta \Bigg\{ &\overline{B}_p\left[\mathrm{i}\, k_p f_{ij} + \frac{1}{2}\nabla_p f_{ij} - \eta k^2 G_\eta\, k_{sp} \nabla_s f_{ij}\right] \\ &-(\nabla_s \overline{B}_p)\left[\frac{1}{2}k_p \frac{\partial f_{ij}}{\partial k_s} + \eta k^2 G_\eta\, k_{sp} f_{ij}\right]\Bigg\} + O(\nabla^2). \end{aligned} \tag{3.190}$$

Using Eqs. (3.189) and (3.190) we obtain an equation for the cross-helicity tensor $g_{ij}(\boldsymbol{k}, \omega)$ as

$$g_{ij}(\boldsymbol{k}, \omega) = G_\eta \left\{ \overline{B}_p \left[\mathrm{i}\, k_p f_{ij} + \frac{1}{2} \nabla_p f_{ij} - \eta k^2 G_\eta \, k_{sp} \nabla_s f_{ij} \right] - \overline{B}_i \left[\mathrm{i}\, k_p f_{pj} \right.\right.$$
$$\left. + \frac{1}{2} \nabla_p f_{pj} - \eta k^2 G_\eta \, k_{sp} \nabla_s f_{pj} \right] - (\nabla_s \overline{B}_p) \left[\frac{1}{2} k_p \frac{\partial f_{ij}}{\partial k_s} + \eta k^2 G_\eta \, k_{sp} f_{ij} \right]$$
$$\left. + \frac{1}{2} (\nabla_s \overline{B}_i) \left[- f_{sj} + k_p \frac{\partial f_{pj}}{\partial k_s} + 2\eta k^2 G_\eta \, k_{sp} f_{pj} \right] \right\}, \qquad (3.191)$$

where $k_{ij} = k_i k_j / k^2$ and $G_\eta \equiv G_\eta(\boldsymbol{k}, \omega) = (\eta \boldsymbol{k}^2 + \mathrm{i}\,\omega)^{-1}$. When $\overline{\boldsymbol{B}}^2/4\pi \ll \overline{\rho} \langle \boldsymbol{u}^2 \rangle$, the correlation function $f_{ij}(\boldsymbol{k}, \omega)$ in Eq. (3.191) should be replaced with the correlation function $f_{ij}^{(0)}(\boldsymbol{k}, \omega)$ for the random flow with a zero mean magnetic field, called the background flow.

Similar derivations allow us to determine the level of magnetic fluctuations $\langle \boldsymbol{b}^2 \rangle = \int E_b(\boldsymbol{k}, \omega, \boldsymbol{R}, t)\, d\boldsymbol{k}\, d\omega$, where the function $E_b(\boldsymbol{k}, \omega, \boldsymbol{R}, t)$ is

$$E_b(\boldsymbol{k}, \omega, \boldsymbol{R}, t) = \int \langle b_i(\boldsymbol{k}_1, \omega_1)\, b_i(\boldsymbol{k}_2, \omega_2) \rangle \, \exp[\mathrm{i}\, \boldsymbol{K} \cdot \boldsymbol{R} + \mathrm{i}\, \Omega t] \, d\boldsymbol{K}\, d\Omega.$$

In particular, using Eq. (3.171), we obtain

$$E_b = \frac{\mathrm{i}}{2} \int \left\{ k_p^{(1)} G_\eta(\boldsymbol{k}_1) \left[\overline{B}_p(\boldsymbol{Q}) \, \langle b_i(\boldsymbol{k}_2)\, u_i(\boldsymbol{k}_1 - \boldsymbol{Q}) \rangle \right.\right.$$
$$\left. - \overline{B}_i(\boldsymbol{Q}) \, \big\langle b_i(\boldsymbol{k}_2)\, u_p(\boldsymbol{k}_1 - \boldsymbol{Q}, \omega) \big\rangle \right]$$
$$+ k_p^{(2)} G_\eta(\boldsymbol{k}_2) \left[\overline{B}_p(\boldsymbol{Q}) \, \langle b_i(\boldsymbol{k}_1)\, u_i(\boldsymbol{k}_2 - \boldsymbol{Q}) \rangle \right.$$
$$\left.\left. - \overline{B}_i(\boldsymbol{Q}) \, \big\langle b_i(\boldsymbol{k}_1)\, u_p(\boldsymbol{k}_2 - \boldsymbol{Q}, \omega) \big\rangle \right] \right\} \exp(\mathrm{i}\, \boldsymbol{K} \cdot \boldsymbol{R})\, d\boldsymbol{K}\, d\boldsymbol{Q}, \qquad (3.192)$$

where $k_i^{(1)} \equiv (\boldsymbol{k}_1)_i = k_i + K_i/2$ and $k_i^{(2)} \equiv (\boldsymbol{k}_2)_i = -k_i + K_i/2$. Here the functions E_b, G_η, b_i and u_i depend also on ω, and the mean magnetic field $\overline{B}_i$ depends on t as well. To simplify the notation, we do not show these dependencies. Next, we neglect the small terms $\sim \mathrm{O}[\ell_0/L_B]$, $\mathrm{O}[\ell_0/L_u]$, which contain the large-scale spatial derivatives of both the mean magnetic field and the turbulent intensity. Therefore, the level of magnetic fluctuations is given by

$$\langle \boldsymbol{b}^2 \rangle = \frac{\mathrm{i}}{2} \int G_\eta(\boldsymbol{k}, \omega)\, k_j \, \left(\overline{B}_i \, g_{ij}^{(-)} - \overline{B}_j \, g_{ii}^{(-)} \right) \, d\boldsymbol{k}\, d\omega, \qquad (3.193)$$

where $g_{ij}^{(-)} = g_{ij}(\boldsymbol{k}, \omega) - g_{ij}(-\boldsymbol{k}, \omega)$. Substituting Eq. (3.191) into Eq. (3.193) and neglecting small terms $\sim \mathrm{O}[\ell_0/L_B]$, $\mathrm{O}[\ell_0/L_u]$, we obtain the level of magnetic fluctuations for small magnetic Reynolds numbers as

$$\langle b^2 \rangle = \frac{1}{2} \int G_\eta^2(\boldsymbol{k}, \omega)\, k_j\, k_p \Big[\overline{B}_j\, \overline{B}_p\, f_{ii}^{(+)} + \overline{\boldsymbol{B}}^2\, f_{pj}^{(+)} - \overline{B}_i\, \overline{B}_p\, f_{ij}^{(+)} - \overline{B}_i\, \overline{B}_j\, f_{pi}^{(+)} \Big]\, d\boldsymbol{k}\, d\omega, \tag{3.194}$$

where $f_{ij}^{(+)} = f_{ij}^{(0)}(\boldsymbol{k}, \omega) + f_{ij}^{(0)}(-\boldsymbol{k}, \omega)$. Here we take into account that for $\overline{\boldsymbol{B}}^2/4\pi \ll \overline{\rho}\langle \boldsymbol{u}^2 \rangle$, the correlation function $f_{ij}(\boldsymbol{k}, \omega)$ in Eq. (3.191) is replaced by the correlation function $f_{ij}^{(0)}(\boldsymbol{k}, \omega)$ for the random flow with a zero mean magnetic field.

3.6.3 Model for Compressible Helical Random Background Flow

In the next step of the derivation, we need a model for the background random flow with a zero mean magnetic field. We consider a random, statistically stationary, density-stratified, inhomogeneous, compressible and helical background flow, which is determined by the following correlation function in the Fourier space:

$$f_{ij}^{(0)}(\boldsymbol{k}, \omega, \boldsymbol{R}) = \frac{\Phi(\omega)}{8\pi\, k^2\, (1 + \sigma_c)} \Bigg\{ \Bigg[E(k) \bigg(\delta_{ij} - k_{ij} + \frac{\mathrm{i}}{k^2}\, (k_j \lambda_i - k_i \lambda_j) + \frac{\mathrm{i}}{2k^2}\, (k_i \nabla_j - k_j \nabla_i) \bigg) + 2\sigma_c\, E_c(k) \bigg(k_{ij} + \frac{\mathrm{i}}{4k^2} (k_i \nabla_j - k_j \nabla_i) + \frac{\mathrm{i}}{2k^2} (k_j \lambda_i - k_i \lambda_j) \bigg) \Bigg] \langle \boldsymbol{u}^2 \rangle - \frac{\mathrm{i}\, E(k)}{k^2}\, \varepsilon_{ijp}\, k_p\, \langle \boldsymbol{u} \cdot (\boldsymbol{\nabla} \times \boldsymbol{u}) \rangle \Bigg\}, \tag{3.195}$$

[see Eq. (2.161)]. Here $\boldsymbol{\lambda} = -\boldsymbol{\nabla} \ln \overline{\rho}$ characterizes the fluid density stratification, δ_{ij} is the Kronecker tensor, ε_{ijn} is the fully antisymmetric Levi-Civita tensor, $\chi = \langle \boldsymbol{u} \cdot (\boldsymbol{\nabla} \times \boldsymbol{u}) \rangle$ is the kinetic helicity and the parameter

$$\sigma_c = \frac{\langle (\boldsymbol{\nabla} \cdot \boldsymbol{u})^2 \rangle}{\langle (\boldsymbol{\nabla} \times \boldsymbol{u})^2 \rangle} \tag{3.196}$$

is the degree of compressibility of the turbulent velocity field. In Eq. (3.195), we neglect higher order effects $\sim \mathrm{O}(\lambda^2 \langle \boldsymbol{u}^2 \rangle, \nabla^2 \langle \boldsymbol{u}^2 \rangle, \lambda_i \nabla_i \langle \boldsymbol{u}^2 \rangle)$, i.e., we assume that $\ell_0 \ll H_\rho$ and $\ell_0 \ll L_u$, where ℓ_0 is the maximum scale of random motions, $L_u = \left| \boldsymbol{\nabla} \ln \langle \boldsymbol{u}^2 \rangle \right|^{-1}$ is the characteristic scale of the inhomogeneity of turbulence and $H_\rho = |\boldsymbol{\lambda}|^{-1}$ is the mean density variation scale, which is assumed to be constant. This means that in Eq. (3.195) we only take into account leading effects, which are linear in stratification ($\propto \ell_0/H_\rho$) and inhomogeneity of turbulence ($\propto \ell_0/L_u$).

Generally, stratification also contributes to div $\boldsymbol{u}$, and therefore it contributes to the parameter σ_c. However, this contribution is small [$\sim \mathrm{O}(\lambda^2 \langle \boldsymbol{u}^2 \rangle)$], and it is neglected in Eq. (3.195). Note that the degree of compressibility, σ_c, depends on the Mach number, but this dependence is not known for arbitrary Mach numbers and has to be determined in direct numerical simulations. The above conditions imply that we separate effects of the arbitrary Mach number, characterized by the parameter σ_c, and density stratification, described by $\boldsymbol{\lambda}$. We also do not consider here effects related to the nonuniformity of kinetic helicity ($\propto \nabla\chi$) or the combined effect of stratification and kinetic helicity ($\propto \boldsymbol{\lambda}\chi$), because their contributions to the turbulent electromotive force are much smaller in comparison with the leading contributions to the turbulent electromotive force which we take into account here.

In Eq. (3.195), we assume that the random flow has a power-law spectrum for incompressible $E(k) = (q-1)\,(k/k_0)^{-q}\,k_0^{-1}$ and compressible $E_c(k) = (q_c - 1)\,(k/k_0)^{-q_c}\,k_0^{-1}$ parts, where the wave number varies in the range, $k_0 \le k \le k_\nu$. Here $k_\nu = 1/\ell_\nu$ is the wave number based on the viscous scale ℓ_ν, and $k_0 = 1/\ell_0 \ll k_\nu$. We assume that there are no random motions for $k < k_0$. This implies that ℓ_0 is the maximum scale of random motions, and $E(k) = E_c(k) = 0$ for $k < k_0$.

We consider the frequency function $\Phi(\omega)$ in the form of a Lorentz profile: $\Phi(\omega) = [\pi\,\tau_0\,(\omega^2 + \tau_0^{-2})]^{-1}$, where $\tau_0 = \ell_0/\sqrt{\langle \boldsymbol{u}^2\rangle}$ is the correlation time and $\sqrt{\langle \boldsymbol{u}^2\rangle}$ is the characteristic turbulent velocity at scale ℓ_0. This model for the frequency function corresponds to the correlation function $\langle u_i(t)u_j(t+\tau)\rangle \propto \exp(-\tau/\tau_0)$. We take into account that for small magnetic Reynolds numbers, $\tau_0 \gg (\eta k^2)^{-1}$ for all turbulent scales.

In spite of the large-scale effects caused by stratification $\boldsymbol{\lambda}$ and inhomogeneous turbulence $\nabla\langle \boldsymbol{u}^2\rangle$, the model (3.195) describes a weakly anisotropic approximation. Indeed, the contributions of these effects to $f_{ij}^{(0)}$ are small because $\ell_0 \ll H_\rho$ and $\ell_0 \ll L_u$. Since we consider only linear effects in $\boldsymbol{\lambda}$ and $\nabla\langle \boldsymbol{u}^2\rangle$, the second rank tensor $f_{ij}^{(0)}$ is constructed as a linear combination of

- symmetric tensors, δ_{ij} and k_{ij}, with respect to the indexes i and j;
- non-symmetric tensors, $k_i\lambda_j$, $k_j\lambda_i$ and $k_i\nabla_j\langle \boldsymbol{u}^2\rangle$, $k_j\nabla_i\langle \boldsymbol{u}^2\rangle$, describing a non-helical part of turbulence;
- fully antisymmetric tensor $\varepsilon_{ijp}\,k_p$ describing a helical part of turbulence.

To determine unknown coefficients multiplying by these tensors, we use the following conditions in the derivation of Eq. (3.195):

$$\langle \boldsymbol{u}^2\rangle = \int f_{ii}^{(0)}(\boldsymbol{k},\omega,\boldsymbol{K})\,\exp(\mathrm{i}\,\boldsymbol{K}\cdot\boldsymbol{R})\,d\boldsymbol{k}\,d\omega\,d\boldsymbol{K}, \tag{3.197}$$

$$\langle \boldsymbol{u}\cdot(\nabla\times\boldsymbol{u})\rangle = \mathrm{i}\,\varepsilon_{ipj}\int (-k_p + K_p/2)\,f_{ij}^{(0)}(\boldsymbol{k},\omega,\boldsymbol{K})\,\exp(\mathrm{i}\,\boldsymbol{K}\cdot\boldsymbol{R})\,d\boldsymbol{k}\,d\omega\,d\boldsymbol{K}, \tag{3.198}$$

$$\langle(\operatorname{div} \boldsymbol{u})^2\rangle = \int (k_i + K_i/2)\,(k_j - K_j/2)\, f_{ij}^{(0)}(\boldsymbol{k}, \omega, \boldsymbol{K}) \exp(\mathrm{i}\,\boldsymbol{K} \cdot \boldsymbol{R})\, d\boldsymbol{k}\, d\omega\, d\boldsymbol{K}, \tag{3.199}$$

$$f_{ij}^{(0)}(\boldsymbol{k}, \omega, \boldsymbol{K}) = f_{ji}^{*(0)}(\boldsymbol{k}, \omega, \boldsymbol{K}) = f_{ji}^{(0)}(-\boldsymbol{k}, \omega, \boldsymbol{K}). \tag{3.200}$$

For very low Mach numbers and when the parameter σ_c is very small, the continuity equation can be written in the anelastic approximation, $\operatorname{div}(\overline{\rho}\,\boldsymbol{u}) = 0$, which implies that $(\mathrm{i}\,k_i + \mathrm{i}\,K_i/2 - \lambda_i) f_{ij}^{(0)}(\boldsymbol{k}, \omega, \boldsymbol{K}) = 0$ and $(-\mathrm{i}\,k_j + \mathrm{i}\,K_j/2 - \lambda_j) f_{ij}^{(0)}(\boldsymbol{k}, \omega, \boldsymbol{K}) = 0$.

3.6.4 Turbulent Electromotive Force and Level of Magnetic Fluctuations for Rm $\ll 1$

The turbulent electromotive force $\overline{\boldsymbol{\mathcal{E}}} = \langle \boldsymbol{u} \times \boldsymbol{b} \rangle$ is given by

$$\overline{\mathcal{E}}_i = \varepsilon_{inm} \int g_{mn}(\boldsymbol{k}, \omega)\, d\boldsymbol{k}\, d\omega. \tag{3.201}$$

For integration over ω and $\boldsymbol{k}$, we use the integrals given by Eqs. (1.133)–(1.137). Integrating in the ω space and $\boldsymbol{k}$ space in Eq. (3.201) and neglecting the higher-order effects $\sim\mathrm{O}(\lambda^2 \langle \boldsymbol{u}^2 \rangle,\ \nabla^2 \langle \boldsymbol{u}^2 \rangle, \lambda_i \nabla_i \langle \boldsymbol{u}^2 \rangle)$, we arrive at an equation for the turbulent electromotive force for compressible, density-stratified, inhomogeneous and helical background turbulence:

$$\boldsymbol{\mathcal{E}} = \alpha \overline{\boldsymbol{B}} + \boldsymbol{V}^{\rm eff} \times \overline{\boldsymbol{B}} - \eta_{_T}\, \boldsymbol{\nabla} \times \overline{\boldsymbol{B}}, \tag{3.202}$$

where the α effect, the turbulent magnetic diffusivity $\eta_{_T}$, and the pumping velocity $\boldsymbol{V}^{\rm eff}$ for Rm $\ll 1$ are given by

$$\alpha = -\frac{(q-1)}{3(q+1)} \frac{\tau_0\, \langle \boldsymbol{u} \cdot (\boldsymbol{\nabla} \times \boldsymbol{u}) \rangle}{1 + \sigma_c}\, \mathrm{Rm}, \tag{3.203}$$

$$\eta_{_T} = \frac{(q-1)}{3(q+1)}\, \tau_0 \left\langle \boldsymbol{u}^2 \right\rangle \left(1 - \frac{(1 + C_\sigma)\, \sigma_c}{1 + \sigma_c}\right) \mathrm{Rm}, \tag{3.204}$$

$$\boldsymbol{V}^{\rm eff} = -\frac{1}{2}\, \boldsymbol{\nabla} \eta_{_T}, \tag{3.205}$$

$\tau_0 = \ell_0 / \sqrt{\langle \boldsymbol{u}^2 \rangle}$, and

$$C_\sigma = \frac{(q_c - 1)\,(q+1)}{(q_c + 1)\,(q-1)}. \tag{3.206}$$

Since $\tau_0\, \mathrm{Rm} = \ell_0^2/\eta$, the turbulent transport coefficients given by Eqs. (3.203)–(3.205) are determined only by the resistive time scale, ℓ_0^2/η.

The effective pumping velocity $\boldsymbol{V}^{\rm eff}$ of the large-scale magnetic field is independent of $\boldsymbol{\lambda}$. Indeed, the contribution of $\boldsymbol{\lambda}$ to the effective pumping velocity $\boldsymbol{V}^{\rm eff}$ can only arise from the term

$$g_{ij}^{(1)}(\boldsymbol{k},\omega) = {\rm i}\, k_p G_\eta(k,\omega) \left[\overline{B}_p f_{ij}^{(0)}(\boldsymbol{k},\omega) - \overline{B}_i f_{pj}^{(0)}(\boldsymbol{k},\omega)\right]. \qquad (3.207)$$

Integration over solid angles in the $\boldsymbol{k}$ space in $g_{ij}^{(1)} = \int g_{ij}^{(1)}(\boldsymbol{k},\omega)\, d\boldsymbol{k}\, d\omega$ yields

$$g_{ij}^{(1)} = -\frac{1}{6\,(1+\sigma_c)} \int dk E(k) \int d\omega\, \Phi(\omega)\, G_\eta(k,\omega) \\ \times \left[\overline{B}_i \lambda_j + \overline{B}_j \lambda_i - \left(\sigma_c + \frac{1}{2}\right) (\overline{B}_i \nabla_j + \overline{B}_j \nabla_i) \overline{\boldsymbol{u}^2}\right], \qquad (3.208)$$

which is a symmetric tensor with respect to the indexes i and j. Therefore, it cannot contribute to the turbulent electromotive force nor to $\boldsymbol{V}^{\rm eff}$. For integration over solid angles in the $\boldsymbol{k}$ space in the expression for $g_{ij}^{(1)}$, we take into account that $\int k_{ij} \sin\theta\, d\theta\, d\varphi = (4\pi/3)\delta_{ij}$, where we use spherical coordinates (k, θ, φ) in the $\boldsymbol{k}$ space.

The remaining contribution, which is proportional to the mean magnetic field, is

$$g_{ij}^{(2)}(\boldsymbol{k},\omega) = G_\eta \left[\overline{B}_p \left(\frac{1}{2}\nabla_p f_{ij}^{(0)} - \eta k^2 G_\eta\, k_{sp} \nabla_s f_{ij}^{(0)}\right) \right. \\ \left. -\overline{B}_i \left(\frac{1}{2}\nabla_s - \eta k^2 G_\eta\, k_{sp} \nabla_p\right) f_{sj}^{(0)}\right]. \qquad (3.209)$$

Since we neglect the effects $\sim O\big(\lambda_i \nabla_i \langle \boldsymbol{u}^2\rangle, \lambda^2 \langle \boldsymbol{u}^2\rangle, \nabla^2 \langle \boldsymbol{u}^2\rangle\big)$, the term $g_{ij}^{(2)}(\boldsymbol{k},\omega)$ describes only the contribution to $\boldsymbol{V}^{\rm eff}$ caused by $\boldsymbol{\nabla} \langle \boldsymbol{u}^2\rangle$, i.e., the effect of inhomogeneity of turbulence, rather than the effect of stratification on $\boldsymbol{V}^{\rm eff}$. A non-zero contribution to $g_{ij}^{(2)}$ (and to the turbulent electromotive force) is only from the last two terms in Eq. (3.209), which are proportional to $f_{sj}^{(0)}$. Indeed, the first two terms in Eq. (3.209) are proportional to $f_{ij}^{(0)}$, which is a symmetric tensor with respect to the indexes i and j. Therefore, these terms in Eq. (3.209) cannot contribute to the turbulent electromotive force. Here we also take into account that $\int k_i \sin\theta\, d\theta\, d\varphi = 0$. After integration in the $\boldsymbol{k}$ and ω space, we arrive at Eq. (3.205) for the effective pumping velocity $\boldsymbol{V}^{\rm eff}$ for small magnetic Reynolds numbers.

Equations (3.203)–(3.205) imply that for small magnetic Reynolds numbers, compressibility effects characterized by the parameter σ_c decrease the α effect, turbulent magnetic diffusion and effective pumping velocity. However, the total magnetic diffusivity $\eta + \eta_{_T}$ cannot be negative, because for ${\rm Rm} \ll 1$ the microscopic magnetic diffusivity η caused by the electrical conductivity of the fluid is much larger than the turbulent magnetic diffusivity, i.e., $\eta \gg |\eta_{_T}|$.

As follows from Eqs. (3.202) to (3.204), the resulting turbulent electromotive force is determined by the isotropic α effect and turbulent diffusivity. The reason for that is as follows. We consider the case of weak mean magnetic fields, where the energy of the mean magnetic field is much smaller than the turbulent kinetic energy. For a large mean magnetic field, when the energy of the mean magnetic field is of the order of the turbulent kinetic energy, the α tensor and turbulent diffusivity are anisotropic. We do not consider here the effects of uniform rotation or large-scale shear on the turbulent electromotive force. In the case of rotating turbulence or turbulence with large-scale shear, the α tensor and turbulent diffusivity are anisotropic. For anisotropic background turbulence, e.g., for turbulent convection, or for magnetically driven turbulence with relativistic particles, the α tensor and turbulent diffusivity are also anisotropic.

Now we determine the level of magnetic fluctuations for small magnetic Reynolds numbers using Eq. (3.194). For the integration over ω and $\boldsymbol{k}$ we use the integrals given by Eqs. (1.134)–(1.136). After integration in Eq. (3.194) over ω and over angles in the $\boldsymbol{k}$ space, we obtain

$$\langle \boldsymbol{b}^2 \rangle = \frac{\overline{\boldsymbol{B}}^2 \, \langle \boldsymbol{u}^2 \rangle}{3\, \eta^2 \, (1+\sigma_c)} \int_{k0}^{k_\nu} k^{-2} \, [E(k) + 2\sigma_c E_c(k)] \, dk. \tag{3.210}$$

Integrating over k in Eq. (3.210), we arrive at the expression for the level of magnetic fluctuations for small magnetic Reynolds numbers

$$\langle \boldsymbol{b}^2 \rangle = \frac{q-1}{3(q+1)} \left(\frac{1 + 2C_\sigma \, \sigma_c}{1+\sigma_c} \right) \mathrm{Rm}^2 \, \overline{\boldsymbol{B}}^2. \tag{3.211}$$

Since $\langle \boldsymbol{b}^2 \rangle \sim \mathrm{Rm}^2 \, \overline{\boldsymbol{B}}^2$, the energy of magnetic fluctuations is much smaller than the energy of the mean magnetic field for small magnetic Reynolds numbers.

3.6.5 Derivation of Equations for Second Moments for Rm $\gg 1$

In this section, we consider the case of large fluid and magnetic Reynolds numbers, so that turbulence is fully developed and the turbulent correlation time is scale-dependent, like in Kolmogorov type turbulence. We take into account the nonlinear terms in Eqs. (3.169) and (3.170) for velocity and magnetic fluctuations and apply the spectral τ approach.

Using Eq. (3.170) for magnetic fluctuations $\boldsymbol{b}$ and the Navier-Stokes equation (3.169) for velocity fluctuations $\boldsymbol{u}$ in Fourier space, we derive an evolution equation for the cross-helicity tensor,

$$g_{ij}(\boldsymbol{k}, \boldsymbol{R}, t) = \int \langle b_i(\boldsymbol{k} + \boldsymbol{K}/2, t) \, u_j(-\boldsymbol{k} + \boldsymbol{K}/2, t) \rangle \exp[\mathrm{i}\, \boldsymbol{K} \cdot \boldsymbol{R}] \, d\boldsymbol{K},$$

which depends on the velocity correlation function:

$$f_{ij}(\boldsymbol{k}, \boldsymbol{R}, t) = \int \langle u_i(\boldsymbol{k} + \boldsymbol{K}/2, t)\, u_j(-\boldsymbol{k} + \boldsymbol{K}/2, t)\rangle \exp[\mathrm{i}\boldsymbol{K} \cdot \boldsymbol{R}]\, d\boldsymbol{K}.$$

The evolution equation for the cross-helicity tensor reads:

$$\frac{\partial g_{ij}(\boldsymbol{k})}{\partial t} = \mathrm{i}\,(\boldsymbol{k} \cdot \overline{\boldsymbol{B}})\, f_{ij}(\boldsymbol{k}) - \mathrm{i}\, k_m \overline{B}_i\, f_{mj}(\boldsymbol{k}) + I_{ij}(\boldsymbol{k}) + g_{ij}^{(\mathrm{N})}(\boldsymbol{k}), \quad (3.212)$$

where for brevity of notation, we omit the large-scale variables t and $\boldsymbol{R}$ in the functions $f_{ij}(\boldsymbol{k}, \boldsymbol{R}, t)$, $g_{ij}(\boldsymbol{k}, \boldsymbol{R}, t)$, $g_{ij}^{(\mathrm{N})}(\boldsymbol{k}, \boldsymbol{R}, t)$ and $\overline{\boldsymbol{B}}(\boldsymbol{R}, t)$. Here the term $g_{ij}^{(\mathrm{N})}$

$$g_{ij}^{(\mathrm{N})}(\boldsymbol{k}) = \int \left[\left\langle b_i(\boldsymbol{k}_1)\, \frac{\partial u_j(\boldsymbol{k}_2)}{\partial t}\right\rangle + \left\langle b_i^{(\mathrm{N})}(\boldsymbol{k}_1)\, u_j(\boldsymbol{k}_2)\right\rangle\right] \exp[\mathrm{i}\boldsymbol{K} \cdot \boldsymbol{R}]\, d\boldsymbol{K} \quad (3.213)$$

is determined by the third-order moments appearing due to the nonlinear terms. Note that since the Navier-Stokes equation is nonlinear, it appears only in the third-order moments $g_{ij}^{(\mathrm{N})}$. Here we take into account that the fluid pressure is nonlinear in velocity fluctuations. For large magnetic and fluid Reynolds numbers, the dissipative terms caused by the kinematic viscosity ν and magnetic diffusivity η in Eq. (3.212) are negligibly small in comparison with the nonlinear terms. The term I_{ij}, which contains the large-scale spatial derivatives of both the mean magnetic field and the turbulent intensity, is given by

$$I_{ij}(\boldsymbol{k}) = \frac{1}{2}\left[(\overline{\boldsymbol{B}} \cdot \nabla)\, f_{ij} - k_n \left(\frac{\partial f_{ij}}{\partial k_s}\, \nabla_s \overline{B}_n - \frac{\partial f_{nj}}{\partial k_s}\, \nabla_s \overline{B}_i\right) - \nabla_n \left(\overline{B}_i\, f_{nj}\right)\right]. \quad (3.214)$$

This term determines the turbulent magnetic diffusivity and effects of inhomogeneous turbulence.

For the derivation of Eq. (3.212), we perform the following calculation. We use the identity

$$\frac{\partial}{\partial t}\langle b_i(\boldsymbol{k}_1, t)\, u_j(\boldsymbol{k}_2, t)\rangle = \left\langle \frac{\partial b_i(\boldsymbol{k}_1, t)}{\partial t}\, u_j(\boldsymbol{k}_2, t)\right\rangle + \left\langle b_i(\boldsymbol{k}_1, t)\, \frac{\partial u_j(\boldsymbol{k}_2, t)}{\partial t}\right\rangle. \quad (3.215)$$

We rewrite the induction equation (3.170) for magnetic fluctuations in the $\boldsymbol{k}$ space,

$$\frac{\partial b_i(\boldsymbol{k})}{\partial t} = \mathrm{i}\, k_p \left(\int \overline{B}_p(\boldsymbol{Q})\, u_i(\boldsymbol{k} - \boldsymbol{Q})\, d\boldsymbol{Q} - \int \overline{B}_i(\boldsymbol{Q})\, u_p(\boldsymbol{k} - \boldsymbol{Q})\, d\boldsymbol{Q}\right) + b_i^{(\mathrm{N})}(\boldsymbol{k}), \quad (3.216)$$

using Eqs. (3.174)–(3.176). Here $b_i^{(\mathrm{N})}(\boldsymbol{k})$ includes the nonlinear terms. For brevity of notation, we omit below the variable t in the functions $\overline{B}_i(\boldsymbol{Q}, t)$, $b_i(\boldsymbol{k}, t)$, $b_i^{(\mathrm{N})}(\boldsymbol{k}, t)$ and $u_i(\boldsymbol{k}, t)$. We use Eqs. (3.215) and (3.216), which yield

$$\frac{\partial g_{ij}(\boldsymbol{k}, \boldsymbol{R})}{\partial t} = \tilde{J}_{ij}^{(1)}(\boldsymbol{k}, \boldsymbol{R}) - \tilde{J}_{ij}^{(2)}(\boldsymbol{k}, \boldsymbol{R}) + g_{ij}^{(\mathrm{N})}(\boldsymbol{k}, \boldsymbol{R}), \tag{3.217}$$

where

$$\tilde{J}_{ij}^{(1)}(\boldsymbol{k}, \boldsymbol{R}) = \mathrm{i} \int (k_p + K_p/2)\, \overline{B}_p(\boldsymbol{Q}) \left\langle u_i(\boldsymbol{k} + \boldsymbol{K}/2 - \boldsymbol{Q}) u_j(-\boldsymbol{k} + \boldsymbol{K}/2) \right\rangle \times \exp(\mathrm{i}\, \boldsymbol{K} \cdot \boldsymbol{R})\, d\boldsymbol{K}\, d\boldsymbol{Q}, \tag{3.218}$$

$$\tilde{J}_{ij}^{(2)}(\boldsymbol{k}, \boldsymbol{R}) = \mathrm{i} \int (k_p + K_p/2)\, \overline{B}_i(\boldsymbol{Q}) \left\langle u_p(\boldsymbol{k} + \boldsymbol{K}/2 - \boldsymbol{Q}) u_j(-\boldsymbol{k} + \boldsymbol{K}/2) \right\rangle \times \exp(\mathrm{i}\, \boldsymbol{K} \cdot \boldsymbol{R})\, d\boldsymbol{K}\, d\boldsymbol{Q}. \tag{3.219}$$

Next, we perform calculations that are similar to Eqs. (3.181), (3.183)–(3.184). In particular, we introduce new variables, $\tilde{\boldsymbol{k}} = (\tilde{\boldsymbol{k}}_1 - \tilde{\boldsymbol{k}}_2)/2 = \boldsymbol{k} - \boldsymbol{Q}/2$ and $\tilde{\boldsymbol{K}} = \tilde{\boldsymbol{k}}_1 + \tilde{\boldsymbol{k}}_2 = \boldsymbol{K} - \boldsymbol{Q}$, and use the Taylor expansion for $|\boldsymbol{Q}| \ll |\boldsymbol{k}|$ and $|\boldsymbol{K}| \ll |\boldsymbol{k}|$, which yield

$$\tilde{J}_{ij}^{(1)}(\boldsymbol{k}, \boldsymbol{R}) = \overline{B}_p \left(\mathrm{i}\, k_p f_{ij} + \frac{1}{2} \nabla_p f_{ij} \right) - \frac{1}{2} k_m \frac{\partial f_{ij}}{\partial k_p} \nabla_p \overline{B}_m + O(\nabla^2), \tag{3.220}$$

$$\tilde{J}_{ij}^{(2)}(\boldsymbol{k}, \boldsymbol{R}) = \overline{B}_i \left(\mathrm{i}\, k_p f_{pj} + \frac{1}{2} \nabla_p f_{pj} \right) + \frac{1}{2} \left(f_{pj} - k_m \frac{\partial f_{mj}}{\partial k_p} \right) (\nabla_p \overline{B}_i) + O(\nabla^2). \tag{3.221}$$

Therefore, Eqs. (3.216)–(3.221) yield Eq. (3.212) for $\mathrm{Rm} \gg 1$.

Similar derivations allow us to determine the level of magnetic fluctuations $\langle \boldsymbol{b}^2 \rangle = \int E_b(\boldsymbol{k}, \boldsymbol{R}, t)\, d\boldsymbol{k}$, where the function $E_b(\boldsymbol{k}, \boldsymbol{R}, t)$ is

$$E_b(\boldsymbol{k}, \boldsymbol{R}, t) = \int \langle b_i(\boldsymbol{k}_1, t)\, b_i(\boldsymbol{k}_2, t) \rangle \exp[\mathrm{i}\, \boldsymbol{K} \cdot \boldsymbol{R}]\, d\boldsymbol{K}.$$

To derive an evolution equation for the function $E_b(\boldsymbol{k}, \boldsymbol{R}, t)$, we use the identity

$$\frac{\partial}{\partial t} \langle b_i(\boldsymbol{k}_1, t)\, b_i(\boldsymbol{k}_2, t) \rangle = \left\langle \frac{\partial b_i(\boldsymbol{k}_1, t)}{\partial t}\, b_i(\boldsymbol{k}_2, t) \right\rangle + \left\langle b_i(\boldsymbol{k}_1, t)\, \frac{\partial b_i(\boldsymbol{k}_2, t)}{\partial t} \right\rangle. \tag{3.222}$$

Equations (3.216) and (3.222) yield the evolution equation for the function $E_b(\boldsymbol{k})$ as

$$\frac{\partial E_b(\boldsymbol{k})}{\partial t} = -\mathrm{i}\, (\boldsymbol{k} \cdot \overline{\boldsymbol{B}})\, g_{ii}^{(-)} + \mathrm{i}\, k_j \overline{B}_i\, g_{ij}^{(-)} + \tilde{I}(\boldsymbol{k}) + E_b^{(\mathrm{N})}(\boldsymbol{k}), \tag{3.223}$$

where $g_{ij}^{(-)} = g_{ij}(\boldsymbol{k}) - g_{ij}(-\boldsymbol{k})$. For brevity of notation, we omit the large-scale variables t and $\boldsymbol{R}$ in the functions $g_{ij}(\boldsymbol{k}, \boldsymbol{R}, t)$, $\tilde{I}(\boldsymbol{k}, \boldsymbol{R}, t)$, $E_b^{(\mathrm{N})}(\boldsymbol{k}, \boldsymbol{R}, t)$ and $\overline{\boldsymbol{B}}(\boldsymbol{R}, t)$. Here the term $E_b^{(\mathrm{N})}$

$$E_b^{(\mathrm{N})}(\boldsymbol{k}) = \int \left[\left\langle b_i(\boldsymbol{k}_1)\, b_i^{(\mathrm{N})}(\boldsymbol{k}_2) \right\rangle + \left\langle b_i^{(\mathrm{N})}(\boldsymbol{k}_1)\, b_j(\boldsymbol{k}_2) \right\rangle \right] \exp[\mathrm{i}\, \boldsymbol{K} \cdot \boldsymbol{R}]\, d\boldsymbol{K} \tag{3.224}$$

is determined by the third-order moments appearing due to the nonlinear terms. For large magnetic and fluid Reynolds numbers, the dissipative terms caused by the magnetic diffusivity η in Eq. (3.223) are negligibly small in comparison with the nonlinear terms. The term $\tilde{I}$, which contains the large-scale spatial derivatives of both the mean magnetic field and the turbulent intensity, is very small ($\sim \mathrm{O}[\ell_0/L_B]$, $\mathrm{O}[\ell_0/L_u]$), and it is neglected.

We consider the case of weak mean magnetic fields, where the energy of the mean magnetic field is much less than the turbulent kinetic energy. This implies that here we do not investigate quenching of the turbulent transport coefficients. Therefore, we do not need evolution equations for the second moments of velocity $\langle u_i u_j \rangle$ and magnetic field $\langle b_i b_j \rangle$ to determine the turbulent electromotive force. This implies that we consider linear effects in the mean magnetic field.

Equations (3.212) and (3.223) for the cross-helicity tensor g_{ij} and for the function $E_b(\boldsymbol{k})$ include first-order spatial differential operators $\hat{\mathcal{M}}$ applied to the third-order moments $F^{(\mathrm{III})}$. The problem arises of how to close these equations, i.e., how to express the third-order term $\hat{\mathcal{M}} F^{(\mathrm{III})}$ through the lower moments $F^{(\mathrm{II})}$. We use the spectral τ approximation, which postulates that the deviations of the third-moment terms, $\hat{\mathcal{M}} F^{(\mathrm{III})}(\boldsymbol{k})$, from the contributions to these terms afforded by the background turbulence, $\hat{\mathcal{M}} F^{(\mathrm{III},0)}(\boldsymbol{k})$, can be expressed through similar deviations of the second moments, $F^{(\mathrm{II})}(\boldsymbol{k}) - F^{(\mathrm{II},0)}(\boldsymbol{k})$:

$$\hat{\mathcal{M}} F^{(\mathrm{III})}(\boldsymbol{k}) - \hat{\mathcal{M}} F^{(\mathrm{III},0)}(\boldsymbol{k}) = -\frac{1}{\tau_r(k)} \left[F^{(\mathrm{II})}(\boldsymbol{k}) - F^{(\mathrm{II},0)}(\boldsymbol{k}) \right], \tag{3.225}$$

where $\tau_r(k)$ is the scale-dependent relaxation time, which can be identified with the correlation time $\tau(k)$ of the turbulent velocity field for large fluid and magnetic Reynolds numbers. The functions with the superscript (0) correspond to the background turbulence with zero mean magnetic field. We take into account that $g_{ij}^{(0)} = 0$, because when the mean magnetic field is zero, the turbulent electromotive force vanishes. We do not take into account magnetic fluctuations caused by a small-scale dynamo (i.e., a dynamo with zero mean magnetic field). This implies that $E_b^{(0)} = 0$. Therefore, Eq. (3.225) for $g_{ij}(\boldsymbol{k})$ reduces to $\hat{\mathcal{M}} g_{ij}^{(\mathrm{III})}(\boldsymbol{k}) = -g_{ij}(\boldsymbol{k})/\tau(k)$. Similarly, Eq. (3.225) for $E_b(\boldsymbol{k})$ reduces to $\hat{\mathcal{M}} E_b^{(\mathrm{III})}(\boldsymbol{k}) = -E_b(\boldsymbol{k})/\tau(k)$.

3.6.6 Model for Compressible Helical Turbulence

In the next step of the derivation, we need a model for the background turbulence. We use statistically stationary, density-stratified, inhomogeneous, compressible and helical background turbulence, which is determined by the following correlation function in the $\boldsymbol{k}$ space:

$$f_{ij}^{(0)}(\boldsymbol{k},\boldsymbol{R}) = \frac{1}{8\pi\, k^2\,(1+\sigma_c)}\Bigg\{\bigg[E(k)\bigg(\delta_{ij} - k_{ij} + \frac{\mathrm{i}}{k^2}\,(k_j\lambda_i - k_i\lambda_j) + \frac{\mathrm{i}}{2k^2}\,(k_i\nabla_j - k_j\nabla_i)\bigg) + 2\sigma_c\, E_c(k)\,\bigg(k_{ij} + \frac{\mathrm{i}}{4k^2}\,(k_i\nabla_j - k_j\nabla_i) + \frac{\mathrm{i}}{2k^2}(k_j\lambda_i - k_i\lambda_j)\bigg)\bigg]\langle \boldsymbol{u}^2\rangle - \frac{\mathrm{i}E(k)}{k^2}\,\varepsilon_{ijp}\,k_p\,\langle \boldsymbol{u}\cdot(\boldsymbol{\nabla}\times\boldsymbol{u})\rangle\Bigg\} \tag{3.226}$$

[see Eq. (2.188)]. Note that most statements and conditions used for the derivation of the tensorial structure of Eq. (3.195) are also valid for Eq. (3.226). We assume here that the background turbulence is of Kolmogorov type with constant energy flux over the spectrum, i.e., the kinetic energy spectrum for the incompressible part of turbulence in the range of wave numbers $k_0 < k < k_\nu$ is $E(k) = -d\bar\tau(k)/dk$, where the function $\bar\tau(k) = (k/k_0)^{1-q}$ with $1 < q < 3$. The condition $q > 1$ corresponds to finite kinetic energy for very large fluid Reynolds numbers, while $q < 3$ corresponds to finite dissipation of the turbulent kinetic energy at the viscous scale. Similarly, the kinetic energy spectrum for the compressible part of turbulence is $E_c(k) = -d\bar\tau_c(k)/dk$, where the function $\bar\tau_c(k) = (k/k_0)^{1-q_c}$ with $1 < q_c < 3$. For instance, the exponent of the incompressible part of the energy spectrum $q = 5/3$ (the Kolmogorov spectrum), while the exponent of the compressible part of the energy spectrum $q_c = 2$ (like for the Burgers turbulence). The turbulent correlation time in the $\boldsymbol{k}$ space is $\tau(k) = 2\tau_0\left[\bar\tau(k) + \sigma_c\,\bar\tau_c(k)\right]/(1+\sigma_c)$ where $\tau_0 = \ell_0/\sqrt{\langle \boldsymbol{u}^2\rangle}$ and $\sqrt{\langle \boldsymbol{u}^2\rangle}$ is the characteristic turbulent velocity at scale ℓ_0. Note that for fully developed Kolmogorov-like turbulence, $\sigma_c < 1$.

3.6.7 Turbulent Electromotive Force and Level of Magnetic Fluctuations for $Rm \gg 1$

We use the spectral τ approach and assume that the characteristic time of variation of the mean magnetic field $\overline{\boldsymbol{B}}$ is substantially larger than the correlation time $\tau(k)$ for all turbulence scales. This allows us to obtain a stationary solution of Eq. (3.212). We consider linear effects in the mean magnetic field, so that the function $f_{ij}(\boldsymbol{k})$ in Eqs. (3.212) and (3.214) is replaced by $f_{ij}^{(0)}(\boldsymbol{k})$. The turbulent electromotive force is determined by $\overline{\mathcal{E}}_i = \varepsilon_{inm}\int g_{mn}(\boldsymbol{k})\,\mathrm{d}\boldsymbol{k}$. We take into account that the terms with symmetric tensors with respect to the indexes m and n in $g_{mn}(\boldsymbol{k})$

do not contribute to the turbulent electromotive force. Therefore, the equation for the turbulent electromotive force is given by

$$\overline{\mathcal{E}}_m = \varepsilon_{mji} \int \tau(k) \Big\{ \mathrm{i}\, (\boldsymbol{k} \cdot \overline{\boldsymbol{B}})\, f_{ij}^{(0)} - \overline{B}_i \left(\mathrm{i}\, k_n\, f_{nj}^{(0)} + \frac{1}{2} \nabla_n f_{nj}^{(0)} \right) - \left[f_{nj}^{(0)} + \frac{k}{2} \left(\frac{\mathrm{d} \ln \tau}{\mathrm{d}k} \right) k_{np}\, f_{pj}^{(0)} \right] \nabla_n \overline{B}_i \Big\}\, \mathrm{d}\boldsymbol{k}. \tag{3.227}$$

We use the model of background turbulence given by Eq. (3.226). Integrating in the $\boldsymbol{k}$ space of Eq. (3.227), we arrive at an equation for the turbulent electromotive force (3.202), where the α effect, the turbulent magnetic diffusivity $\eta_{_T}$ and the pumping velocity $\boldsymbol{V}^{\rm eff}$ for large magnetic Reynolds numbers are given by

$$\alpha = -\frac{1}{3}\, \tau_0\, \langle \boldsymbol{u} \cdot (\nabla \times \boldsymbol{u}) \rangle \left[1 - \frac{\sigma_c}{1 + \sigma_c} \left(1 + \frac{\tilde{C}_\sigma}{2(1 + \sigma_c)} \right) \right], \tag{3.228}$$

$$\eta_{_T} = \frac{\tau_0\, \langle \boldsymbol{u}^2 \rangle}{3} \left\{ 1 - \frac{\sigma_c}{2(1 + \sigma_c)^2} \left[\tilde{C}_\sigma\, (q - 1) + \sigma_c\, (q_c - 1) \right] \right\}, \tag{3.229}$$

$$\boldsymbol{V}^{\rm eff} = -\frac{1}{6} \nabla \left(\tau_0\, \langle \boldsymbol{u}^2 \rangle \right), \tag{3.230}$$

and

$$\tilde{C}_\sigma = \frac{2(q_c - 1)}{q + q_c - 2}. \tag{3.231}$$

For the derivation of Eqs. (3.228)–(3.230), we use identities (2.193)–(2.196). For irrotational turbulence ($\sigma_c \gg 1$), the α effect, the turbulent magnetic diffusivity $\eta_{_T}$ and the pumping velocity $\boldsymbol{V}^{\rm eff}$ for large magnetic Reynolds numbers are given by

$$\alpha = -\frac{q_c - 1}{3(q + q_c - 2)\, \sigma_c}\, \tau_0\, \langle \boldsymbol{u} \cdot (\nabla \times \boldsymbol{u}) \rangle, \tag{3.232}$$

$$\eta_{_T} = \frac{3 - q_c}{6}\, \tau_0\, \langle \boldsymbol{u}^2 \rangle, \quad \boldsymbol{V}^{\rm eff} = -\frac{1}{6} \nabla \left(\tau_0\, \langle \boldsymbol{u}^2 \rangle \right). \tag{3.233}$$

Note that for a purely irrotational turbulence the kinetic helicity vanishes. Equations (3.228)–(3.233) imply that for large magnetic Reynolds numbers, compressibility decrease both the α effect and turbulent magnetic diffusivity. Since $q < 3$ and $q_c < 3$, the turbulent magnetic diffusivity is always positive for $\mathrm{Rm} \gg 1$, even for very large degrees of compressibility, $\sigma_c \gg 1$ (e.g., for irrotational flows). Note that for an incompressible flow the turbulent magnetic diffusivity is independent of q. On the other hand, the compressibility of fluid flow does not affect the pumping velocity $\boldsymbol{V}^{\rm eff}$ of the mean magnetic field for large magnetic Reynolds numbers.

Now we determine the level of magnetic fluctuations for large magnetic Reynolds numbers. To this end, we use the spectral τ approach in Eqs. (3.212) and (3.223). We assume that the characteristic time of variation of the mean magnetic field $\overline{\boldsymbol{B}}$ is substantially larger than the correlation time $\tau(k)$ for all turbulence scales. This allows us to get stationary solutions of Eqs. (3.212) and (3.223), which yield

$$\langle b^2 \rangle = 2 \int \tau^2(k) \left\{ (\boldsymbol{k} \cdot \overline{\boldsymbol{B}}) \left[(\boldsymbol{k} \cdot \overline{\boldsymbol{B}})\, f_{ii}^{(0)}(\boldsymbol{k}) - k_i \overline{B}_j\, f_{ij}^{(0)}(\boldsymbol{k}) \right] \right.$$
$$\left. - k_j \overline{B}_i \left[(\boldsymbol{k} \cdot \overline{\boldsymbol{B}})\, f_{ij}^{(0)}(\boldsymbol{k}) - k_m \overline{B}_i\, f_{mj}^{(0)}(\boldsymbol{k}) \right] \right\} d\boldsymbol{k}. \qquad (3.234)$$

After integration over angles in the $\boldsymbol{k}$ space, we obtain

$$\langle b^2 \rangle = \frac{2 \overline{\boldsymbol{B}}^2}{3\,(1+\sigma_c)} \int_{k_0}^{k_\nu} \tau^2(k)\, k^2\, \left[E(k) + 2\sigma_c E_c(k) \right] dk. \qquad (3.235)$$

To integrate over k in Eq. (3.235), we use Eq. (2.199) and the following identity obtained for $q = 5/3$, $q_c = 2$ and for large magnetic Prandtl numbers:

$$\int_{k_0}^{k_\nu} \tau^2(k)\, k^2\, E(k)\, dk = 4 \left(\ln \mathrm{Re} + 8\sigma_c + 2\sigma_c^2 \right) \left(\frac{\tau_0}{\ell_0} \right)^2 \left(\frac{\sigma_c}{1+\sigma_c} \right)^2. \qquad (3.236)$$

For small magnetic Prandtl numbers, the integration over k in Eq. (3.236) should be performed up to $k_\eta = \ell_\eta^{-1}$, where $\ell_\eta = \ell_0\, \mathrm{Rm}^{-3/4}$ is the resistive scale. In this case, the fluid Reynolds number Re in Eq. (3.236) should be replaced by the magnetic Reynolds number Rm. Integrating over k in Eq. (3.235), we obtain the expression for the level of magnetic fluctuations for large magnetic Reynolds numbers and large magnetic Prandtl numbers as

$$\langle b^2 \rangle = \frac{4 \overline{\boldsymbol{B}}^2}{3\,(1+\sigma_c)^3} \left[\ln \mathrm{Re} + 20\sigma_c + 2\sigma_c^2 (7 + 2\sigma_c) \right]. \qquad (3.237)$$

For small magnetic Prandtl numbers, the fluid Reynolds number Re in Eq. (3.237) should be replaced by the magnetic Reynolds number Rm. Since fluid and magnetic Reynolds numbers are large, the energy of magnetic fluctuations is larger than the energy of the mean magnetic field. For irrotational turbulence ($\sigma_c \gg 1$), the level of magnetic fluctuations for large magnetic Reynolds numbers is $\langle b^2 \rangle = 16 \overline{\boldsymbol{B}}^2/3$.

3.7 Mean-Field Dynamo

In this section, we consider a theory of generation of the mean magnetic field by turbulent motions of conducting fluid described by the α effect.

3.7.1 Mean-Field Equation

The mean-field equation for the large-scale magnetic field reads

$$\frac{\partial \overline{\boldsymbol{B}}}{\partial t} = \nabla \times \left(\overline{\boldsymbol{U}} \times \overline{\boldsymbol{B}} + \alpha \overline{\boldsymbol{B}} + \boldsymbol{V}^{\rm eff} \times \overline{\boldsymbol{B}} - \eta_{_T} \nabla \times \overline{\boldsymbol{B}} \right) + \eta \, \Delta \overline{\boldsymbol{B}}, \tag{3.238}$$

where α determines the α-effect caused by the kinetic helicity, $\langle \boldsymbol{u} \cdot (\nabla \times \boldsymbol{u}) \rangle$, $\boldsymbol{V}^{\rm eff}$ is the effective pumping velocity that describes the turbulent diamagnetic effect and $\eta_{_T}$ is the coefficient of turbulent magnetic diffusion. These turbulent transport coefficients entering in the turbulent electromotive force, are calculated in previous sections of this chapter.

3.7.2 The α^2 Dynamo

Let us consider homogeneous, isotropic and helical turbulence with α = const. Since turbulence is homogeneous, $\eta_{_T}$ = const and $\boldsymbol{V}^{\rm eff} = 0$. We consider, e.g., a case of large fluid and magnetic Reynolds numbers so that $\eta_{_T} \gg \eta$. We also consider, for simplicity, a case with a zero mean fluid velocity. The mean-field equation for the large-scale magnetic field in this case reads

$$\boxed{\frac{\partial \overline{\boldsymbol{B}}}{\partial t} = \nabla \times \left(\alpha \overline{\boldsymbol{B}} - \eta_{_T} \nabla \times \overline{\boldsymbol{B}} \right),} \tag{3.239}$$

We seek a solution for Eq. (3.239) for perturbations in the following form:

$$\boxed{\overline{\boldsymbol{B}}(t, x, z) = \overline{B}_y(t, x, z) \boldsymbol{e}_y + \nabla \times \left[\overline{A}(t, x, z) \boldsymbol{e}_y \right],} \tag{3.240}$$

where $\boldsymbol{e}_y$ is the unit vector directed along the y-axis. The functions $\overline{B}_y(t, x, z)$ and $\overline{A}(t, x, z)$ are determined by the following equations:

$$\frac{\partial \overline{A}(t, x, z)}{\partial t} = \alpha \overline{B}_y + \eta_{_T} \Delta \overline{A}, \tag{3.241}$$

$$\frac{\partial \overline{B}_y(t, x, z)}{\partial t} = -\alpha \Delta \overline{A} + \eta_{_T} \Delta \overline{B}_y, \tag{3.242}$$

where $\Delta = \nabla_x^2 + \nabla_z^2$ and we take into account that $\nabla \times (\nabla \times \overline{\boldsymbol{B}}) = -\Delta \overline{\boldsymbol{B}}$. The other components of the magnetic field are $\overline{B}_x = -\nabla_z \overline{A}$ and $\overline{B}_z = \nabla_x \overline{A}$. We seek a solution for Eqs. (3.241) and (3.242) in the form

$$\overline{A} = A_0 \exp(\gamma t + {\rm i} \, K_x X + {\rm i} \, K_z Z), \tag{3.243}$$

$$\overline{B}_y = B_0 \exp(\gamma t + {\rm i} \, K_x X + {\rm i} \, K_z Z). \tag{3.244}$$

Substituting these solutions into Eqs. (3.241) and (3.242), we obtain the following system of algebraic equations:

$$(\gamma + \eta_{_T} K^2)\, A_0 - \alpha\, B_0 = 0, \tag{3.245}$$

$$\alpha\, K^2\, A_0 - (\gamma + \eta_{_T} K^2)\, B_0 = 0, \tag{3.246}$$

where $K^2 = K_x^2 + K_z^2$. The solution of Eqs. (3.245) and (3.246) yields the growth rate γ of the dynamo instability,

$$\boxed{\gamma = |\alpha K| - \eta_{_T} K^2,} \tag{3.247}$$

and the ratio A_0/B_0,

$$\frac{A_0}{B_0} = \frac{\gamma + \eta_{_T} K^2}{\alpha K^2} = \frac{\text{sgn}(\alpha)}{|K|}. \tag{3.248}$$

The components of the magnetic field are

$$\boxed{\overline{B}_x = \text{sgn}(\alpha)\, B_0\, \frac{K_z}{|K|}\, \exp(\gamma t)\, \sin(K_x X + K_z Z),} \tag{3.249}$$

$$\boxed{\overline{B}_y = B_0\, \exp(\gamma t)\, \cos(K_x X + K_z Z),} \tag{3.250}$$

$$\boxed{\overline{B}_z = -\text{sgn}(\alpha)\, B_0\, \frac{K_x}{|K|}\, \exp(\gamma t)\, \sin(K_x X + K_z Z).} \tag{3.251}$$

The maximum growth rate of the dynamo instability that is attained at $|K| \equiv K^{\max} = |\alpha|/2\eta_{_T}$ is given by

$$\boxed{\gamma^{\max} = \alpha^2/4\eta_{_T}.} \tag{3.252}$$

3.7.3 The α–Shear or α Ω Dynamo

Let us consider the following equilibrium: $\alpha = $ const and $\overline{\boldsymbol{U}} = (0, Sx, 0)$. We seek a solution of Eq. (3.239) for perturbations in the form (3.240), where the functions $\overline{B}_y(t, x, z)$ and $\overline{A}(t, x, z)$ are determined by the following equations:

$$\frac{\partial \overline{A}(t, x, z)}{\partial t} = \alpha \overline{B}_y + \eta_{_T} \Delta \overline{A}, \tag{3.253}$$

$$\frac{\partial \overline{B}_y(t, x, z)}{\partial t} = -\alpha \Delta \overline{A} - S \nabla_z A + \eta_{_T} \Delta \overline{B}_y. \tag{3.254}$$

The second term, $-S\nabla_z A$, on the right-hand side of Eq. (3.254) is originated from the term $(\overline{\boldsymbol{B}} \cdot \boldsymbol{\nabla})\overline{U}_y$. In this section, we consider the case when $|\alpha \Delta \overline{A}| \ll |S \nabla_z A|$, so that we can neglect the first term on the right-hand side of Eq. (3.254). We seek a solution for Eqs. (3.253) and (3.254) as a real part of the following functions:

$$\overline{A} = A_0 \exp(\tilde{\gamma} t + \mathrm{i}\, K_x X + \mathrm{i}\, K_z Z), \tag{3.255}$$

$$\overline{B}_y = B_0 \exp(\tilde{\gamma} t + \mathrm{i}\, K_x X + \mathrm{i}\, K_z Z). \tag{3.256}$$

Substituting solutions (3.255)–(3.256) into Eqs. (3.253) and (3.254), we obtain the following system of algebraic equations:

$$(\tilde{\gamma} + \eta_{_T} K^2) A_0 - \alpha B_0 = 0, \tag{3.257}$$

$$\mathrm{i}\, S K_z A_0 + (\tilde{\gamma} + \eta_{_T} K^2) B_0 = 0. \tag{3.258}$$

Equations (3.257) and (3.258) yield the growth rate of the dynamo instability and the frequency of the dynamo waves as

$$\boxed{\gamma = \left(\frac{|\alpha\, S\, K_z|}{2}\right)^{1/2} - \eta_{_T} K^2,} \tag{3.259}$$

$$\boxed{\omega = \left(\frac{|\alpha\, S\, K_z|}{2}\right)^{1/2},} \tag{3.260}$$

and

$$\frac{A_0}{B_0} = \frac{\alpha}{|\alpha\, S\, K_z|^{1/2}} \exp(\mathrm{i}\, \pi/4), \tag{3.261}$$

where $\tilde{\gamma} = \gamma - \mathrm{i}\, \omega$, and we use $(-\mathrm{i})^{1/2} = \exp(-\mathrm{i}\, \pi/4) = (1 - \mathrm{i})/\sqrt{2}$. The components of the magnetic field are:

$$\overline{B}_x = \frac{\alpha\, K_z}{|\alpha\, S\, K_z|^{1/2}} B_0 \exp(\gamma t) \cos(K_x X + K_z Z - \omega t - \pi/4), \tag{3.262}$$

$$\overline{B}_y = B_0 \exp(\gamma t) \cos(K_x X + K_z Z - \omega t), \tag{3.263}$$

$$\overline{B}_z = \frac{\alpha\, K_x}{|\alpha\, S\, K_z|^{1/2}} B_0 \exp(\gamma t) \cos(K_x X + K_z Z - \omega t + 3\pi/4). \tag{3.264}$$

The maximum growth rate of the dynamo instability and the maximum frequency of the dynamo waves are given by

$$\boxed{\gamma^{\max} \approx \frac{3}{8} \left(\frac{\alpha^2\, S^2}{2\eta_{_T}}\right)^{1/3},} \tag{3.265}$$

$$\boxed{\omega^{\max} = \frac{1}{2} \left(\frac{\alpha^2\, S^2}{2\eta_{_T}}\right)^{1/3}.} \tag{3.266}$$

The maximum growth rate of the dynamo instability and the maximum frequency of the dynamo waves are attained at $K_x^{\max} = 0$ and

$$\boxed{K_z^{\max} = \frac{1}{2}\left(\frac{|\alpha\, S|}{4\eta_T^2}\right)^{1/3}.} \tag{3.267}$$

3.7.4 Physics of the α Effect

The main reason for the excitation of the large-scale magnetic field is the α-effect caused by the kinetic helicity, $\langle \boldsymbol{u} \cdot (\nabla \times \boldsymbol{u})\rangle$. A non-zero kinetic helicity implies that a number of the left-handed rotating turbulent eddies at a given instant does not exactly equal the number of right-handed rotating eddies (i.e., the number of eddies rotating in the clockwise direction does not equal to that rotating in the counterclockwise direction). A mechanism of the α-effect is as follows. Deformation of the original magnetic field is caused by both the left-handed rotating eddies and the right-handed rotating eddies. Since a number of the left-handed rotating turbulent eddies at a given instant does not exactly equal the number of right-handed rotating eddies, the total effect of deformation of the original magnetic field line is not zero. This causes the generation of a large-scale magnetic field.

In real turbulent systems, the kinetic helicity is created in rotating inhomogeneous turbulent convection (or rotating density-stratified turbulent convection). The deformation of the magnetic field lines is caused by upward and downward rotating turbulent eddies. Since the turbulence is inhomogeneous (or turbulence is density stratified), it breaks a symmetry between the upward and downward eddies, so that their total effect on the mean magnetic field does not vanish, and it creates the mean electric current along the original mean magnetic field.

The $\alpha-$shear (or $\alpha\Omega$) mean-field dynamo mechanism acts in the following way. A nonuniform shear velocity $\overline{U}_y = Sx$ stretches the $\overline{B}_x$ component in the y-direction, creating the $\overline{B}_y$ component of the mean magnetic field [see Eq. (3.254)]. The α effect closes the loop, creating the $\overline{B}_x$ component from the $\overline{B}_y$ component of the mean magnetic field [see Eq. (3.253)]. This creates the dynamo waves predicted by Parker (1955b).

3.7.5 The α^2–Shear or $\alpha^2\Omega$ Dynamo

Now we consider the case when $|\alpha \Delta \overline{A}| \sim |S \nabla_z A|$. We seek a solution for Eqs. (3.253) and (3.254) as a real part of the functions (3.255) and (3.256). Substituting solutions (3.255) and (3.256) into Eqs. (3.253) and (3.254), we obtain the following system of algebraic equations:

$$(\tilde{\gamma} + \eta_{_T} K^2) A_0 - \alpha B_0 = 0, \tag{3.268}$$

$$(\mathrm{i}\, S K_z - \alpha K^2) A_0 + (\tilde{\gamma} + \eta_{_T} K^2) B_0 = 0, \tag{3.269}$$

where $\tilde{\gamma} = \gamma - \mathrm{i}\,\omega$. Equations (3.268) and (3.269) yield the growth rate of the dynamo instability and the frequency of the dynamo waves as

$$\boxed{\gamma = \frac{1}{\sqrt{2}} \left\{ \left[(\alpha\, K_z\, S)^2 + (\alpha\, K)^4 \right]^{1/2} + (\alpha\, K)^2 \right\}^{1/2} - \eta_{_T} K^2,} \tag{3.270}$$

$$\boxed{\omega = \frac{|\alpha\, K_z\, S|}{\sqrt{2}} \left\{ \left[(\alpha\, K_z\, S)^2 + (\alpha\, K)^4 \right]^{1/2} + (\alpha\, K)^2 \right\}^{-1/2},} \tag{3.271}$$

and

$$\frac{A_0}{B_0} = \frac{\alpha}{\gamma + \eta_{_T} K^2 - \mathrm{i}\,\omega}. \tag{3.272}$$

The components of the magnetic field, e.g., for $K_x = 0$ are $\overline{B}_z = 0$,

$$\overline{B}_x = \frac{\alpha k_z\, B_0 \exp(\gamma t)}{(\gamma + \eta_{_T} K_z^2)^2 + \omega^2} \left[(\gamma + \eta_{_T} K_z^2) \sin(K_z Z - \omega t) + \omega \cos(K_z Z - \omega t) \right], \tag{3.273}$$

$$\overline{B}_y = B_0 \exp(\gamma t) \cos(K_z Z - \omega t). \tag{3.274}$$

The dynamo instability is saturated by nonlinear effects (the back-reaction of the generated mean magnetic field on the turbulent motions of the conducting fluids or the nonlinear evolution of magnetic helicity); see the next section.

3.8 Further Reading

3.8.1 Large-Scale Dynamo

The mean-field dynamo approach has been formulated in a seminal paper by M. Steenbeck, F. Krause and K.-H. Rädler in 1966 (Steenbeck et al., 1966; Krause and Rädler, 1980; Roberts and Stix, 1971). Since the formulation of the mean-field dynamo approach, the theories of solar, stellar and galactic magnetic fields have been actively developing over the last 50 years (Moffatt, 1978; Parker, 1979; Krause and Rädler, 1980; Rädler, 1980, 1982; Zeldovich et al., 1983; Ruzmaikin et al., 1988; Rüdiger et al., 2013; Moffatt and Dormy, 2019).

A well-known point of view on the origin of the large-scale solar and galactic magnetic fields is that the magnetic field is generated in the solar convective zone or in the galactic disk by a combined action of helical turbulent motions and large-scale differential (nonuniform) rotation. A non-zero mean kinetic helicity produced by a rotating density-stratified or inhomogeneous turbulence or turbulent convection causes the α effect. An origin of the differential rotation is related to an anisotropic eddy viscosity (Kippenhahn, 1963; Rüdiger, 1989; Rüdiger et al., 2013) and an effect of the turbulent heat flux on the Reynolds stress in a rotating density-stratified turbulent convection (Kleeorin and Rogachevskii, 2006; Rogachevskii and Kleeorin, 2018). The combined effect of helical turbulent motions and differential rotation, referred to as $\alpha\Omega$ or $\alpha^2\Omega$ mean-field dynamo, results in a dynamo instability, i.e., it causes an exponential growth of the large-scale magnetic field. In the context of the solar dynamo, the instability is oscillatory, i.e., it creates the dynamo waves predicted by E. Parker (Parker, 1955b). In the context of the galactic dynamo, the instability is non-oscillatory.

The α effect describes a contribution to the turbulent electromotive force that is given by a pseudo-scalar α multiplying the mean magnetic field. Such a pseudo-scalar can be the result of rotation $\boldsymbol{\Omega}$ combined with stratification of the mean density $\nabla\overline{\rho}$ and/or inhomogeneity of turbulence intensity $\nabla u_{\rm rms}$, respectively. Here, $\overline{\rho}$ is the mean gas density and $u_{\rm rms} = \langle \boldsymbol{u}^2\rangle^{1/2}$ is the rms value of the turbulent velocity. There have been a number of analytic studies quantifying the effects of rotating stratified turbulence on the α effect by means of different approaches (Krause and Rädler, 1980; Rüdiger and Kichatinov, 1993; Rädler et al., 2003; Kleeorin and Rogachevskii, 2003). In particular, it was found using the quasi-linear approach (Steenbeck et al., 1966; Krause and Rädler, 1980) that the diagonal components of the α tensor for slow rotation rate are given by

$$\alpha \approx -\ell_0^2\, (\boldsymbol{\Omega}\cdot\nabla)\ln\left(\overline{\rho}\, u_{\rm rms}\right). \tag{3.275}$$

In the solar convective zone, the mean fluid density changes by seven orders of magnitude, while the turbulent kinetic energy changes by three orders of magnitudes and, according to stellar mixing length theory (Spruit, 1974), $\overline{\rho}\, u_{\rm rms}^3$ would be approximately constant in the solar convective zone. It has been found analytically as well as numerically for forced turbulence and turbulent convection with a stronger temperature stratification and overshoot layer that the α effect for slow rotation and large fluid and magnetic Reynolds numbers is given by

$$\alpha \approx -\ell_0^2\, (\boldsymbol{\Omega}\cdot\nabla)\ln\left(\overline{\rho}^{\zeta}\, u_{\rm rms}\right), \tag{3.276}$$

where $\zeta = 1/2$. Note, however, that several simulations are consistent also with $\zeta = 1/3$ (Brandenburg et al., 2013a).

The dynamo instability is saturated by nonlinear effects, i.e., in the nonlinear dynamo the turbulent transport coefficients, α, η_T and $\boldsymbol{V}^{\rm eff}$ depend on the mean magnetic field $\overline{\boldsymbol{B}}$. The simplest nonlinear saturation mechanism of the dynamo instability is related to α quenching, which prescribes the α effect to be a decreasing function of the mean magnetic field strength, e.g.,

$$\alpha(\overline{\boldsymbol{B}}) = \alpha_0 \left(1 + \frac{\overline{\boldsymbol{B}}^2}{\overline{B}_{\rm eq}^2}\right)^{-1}, \tag{3.277}$$

where $\overline{B}_{\rm eq}^2 = 4\pi\,\overline{\rho}\,\langle \boldsymbol{u}^2 \rangle$ is the squared equipartition mean magnetic field. This implies that the mean magnetic field strength at which quenching becomes significant is estimated from the equipartition between the energy density of the mean magnetic field and the turbulent kinetic energy density. When applied to galactic dynamos, this picture results in robust magnetic field models that are compatible with observations (Ruzmaikin et al., 1988).

However, this picture is obviously oversimplified, and various attempts to suggest a more developed version of nonlinear dynamo theory have been undertaken by different authors [see the review by Brandenburg and Subramanian (2005a)]. The above-mentioned nonlinearity is referred to as algebraic nonlinearity. The algebraic nonlinearity is caused by the feedback of the growing large-scale magnetic field on the fluid motion. The quantitative theories of the algebraic nonlinearities of the α effect, the turbulent magnetic diffusion and the effective pumping velocity have been developed using the quasi-linear approach for small fluid and magnetic Reynolds numbers (Rüdiger and Kichatinov, 1993; Kitchatinov et al., 1994; Rüdiger et al., 2013) and the tau approach for large fluid and magnetic Reynolds numbers (Field et al., 1999; Rogachevskii and Kleeorin, 2000, 2001, 2004, 2006).

Validations of the quasi-linear approach and the τ approach for different situations have been performed in various direct numerical simulations (Brandenburg et al., 2004, 2012a; Brandenburg and Subramanian, 2005a,b; Rogachevskii et al., 2011, 2012, 2018; Haugen et al., 2012). In magnetohydrodynamics, the τ approach reproduces many well-known phenomena found by different methods, e.g., the τ approach yields correct formulae for the α effect (Krause and Rädler, 1980; Rüdiger and Kichatinov, 1993; Rogachevskii and Kleeorin, 2000; Rädler et al., 2003; Kleeorin and Rogachevskii, 2003, 2007), the turbulent diamagnetic and paramagnetic velocities (Krause and Rädler, 1980; Kitchatinov et al., 1994; Rogachevskii and Kleeorin, 2001, 2004; Rädler et al., 2003; Kleeorin and Rogachevskii, 2007), the turbulent magnetic diffusion, the $\boldsymbol{\Omega} \times \boldsymbol{J}$ effect and the κ effect (Krause and Rädler, 1980; Rogachevskii and Kleeorin, 2001, 2004; Rädler et al., 2003).

An effect of kinetic helicity on turbulent magnetic diffusivity has been recently studied using direct numerical simulations of forced helical turbulence (Brandenburg et al., 2017a). In particular, for large magnetic Reynolds numbers, the turbulent magnetic diffusivity without helicity is about twice the value with helicity. This effect has not yet been explained. Turbulent electromotive force, including the cross-helicity effect of inhomogeneous turbulence, also has been calculated in the framework of the multiple-scale direct-interaction approximation (Yoshizawa, 1990; Yokoi, 2013). This theory confirms the results for the turbulent electromotive force obtained using other analytical approaches.

Compressibility effects on turbulent electromotive force have been also investigated using the quasi-linear approach, the spectral tau approach and the multiple-scale direct-interaction approximation in a number of studies (Krause and Rädler, 1980; Rädler et al., 2011; Rogachevskii et al., 2018; Yokoi, 2018a,b). In particular, the compressibility of the fluid flow reduces the turbulent magnetic diffusivity both at small (Krause and Rädler, 1980) and large magnetic Reynolds numbers (Rogachevskii et al., 2018). The reduction of turbulent magnetic diffusivity by the compressibility of fluid flow has been also confirmed by the test-field method in direct numerical simulations for an irrotational homogeneous deterministic flow (Rädler et al., 2011). The compressibility of fluid flow decreases the α effect for arbitrary Rm and it reduces the effective pumping velocity of the mean magnetic field for small Rm, but it does not affect the effective pumping velocity for large magnetic Reynolds numbers (Rogachevskii et al., 2018).

In addition to the algebraic nonlinearity, there is also a dynamic nonlinearity caused by an evolution of magnetic helicity density during the nonlinear stage of the dynamo. In particular, in a nonlinear stage of mean magnetic field evolution, the α effect is the sum of the kinetic and magnetic parts, where the kinetic part of the α effect is proportional to the kinetic helicity $H_{\rm u} = \langle \boldsymbol{u} \cdot (\boldsymbol{\nabla} \times \boldsymbol{u}) \rangle$ and the magnetic part of the α effect is proportional to the current helicity $H_{\rm c} = \langle \boldsymbol{b} \cdot (\boldsymbol{\nabla} \times \boldsymbol{b}) \rangle$ (Pouquet et al., 1976). The dynamics of the current helicity H_c are determined by the evolution of the small-scale magnetic helicity density $H_{\rm m} = \langle \boldsymbol{a} \cdot \boldsymbol{b} \rangle$, where $\boldsymbol{b} = \boldsymbol{\nabla} \times \boldsymbol{a}$. The total magnetic helicity, i.e., the sum of the magnetic helicity densities of the large-scale and small-scale fields, $H_{\rm M} + H_{\rm m}$, integrated over the volume, $\int (H_{\rm M} + H_{\rm m}) \, dr^3$, is conserved for very small microscopic magnetic diffusivity η. Here $H_{\rm M} = \overline{\boldsymbol{A}} \cdot \overline{\boldsymbol{B}}$ is the magnetic helicity density of the large-scale field $\overline{\boldsymbol{B}} = \boldsymbol{\nabla} \times \overline{\boldsymbol{A}}$.

The governing equation for magnetic helicity density $H_{\rm m}$ has been derived for an isotropic turbulence by Kleeorin and Ruzmaikin (1982) and for an arbitrary anisotropic turbulence by Kleeorin and Rogachevskii (1999). This equation has been used for the analytical study of solar dynamos (Kleeorin et al., 1994b, 1995), as well as for mean-field numerical modeling of solar and galactic dynamos (Covas et al., 1997, 1998; Kleeorin et al., 2000, 2002b, 2003a,b; Brandenburg and Subramanian, 2005a).

As the dynamo amplifies the large-scale magnetic field, the large-scale magnetic helicity density $H_{\rm M}$ grows in time. In particular, the evolution of the large-scale magnetic helicity density $H_{\rm M}$ is determined by the following equation:

$$\frac{\partial H_{\rm M}}{\partial t} + \nabla \cdot \boldsymbol{F}_{\rm M} = 2\boldsymbol{\mathcal{E}} \cdot \overline{\boldsymbol{B}} - 2\eta H_C \ , \tag{3.278}$$

where $\boldsymbol{\mathcal{E}} = \langle \boldsymbol{u} \times \boldsymbol{b} \rangle$ is the turbulent electromotive force that determines generation and dissipation of the large-scale magnetic field, $2\boldsymbol{\mathcal{E}} \cdot \overline{\mathbf{B}}$ is the source of $H_{\rm M}$ due to the dynamo generated large-scale magnetic field, and $\mathbf{F}_{\rm M}$ is the flux of magnetic helicity density of the large-scale field that determines its transport. Since the total magnetic helicity $\int (H_{\rm M} + H_{\rm m})\, dr^3$ is conserved for very small magnetic diffusivity, the magnetic helicity density $H_{\rm m}$ of the small-scale field changes during the dynamo action, and its evolution is determined by the dynamic equation (Kleeorin and Ruzmaikin, 1982; Zeldovich et al., 1983; Kleeorin et al., 1994b, 1995; Kleeorin and Rogachevskii, 1999)

$$\frac{\partial H_{\rm m}}{\partial t} + \nabla \cdot \boldsymbol{F}_{\rm m} = -2\boldsymbol{\mathcal{E}} \cdot \overline{\boldsymbol{B}} - 2\eta H_c \ , \tag{3.279}$$

where $-2\boldsymbol{\mathcal{E}} \cdot \overline{\boldsymbol{B}}$ is the source of $H_{\rm m}$ due to the dynamo generated large-scale magnetic field, $\mathbf{F}_{\rm m}$ is the flux of magnetic helicity density of the small-scale field that determines its transport and $2\eta H_c$ is the dissipation rate of $H_{\rm m}$. The source of the small-scale and large-scale magnetic helicity densities is only located in the turbulent region [see Eqs. (3.278) and (3.279)].

The characteristic decay time of the magnetic helicity density $H_{\rm m}$ of the small-scale field is of the order of $T_{\rm m} = \tau_0\, {\rm Rm}$, while the characteristic time for the decay of kinetic helicity is of the order of the turn over time $\tau_0 = \ell_0/u_0$ of turbulent eddies in the integral turbulence scale ℓ_0, where ${\rm Rm} = \ell_0\, u_0/\eta$ is the magnetic Reynolds number. The current helicity $H_{\rm c}$ is not an integral of motion and the characteristic decay time of $H_{\rm c}$ varies from a short timescale τ_0 to much larger timescales. On the other hand, the characteristic decay times of the current helicity of large-scale field, $H_C = \overline{\boldsymbol{B}} \cdot (\nabla \times \overline{\boldsymbol{B}})$, and of the large-scale magnetic helicity $H_{\rm M}$ are of the order of the turbulent diffusion time. For weakly inhomogeneous turbulence, the small-scale current helicity density $H_{\rm c}$ is proportional to the small-scale magnetic helicity density $H_{\rm m}$ (Kleeorin and Rogachevskii, 1999).

Using the steady-state solution of Eq. (3.279) with a zero flux $\mathbf{F}_{\rm m} = 0$ of the magnetic helicity density of a small-scale field, it has been concluded that the critical mean magnetic field strength, $\overline{B}_{\rm cr}$, at which the dynamic α quenching becomes significant, in fact is much lower than the equipartition value, e.g., $\overline{B}_{\rm cr} = \overline{B}_{\rm eq}\, {\rm Rm}^{-1/2}$ (Vainshtein and Cattaneo, 1992; Gruzinov and Diamond, 1994). In astrophysics, e.g., in galactic disks and in the convective zone of the sun, magnetic Reynolds numbers are very large, so it implies that dynamo action saturates at a magnetic field strength that is much lower than the equipartition value. This effect is referred to as a catastrophic quenching of the α effect. On the other hand, the observed large-scale field strengths in spiral galaxies is of the order of the equipartition value (Ruzmaikin et al., 1988).

The evolution of magnetic helicity appears, however, to be a more complicated process than can simply be described by a balance of magnetic helicity in a given volume. It is necessary to take into account fluxes of magnetic helicity (Kleeorin et al., 2000). This implies that the turbulent transport of magnetic helicity through the boundaries (the open boundary conditions in simulations) should be taken into account (Blackman and Field, 2000). Using this idea, it has been shown that a non-linear galactic dynamo governed by a dynamic equation for the magnetic helicity density $H_{\rm m}$ of a small-scale field with non-zero fluxes (the dynamical nonlinearity) results in a steady mean magnetic field comparable with the equipartition magnetic field (Kleeorin et al., 2000, 2002b, 2003a,b; Blackman and Brandenburg, 2002; Brandenburg and Subramanian, 2005a; Del Sordo et al., 2013). Different forms of magnetic helicity fluxes have been derived in various studies (Kleeorin and Rogachevskii, 1999; Kleeorin et al., 2000, 2002b; Vishniac and Cho, 2001; Subramanian and Brandenburg, 2004; Brandenburg and Subramanian, 2005a; Shukurov et al., 2006; Hubbard and Brandenburg, 2010a,b; Mitra et al., 2010). Numerical simulations demonstrate that the dynamics of magnetic helicity plays a crucial role in solar dynamo as well (Kleeorin et al., 2003a, 2016, 2020; Sokoloff et al., 2006; Zhang et al., 2006, 2012; Hubbard and Brandenburg, 2012; Del Sordo et al., 2013; Safiullin et al., 2018).

In a weakly ionized plasma, the Hall effect can be a dominant nonlinearity that causes a dynamo saturation. Examples of such flows include ionospheric plasma at low altitudes and laboratory plasma (Akasofu and Chapman, 1972; Galeev et al., 1983; Chen, 1984; Diamond et al., 2010). In particular, magnetic field observations in the ionosphere of Venus have been explained on the basis of the theory of magnetic field evolution where the Hall nonlinearity is a key mechanism of the dynamo saturation (Kleeorin et al., 1994a; Kleeorin and Rogachevskii, 1994b). The effect of different nonlinearities (Lorentz force and Hall effect) on the saturation of a magnetic field generated by a three-component plasma flow (ions, electrons and neutral particles) has been studied by means of numerical simulations (Galanti et al., 1994, 1995). The velocity field of the neutral particles is forced by a deterministic incompressible three-dimensional Arnold–Beltrami–Childress (ABC) flow with chaotic stream lines. Various properties of the ABC flow capable of generating a magnetic field have been studied in a number of publications (Arnol'd, 1965; Childress, 1970; Childress and Gilbert, 1995; Dombre et al., 1986; Galanti et al., 1992). It is notable that the Lorentz force yields a cubic nonlinearity in terms of the magnetic field in the induction equation, whereas the Hall effect is a quadratic nonlinearity. This basic difference is essential for the energy level at saturation of the magnetic field as well as its spatial structure. Four typical regimes of the nonlinear evolution of the magnetic field, corresponding to different types of nonlinearities (Lorentz force or Hall effect) and different types of collisions (ion-ion collisions or ion-neutral particle collisions), have been identified in numerical simulations (Galanti et al., 1994, 1995). The considered model represents weakly ionized plasma, where the effect of the charged component of the plasma on the motion of the neutral particles is negligibly small. In the numerical simulations, the structure of the saturated magnetic field, the transitions between various regimes and the evolution of the magnetic field have been studied (Galanti et al., 1994, 1995). Analytical scalings for the level of the saturated magnetic field and derived conditions for the different regimes of the magnetic field evolution are in agreement with the results of the numerical simulations. Note also that dynamo action in various deterministic flows with chaotic stream lines has been studied in a number of publications (Childress and Gilbert, 1995; Galloway and Frisch, 1986; Finn and Ott, 1988, 1990; Gilbert and Childress, 1990; Galloway and Proctor, 1992; Courvoisier et al., 2006; Rädler and Brandenburg, 2009; Devlen et al., 2013).

The mean-field approach can be also used for flows with fluctuating in time governing parameters (Pétrélis and Fauve, 2006; Kleeorin et al., 2009). In particular, the effect of velocity fluctuations on the generation of a magnetic field by a cellular flow of an electrically conducting fluid has been studied analytically by Pétrélis and Fauve (2006). Two cases, a Roberts flow subject to an amplitude modulation and the effect of phase fluctuations, are considered. The magnetic field can grow at large scale compared to the scale of the flow. This requires scale separations. In the case of a modulation of the flow amplitude, the dynamo onset shift is proportional to the product of the axial velocity modulation with the azimuthal one. The sign of the displacement depends on the phase difference between the modulations: if they act in phase, the dynamo threshold is lowered, while it is increased if they act out of phase. Fluctuations of the phase of the cellular flow always inhibit the dynamo process. However, the part of the alpha effect that drives the instability is lowered by phase fluctuations (Pétrélis and Fauve, 2006).

Amplitude fluctuating ABC and Roberts flows also have been studied analytically by Kleeorin et al. (2009). The derived mean-field dynamo equations for fluctuating ABC and Roberts flows have growing solutions with a growth rate of the large-scale magnetic field that is not controlled by molecular magnetic diffusivity. Three approaches are used in the derivation of the mean-field dynamo equations: (i) the quasi-linear approach for small Reynolds numbers (with the correlation time $\tau = \ell_0^2/\eta$); (ii) the path integral approach for the delta-correlated in time random velocity field; (iii) the tau approach for large fluid and magnetic Reynolds numbers. All these studies confirm the Zeldovich idea that the nonstationarity of the fluid flow may remove the obstacle in the large-scale dynamo action of classic stationary flows (Zeldovich et al., 1990). Note that stationary deterministic flows like the ABC flow with constant coefficients cannot cause a dynamo of the large-scale magnetic field, i.e., stationary flows cannot generate a magnetic field in scales that are larger than the characteristic scales of the fluid flow. In particular, the ABC flow with constant coefficients excites a magnetic field in the scales which are smaller than the scales of the fluid flow. The generation of the magnetic field by the stationary ABC flow requires the magnetic Reynolds numbers, which are larger than 10. On the other hand, stationary Roberts flow cannot generate magnetic field. In contrast, the generation of large-scale magnetic fields by random ABC or Roberts flows can be possible even for small magnetic Reynolds numbers (Kleeorin et al., 2009).

In a turbulence with a large-scale shear and vanishing mean kinetic helicity, there can be also a large-scale non-helical shear dynamo. Different mechanisms of the large-scale shear dynamos in turbulent flows have been investigated in a number of analytical studies (Sokolov, 1997; Silant'ev, 2000; Rogachevskii and Kleeorin, 2003, 2004; Kleeorin and Rogachevskii, 2008; Sridhar and Subramanian, 2009a,b; Sridhar and Singh, 2010; Singh and Sridhar, 2011; Sridhar and Singh, 2014; Singh, 2016) and numerical simulations (Vishniac and Brandenburg, 1997; Yousef et al., 2008a,b; Brandenburg et al., 2008; Singh and Jingade, 2015).

The kinetic helicity not only causes the generation of a large-scale magnetic field but also can result in the generation of large-scale vorticity in rotating inhomogeneous or density-stratified turbulence (Kleeorin and Rogachevskii, 2018). The large-scale nonuniform flow is also produced by a combined effect of a rotating incompressible turbulence and inhomogeneous kinetic helicity (Yokoi and Yoshizawa, 1993; Yokoi and Brandenburg, 2016). These effects result in the formation of a large-scale shear, and in turn, its interaction with the small-scale turbulence causes an excitation of the large-scale instability, known as a vorticity dynamo (Elperin et al., 2003, 2007b; Kleeorin and Rogachevskii, 2018; Yousef et al., 2008a; Käpylä et al., 2009). The latter is due to a combined effect of the large-scale shear and Reynolds stress-induced generation of the mean vorticity. The Reynolds stress-induced generation of the mean vorticity is caused by the effect of large-scale shear on the Reynolds stress (Elperin et al., 2003, 2007b). A fast rotation suppresses this large-scale instability.

Large-scale mean-field dynamo is also possible in plasmas with relativistic particles. In particular, an α effect is driven by the cosmic ray Bell instability (Bell, 2004, 2005) exciting left–right asymmetric turbulence. Theoretical and numerical studies (Rogachevskii et al., 2012) show that the dynamics of magnetic fields in plasmas with relativistic particles involve the following stages: (i) in the early stage, the small-scale Bell instability that results in a production of small-scale turbulence is excited; (ii) in the intermediate stage, the produced turbulence causes the formation of larger-scale magnetic structures due to the large-scale mean-field dynamo;

(iii) finally, the large-scale magnetic field is dissipated. These phenomena can be important for large-scale magnetic field amplification associated with cosmic ray production and diffusive shock acceleration in supernova remnants and blast waves from gamma-ray bursts. Magnetic field amplification by Bell turbulence in supernova remnants is found to be significant (Bykov et al., 2011; Rogachevskii et al., 2012; Beresnyak and Li, 2014).

Large-scale mean-field dynamo in plasmas with relativistic particles is also possible due to the chiral magnetic effect (Vilenkin, 1980) caused by an asymmetry between the number densities of left- and right-handed fermions, which can be maintained at high energies (Kharzeev, 2014; Kharzeev et al., 2016). The chiral magnetic effect results in laminar chiral dynamo instability, which is not related to fluid motions. This small-scale dynamo instability causes the production of small-scale turbulence with vanishing kinetic helicity (Rogachevskii et al., 2017; Brandenburg et al., 2017b; Schober et al., 2018, 2019, 2020). The produced turbulence results in large-scale dynamos caused by the chiral alpha effect, which is not related to the kinetic helicity and is dominant for large fluid and magnetic Reynolds numbers. The chiral alpha effect is due to an interaction of the chiral magnetic effect and fluctuations of the small-scale current produced by tangling magnetic fluctuations (Rogachevskii et al., 2017). These dynamo effects may have interesting consequences in the dynamics of the early universe, neutron stars and quark–gluon plasma.

3.8.2 Formation of Inhomogeneous Large-Scale Magnetic Fields

The solar magnetic fields obey the 11-year solar cyclic activity (Moffatt, 1978; Parker, 1979; Krause and Rädler, 1980; Zeldovich et al., 1983; Schrijver and Zwaan, 2008; Stix, 2012; Rüdiger et al., 2013; Ossendrijver, 2003; Brandenburg and Subramanian, 2005a; Jones et al., 2010). The observed solar activity is associated with strongly concentrated magnetic fields as sunspots with the characteristic spatial scale of the order of solar super-granulation, about 10^4 km (Thomas and Weiss, 2008; Weiss and Proctor, 2014). On the other hand, the mean-field dynamo generates smooth large-scale magnetic fields with the characteristic scale of the order of the solar radius (about 10^6 km). So it was not clear for many years how to relate the mean-field dynamo to sunspots.

There are different mechanisms for the formation of inhomogeneous magnetic structures, e.g., the magnetic buoyancy instability (or interchange instability) of a stratified continuous magnetic field (Newcomb, 1961; Parker, 1966; Gilman, 1970; Acheson, 1978), the magnetic flux expulsion (Weiss, 1966; Weiss and Proctor, 2014), the topological magnetic pumping (Drobyshevski and Yuferev, 1974) and the negative effective magnetic pressure instability (Kleeorin et al., 1989, 1990).

Magnetic buoyancy applies in the literature for different situations. The first mechanism corresponds to the magnetic buoyancy instability of stratified continuous magnetic field (Newcomb, 1961; Parker, 1966; Gilman, 1970; Acheson, 1978; Priest, 1982; Tobias and Hughes, 2004; Kersale et al., 2007), and magnetic flux tube concept is not used there. The second mechanism describes the buoyancy of discrete magnetic flux tubes discussed in different contexts in solar physics and astrophysics (Parker, 1955a; Spruit, 1981; Spruit and Van Ballegooijen, 1982; Schussler et al., 1994; Dikpati and Gilman, 2006). This is also related to the problem of the storage of magnetic fields in the overshoot layer near the bottom of the solar convective zone (Spiegel and Weiss, 1980; Tobias et al., 2001).

A source of free energy for the magnetic buoyancy instability is the energy of the gravitational field. The magnetic buoyancy instability is excited when the characteristic scale of original magnetic field variation is smaller than the density stratification height. Therefore, in a strongly density-stratified solar convective zone where the plasma density in radial direction varies by seven orders of magnitude, this instability is only excited when the initial magnetic field is already strongly nonuniform.

A universal mechanism of the formation of the nonuniform distribution of magnetic flux is associated with a magnetic-flux expulsion. The expulsion of magnetic flux from two-dimensional flows (a single vortex and a grid of vortices) has been demonstrated by Weiss (1966). In the context of solar and stellar convection, the topological asymmetry of stationary thermal convection plays a very important role in the magnetic field dynamics. In particular, topological magnetic pumping is caused by the topological asymmetry of the thermal convection (Drobyshevski and Yuferev, 1974). The fluid rises at the centers of the convective cells and falls at their peripheries. The ascending fluid elements (contrary to the descending fluid elements) are disconnected from one to another. This causes a topological magnetic pumping effect, allowing downward transport of the mean horizontal magnetic field to the bottom of a cell but impeding its upward return (Drobyshevski and Yuferev, 1974; Zeldovich et al., 1983; Galloway and Proctor, 1983).

Another mechanism of magnetic spot formation is related to the negative effective magnetic pressure instability (NEMPI), which can be excited even for uniform initial large-scale magnetic field (Kleeorin et al., 1989, 1990). The mechanism of this instability is based on the suppression of total (hydrodynamic and magnetic) turbulent pressure by a large-scale magnetic field. NEMPI can be understood as a negative contribution of turbulence to the effective mean magnetic pressure (the sum of non-turbulent and turbulent contributions). At large fluid and magnetic Reynolds numbers, this turbulent contribution becomes large and a large-scale instability can be excited, redistributing the magnetic flux produced by the mean-field dynamo. A source of free energy for the negative effective magnetic pressure instability is the energy of turbulence or turbulent convection. This phenomenon has been studied analytically using the mean-field approach (Kleeorin et al., 1993, 1996; Kleeorin and Rogachevskii, 1994a; Rogachevskii and Kleeorin, 2007) and numerically via mean-field simulations (Brandenburg et al., 2010, 2016; Kemel et al., 2012b, 2013a; Losada et al., 2012; Kemel et al., 2013b; Jabbari et al., 2013, 2014), large-eddy simulations (Brandenburg et al., 2014; Käpylä et al., 2012, 2016) and direct numerical simulations (Brandenburg et al., 2011a, 2012b, 2013b; Kemel et al., 2012a; Losada et al., 2013; Mitra et al., 2014; Warnecke et al., 2013, 2016; Jabbari et al., 2015, 2016, 2017).

Direct numerical simulations demonstrate that the effective mean magnetic pressure can be negative in any kind of turbulence, e.g., in non-stratified (Brandenburg et al., 2010) and stratified (Brandenburg et al., 2012b) isothermal turbulence, polytropic stably stratified turbulence (Losada et al., 2014), turbulent convection (Käpylä et al., 2012, 2016) and turbulence with an upper non-turbulent coronal layer (Warnecke et al., 2013, 2016; Losada et al., 2019). However, the actual instability (NEMPI) is excited only in stratified turbulence when the initial mean magnetic field is less than the equipartition field and there is enough scale separation between the integral turbulence scale and the characteristic scale of the mean magnetic field variations. For very large fluid and magnetic Reynolds numbers, NEMPI weakly dependent on the level of turbulence. NEMPI causes formation of magnetic spots (Brandenburg et al., 2013b, 2016) and bipolar regions (Warnecke et al., 2013, 2016; Losada et al., 2019).

In direct numerical simulations (Jabbari et al., 2016) in a two-layer model with a non-zero kinetic helicity in the lower layer and nonhelical turbulence in the upper layer, the magnetic field is found to develop strongly concentrated bipolar structures near the surface. They form elongated bands with a sharp interface between opposite polarities, where a current sheet is formed and magnetic reconnection of oppositely oriented field lines occurs. The reconnection rate in the numerical simulations has been determined by measuring either the inflow velocity in the vicinity of the current sheet or by measuring the electric field in the reconnection region. It has been demonstrated (Jabbari et al., 2016) that for large Lundquist numbers, $\mathrm{Lu} > 10^3$, the reconnection rate is nearly independent of the Lundquist number in agreement with results of direct numerical simulations performed in simpler settings where turbulent reconnection has been studied (Loureiro et al., 2009, 2012). Further steps toward more realistic modeling of the formation of magnetic spots and bipolar regions include replacing forced turbulence by self-consistently driven convective motions that are influenced by the radiative cooling at the surface together with partial ionization. Including more realistic physical processes at the solar surface might also help reproduce the surrounding sunspot structures.

The formation of magnetic spots and bipolar regions are related to observed solar activity. In view of observations, an important parameter of the solar activity is the Wolf number $W = 10g_{\rm w} + f_{\rm w}$ (Gibson, 1973; Stix, 2012), where $g_{\rm w}$ is the number of sunspot groups and $f_{\rm w}$ is the total number of sunspots in the visible part of the Sun. This parameter has been measured over the course of three centuries. To predict solar activity, a nonlinear mean-field dynamo model and a budget equation for dynamics of Wolf numbers have been recently used (Kleeorin et al., 2016, 2020; Safiullin et al., 2018). This dynamo model takes into account the algebraic and dynamic nonlinearities of the alpha effect, where the equation for dynamic nonlinearity is derived from the conservation law for the magnetic helicity. The budget equation for the evolution of the Wolf number is based on a formation mechanism of sunspots related to the negative effective magnetic pressure instability. This instability redistributes the magnetic flux produced by the mean-field dynamo. To predict solar activity on the time scale of one month, a method based on a combination of the numerical solution of the nonlinear mean-field dynamo equations and the artificial neural network has been used. A comparison of the results of the prediction of the solar activity with the observed Wolf numbers demonstrates a good agreement between the forecast and observations (Safiullin et al., 2018).

3.8.3 Small-Scale Dynamo

There are two kinds of turbulent dynamos, large-scale and small-scale dynamos. The large-scale mean-field dynamo implies that the amplification of the magnetic field occurs at scales that are much larger than the maximum scale of turbulent motions. Self-excitation of magnetic fluctuations with a zero mean magnetic field is called a small-scale dynamo (or fluctuation dynamo), and generation of magnetic fluctuations occurs at scales that are smaller than the maximum scale of turbulent motions (Zeldovich et al., 1990; Childress and Gilbert, 1995; Kulsrud and Anderson, 1992).

The mechanisms of the small-scale dynamo action are different depending on magnetic Prandtl numbers $\mathrm{Pm} = \nu/\eta$ (Batchelor, 1950; Moffatt, 1978), where ν is the kinematic viscosity of the fluid and η is the magnetic diffusion due to the electrical conductivity of the fluid. For large magnetic Prandtl numbers, the self-excitation of magnetic fluctuations is caused by the random stretching of the magnetic field by smooth velocity field. This type of dynamo has been comprehensively studied analytically (Zeldovich et al., 1990; Kleeorin and Rogachevskii, 1994b; Kleeorin et al., 1994a, 2002a; Bhat and Subramanian, 2014), in direct numerical simulations of forced turbulence (Schekochihin et al., 2004; Haugen et al., 2004; Brandenburg and Subramanian, 2005a; Federrath et al., 2014; Federrath, 2016) and in turbulent convection (Brandenburg et al., 1996; Cattaneo, 1999).

The nature of small-scale dynamo for low magnetic Prandtl numbers is different, e.g., it is driven by velocity fluctuations at the resistive scale that is located in the inertial-subrange of scales. The small-scale dynamo at low magnetic Prandtl numbers has been studied analytically (Kazantsev, 1968; Rogachevskii and Kleeorin, 1997, 1999; Vincenzi, 2002; Boldyrev and Cattaneo, 2004; Arponen and Horvai, 2007; Kleeorin and Rogachevskii, 2012; Afonso et al., 2019) for a Gaussian white-noise velocity field (the so-called Kraichnan-Kazantsev model), and in direct numerical simulations (Schekochihin et al., 2005, 2007; Iskakov et al., 2007; Brandenburg, 2011; Federrath et al., 2014; Federrath, 2016). Since the magnetic energy is not conserved, the second moment of a magnetic field has anomalous scaling (Vergassola, 1996; Rogachevskii and Kleeorin, 1997).

The small-scale dynamo instability is excited when the magnetic Reynolds number, Rm, is larger than the critical magnetic Reynolds number, $\mathrm{Rm}^{\mathrm{cr}}$. Analytical models based on the Kraichnan-Kazantsev model of a homogeneous, isotropic, non-helical and incompressible velocity field yield $\mathrm{Rm}^{\mathrm{cr}} \approx 410$ at very low magnetic Prandtl numbers, and compressibility of fluid flow causes a strong increase of the critical magnetic Reynolds number at $\mathrm{Pm} \ll 1$ (Rogachevskii and Kleeorin, 1997; Afonso et al., 2019). Direct numerical simulations of small-scale dynamo of the Navier-Stokes turbulence (Schekochihin et al., 2007; Iskakov et al., 2007; Brandenburg, 2011) show that $\mathrm{Rm}^{\mathrm{cr}}$ is around 200 for small magnetic Prandtl numbers, and it is at three times larger than for the small-scale dynamo at large and moderate Prandtl numbers (Schekochihin et al., 2004; Haugen et al., 2004; Schekochihin et al., 2005). These numerical results at large and moderate Prandtl numbers are in agreement with different analytical models (Zeldovich et al., 1990; Kleeorin and Rogachevskii, 1994b; Kleeorin et al., 1994a; Boldyrev and Cattaneo, 2004).

Small-scale dynamos are ubiquitous in a broad range of turbulent flows with large-scale shear, ranging from solar and galactic magnetism to accretion disks, cosmology, and structure formation. As follows from high-resolution direct numerical simulations (Singh et al., 2017), large-scale shear supports small-scale dynamo action in nonhelically forced turbulence with zero mean magnetic field, i.e., the dynamo growth rate increases with shear and shear enhances or even produces turbulence, which, in turn, further increases the dynamo growth rate. When the production rates of turbulent kinetic energy due to shear and forcing are comparable, the growth rate γ of the small-scale dynamo and the turbulent rms velocity u_{rms} increase with shear rate S independently of the magnetic Prandtl number as $\gamma \propto |S|$ and $u_{\mathrm{rms}} \propto |S|^{2/3}$. For large fluid and magnetic Reynolds numbers, γ, normalized by its shear-free value, depends only on shear (Singh et al., 2017).

4

Analysis Based on Budget Equations

In this chapter, we derive budget equations for turbulent kinetic and potential energy densities and turbulent heat flux. We perform analyses based on these budget equations to investigate the turbulent transport of the momentum and temperature in non-stratified isotropic homogeneous and incompressible turbulence, as well as in stably stratified shear-produced turbulence, as studied in Boussinesq approximation.

4.1 Budget Equations for Non-Stratified Turbulence

In this section, we derive budget equations for and perform an analysis of a nonstratified isotropic homogeneous and incompressible turbulence.

4.1.1 Budget Equation for Turbulent Kinetic Energy Density

First, we derive a budget equation for turbulent kinetic energy density. The evolution of fluid velocity fluctuations $\boldsymbol{u}$ in incompressible flow is determined by the Navier-Stokes equation:

$$\rho \left(\frac{\partial \boldsymbol{u}}{\partial t} + (\boldsymbol{u} \cdot \nabla)\boldsymbol{u} \right) = -\nabla P + \rho \, \nu \, \Delta \boldsymbol{u} + \rho \boldsymbol{f}, \tag{4.1}$$

where ν is the kinematic viscosity, $\rho \boldsymbol{f}$ is the external force that produces a turbulent flow and P and ρ are the fluid pressure and density, respectively. For simplicity, we consider the case of a zero mean velocity. We multiply Eq. (4.1) by $\boldsymbol{u}$, which yields

$$\rho \left(\boldsymbol{u} \cdot \frac{\partial \boldsymbol{u}}{\partial t} + \boldsymbol{u} \cdot (\boldsymbol{u} \cdot \nabla)\boldsymbol{u} \right) = -(\boldsymbol{u} \cdot \nabla) P + \rho \, \nu \, \boldsymbol{u} \cdot \Delta \boldsymbol{u} + \rho \, \boldsymbol{u} \cdot \boldsymbol{f}. \tag{4.2}$$

To derive the budget equation for the turbulent kinetic energy density, we use the following identities:

$$\frac{\partial}{\partial t}\left(\frac{\rho\, \boldsymbol{u}^2}{2}\right) = \rho \boldsymbol{u} \cdot \frac{\partial \boldsymbol{u}}{\partial t}, \tag{4.3}$$

$$\begin{aligned} \text{div}\left(\boldsymbol{u}\frac{\rho\, \boldsymbol{u}^2}{2}\right) &= \boldsymbol{\nabla} \cdot \left(\boldsymbol{u}\frac{\rho\, \boldsymbol{u}^2}{2}\right) = \frac{\rho\, \boldsymbol{u}^2}{2}\boldsymbol{\nabla} \cdot \boldsymbol{u} + (\boldsymbol{u} \cdot \boldsymbol{\nabla})\left(\frac{\rho\, \boldsymbol{u}^2}{2}\right) \\ &= \frac{\rho}{2}(\boldsymbol{u} \cdot \boldsymbol{\nabla})\boldsymbol{u}^2 = \rho \boldsymbol{u} \cdot (\boldsymbol{u} \cdot \boldsymbol{\nabla})\boldsymbol{u}, \end{aligned} \tag{4.4}$$

$$\text{div}\,(\boldsymbol{u}\, P) = \boldsymbol{\nabla} \cdot (\boldsymbol{u}\, P) = P\boldsymbol{\nabla} \cdot \boldsymbol{u} + (\boldsymbol{u} \cdot \boldsymbol{\nabla})P = (\boldsymbol{u} \cdot \boldsymbol{\nabla})P, \tag{4.5}$$

$$\begin{aligned} \boldsymbol{u} \cdot \Delta \boldsymbol{u} &= u_i \nabla_j \nabla_j u_i = \nabla_j \left(u_i \nabla_j u_i\right) - \left(\nabla_j u_i\right)\left(\nabla_j u_i\right) \\ &= \nabla_j \left(\nabla_j \frac{\boldsymbol{u}^2}{2}\right) - \left(\nabla_j u_i\right)^2 = \text{div}\left(\boldsymbol{\nabla}\frac{\boldsymbol{u}^2}{2}\right) - \left(\nabla_j u_i\right)^2. \end{aligned} \tag{4.6}$$

Using Eqs. (4.3)–(4.6), we rewrite Eq. (4.2) as

$$\frac{\partial}{\partial t}\left(\frac{\rho\, \boldsymbol{u}^2}{2}\right) + \text{div}\left[\boldsymbol{u}\left(\frac{\rho\, \boldsymbol{u}^2}{2} + P\right) - \boldsymbol{\nabla}\left(\nu\frac{\rho\, \boldsymbol{u}^2}{2}\right)\right] = -\nu\, \rho\, \left(\nabla_j u_i\right)^2 + \rho\, \boldsymbol{u} \cdot \boldsymbol{f}. \tag{4.7}$$

We average Eq. (4.7) over an ensemble of turbulent velocity field. This yields the budget equation for the density of turbulent kinetic energy (TKE), $E_K = \rho\, \langle \boldsymbol{u}^2 \rangle/2$ (Kolmogorov, 1942),

$$\frac{\partial E_K}{\partial t} + \text{div}\, \boldsymbol{\Phi}_K = -\varepsilon_K + \Pi_K, \tag{4.8}$$

where

$$\boldsymbol{\Phi}_K = \left\langle \boldsymbol{u}\left(\frac{\rho\, \boldsymbol{u}^2}{2} + P\right)\right\rangle - \nu\, \boldsymbol{\nabla}\left(\frac{\rho\, \langle \boldsymbol{u}^2 \rangle}{2}\right) \tag{4.9}$$

is the flux of TKE,

$$\varepsilon_K = \nu\, \rho\, \langle (\nabla_j u_i)^2 \rangle \tag{4.10}$$

is the dissipation rate of TKE and $\Pi_K = \rho\, \langle \boldsymbol{u} \cdot \boldsymbol{f} \rangle$ is the production rate of TKE caused by the external force.

To calculate the dissipation rate of TKE in homogeneous isotropic and incompressible turbulence, we determine the following correlation function using the multi-scale approach (see Section 1.4):

$$\langle(\nabla_j u_i)^2\rangle = \lim_{x\to y}\langle[\nabla_j u_i(t,x)]\,[\nabla_j u_i(t,y)]\rangle = \int\langle(\nabla_j u_i)_k(\nabla_j u_i)_{-k}\rangle\,dk$$
$$= \int\langle[\mathrm{i}\,k_j u_i(t,k)]\,[-\mathrm{i}\,k_j u_i(t,-k)]\rangle\,dk = \int k^2\langle u_i(t,k)u_i(t,-k)\rangle\,dk. \tag{4.11}$$

We use the following model for the second moments, $\langle u_i(t,\boldsymbol{k})\,u_j(t,-\boldsymbol{k})\rangle$, of the turbulent velocity field for an isotropic homogeneous and incompressible turbulence in a Fourier space:

$$\langle u_i(t,\boldsymbol{k})\,u_j(t,-\boldsymbol{k})\rangle = \frac{\langle \boldsymbol{u}^2\rangle\,E(k)}{8\pi k^2}\left[\delta_{ij} - \frac{k_i\,k_j}{k^2}\right], \tag{4.12}$$

where δ_{ij} is the Kronecker tensor, the Kolmogorov energy spectrum function is $E(k) = k_0^{-1}\,(2/3)\,(k/k_0)^{-5/3}$, $k_0 \le k \le k_\nu$, the wave number $k_0 = 1/\ell_0$, the length ℓ_0 is the integral scale of turbulent motions, the wave number $k_\nu = \ell_\nu^{-1}$ and $\ell_\nu = \ell_0\mathrm{Re}^{-3/4}$ is the Kolmogorov (viscous) scale. The trace $\langle u_i(t,\boldsymbol{k})\,u_i(t,-\boldsymbol{k})\rangle$ of the velocity tensor (4.12) is

$$\langle u_i(t,\boldsymbol{k})\,u_i(t,-\boldsymbol{k})\rangle = \frac{\langle \boldsymbol{u}^2\rangle\,E(k)}{4\pi k^2}, \tag{4.13}$$

where we take into account that $\delta_{ii} = \delta_{11} + \delta_{22} + \delta_{33} = 3$. Substituting Eq. (4.13) into Eq. (4.11) and using Eq. (4.10), we obtain

$$\varepsilon_K = \nu\,\rho\,\frac{\langle \boldsymbol{u}^2\rangle}{4\pi}\int_{k_0}^{k_\nu} k^2\,E(k)\,dk\int_0^{2\pi} d\varphi\int_0^{\pi}\sin\vartheta\,d\vartheta. \tag{4.14}$$

For integration in the $\boldsymbol{k}$ space of Eq. (4.14), we use the following integrals:

$$\int_0^{2\pi} d\varphi\int_0^{\pi}\sin\vartheta\,d\vartheta = 4\pi, \tag{4.15}$$

$$\int_{k_0}^{k_\nu} k^2\,E(k)\,dk = \frac{\mathrm{Re}}{2\ell_0^2}. \tag{4.16}$$

Substituting Eq. (4.15) and (4.16) into Eq. (4.14), we obtain an expression for the dissipation rate of the density of turbulent kinetic energy in homogeneous isotropic and incompressible turbulence:

$$\varepsilon_K = \frac{E_K}{\tau_0}. \tag{4.17}$$

4.1.2 Budget Equation for the Second Moment of Temperature Fluctuations

In this section, we derive a budget equation for the second moment of temperature fluctuations, $E_\theta = \langle\theta^2\rangle/2$. The equation for temperature fluctuations is obtained by subtracting the mean-field equation from the equation for the total temperature:

$$\frac{\partial \theta}{\partial t} + (\boldsymbol{u} \cdot \nabla)\theta - \langle (\boldsymbol{u} \cdot \nabla)\theta \rangle - D^{(\theta)} \Delta \theta = -(\boldsymbol{u} \cdot \nabla)\overline{T}, \tag{4.18}$$

where $\overline{T}$ is the mean fluid temperature, θ are temperature fluctuations and $D^{(\theta)}$ is the molecular diffusion coefficient. We multiply Eq. (4.18) by θ,

$$\theta \frac{\partial \theta}{\partial t} + \theta(\boldsymbol{u} \cdot \nabla)\theta - \theta\langle (\boldsymbol{u} \cdot \nabla)\theta \rangle - D^{(\theta)} \theta \Delta \theta = -\theta(\boldsymbol{u} \cdot \nabla)\overline{T}, \tag{4.19}$$

and use the following identities:

$$\theta \frac{\partial \theta}{\partial t} = \frac{\partial}{\partial t}\left(\frac{\theta^2}{2}\right), \tag{4.20}$$

$$\theta\, (\boldsymbol{u} \cdot \nabla)\theta = (\boldsymbol{u} \cdot \nabla)\frac{\theta^2}{2} = \operatorname{div}\left(\boldsymbol{u}\, \frac{\theta^2}{2}\right), \tag{4.21}$$

$$\begin{aligned} \theta \Delta \theta &= \theta \nabla_j \nabla_j \theta = \nabla_j \left(\theta \nabla_j \theta\right) - \left(\nabla_j \theta\right)^2 = \nabla_j \left(\nabla_j \frac{\theta^2}{2}\right) - \left(\nabla_j \theta\right)^2 \\ &= \operatorname{div}\left(\nabla \frac{\theta^2}{2}\right) - (\nabla \theta)^2 . \end{aligned} \tag{4.22}$$

Using Eqs. (4.20)–(4.22), we rewrite Eq. (4.19) in the following form:

$$\frac{\partial}{\partial t}\left(\frac{\theta^2}{2}\right) + \operatorname{div}\left(\boldsymbol{u}\,\frac{\theta^2}{2} - D^{(\theta)} \nabla \frac{\theta^2}{2}\right) - \theta\langle (\boldsymbol{u} \cdot \nabla)\theta \rangle = -D^{(\theta)}\, (\nabla \theta)^2 - (\boldsymbol{u}\theta \cdot \nabla)\overline{T}. \tag{4.23}$$

We average Eq. (4.23) over an ensemble of turbulent velocity field, which yields the budget equation for the second moment E_θ of temperature fluctuations as

$$\frac{\partial E_\theta}{\partial t} + \operatorname{div} \boldsymbol{\Phi}_\theta = -\varepsilon_\theta + \Pi_\theta, \tag{4.24}$$

where

$$\boldsymbol{\Phi}_\theta = \frac{\langle \boldsymbol{u}\, \theta^2 \rangle}{2} - D^{(\theta)} \nabla \frac{\langle \theta^2 \rangle}{2} \tag{4.25}$$

is the flux of E_θ,

$$\varepsilon_\theta = D^{(\theta)} \left\langle (\nabla \theta)^2 \right\rangle \tag{4.26}$$

is the dissipation rate of E_θ,

$$\Pi_\theta = -\left(\langle \boldsymbol{u}\theta \rangle \cdot \nabla\right) \overline{T} = -\left(\boldsymbol{F}^{(\theta)} \cdot \nabla\right) \overline{T} \tag{4.27}$$

is the production rate of E_θ and $\boldsymbol{F}^{(\theta)} = \langle \boldsymbol{u}\theta \rangle$ is the turbulent heat flux. Here we take into account that $\langle \theta\, \langle (\boldsymbol{u} \cdot \nabla)\theta \rangle \rangle = \langle \theta \rangle\, \langle (\boldsymbol{u} \cdot \nabla)\theta \rangle = 0$.

To calculate the dissipation rate of E_θ in a homogeneous isotropic and incompressible turbulence, we determine the following correlation function using the multi-scale approach:

$$\begin{aligned}\langle(\nabla_j\theta)^2\rangle &= \lim_{x\to y}\Big\langle\Big[\nabla_j\theta(t,\boldsymbol{x})\Big]\Big[\nabla_j\theta(t,\boldsymbol{y})\Big]\Big\rangle = \int \langle(\nabla_j\theta)_{\boldsymbol{k}}(\nabla_j\theta)_{-\boldsymbol{k}}\rangle\,d\boldsymbol{k}\\ &= \int \Big\langle\Big[\mathrm{i}\,k_j\theta(t,\boldsymbol{k})\Big]\Big[-\mathrm{i}\,k_j\theta(t,-\boldsymbol{k})\Big]\Big\rangle d\boldsymbol{k} = \int k^2\langle\theta(t,\boldsymbol{k})\theta(t,-\boldsymbol{k})\rangle\,d\boldsymbol{k}.\end{aligned} \tag{4.28}$$

We use the following model for the second moments, $\langle\theta(t,\boldsymbol{k})\,\theta(t,-\boldsymbol{k})\rangle$, of temperature fluctuations in an isotropic homogeneous and incompressible turbulence in a Fourier space:

$$\langle\theta(t,\boldsymbol{k})\,\theta(t,-\boldsymbol{k})\rangle = \frac{\langle\theta^2\rangle\,\tilde{E}_\theta(k)}{4\pi k^2}, \tag{4.29}$$

where the spectrum function is $\tilde{E}_\theta(k) = k_0^{-1}\,(2/3)\,(k/k_0)^{-5/3}$ (see Section 1.2.2), $k_0 \le k \le k_D$, the wave number $k_0 = 1/\ell_0$, the length ℓ_0 is the integral scale, the wave number $k_D = \ell_D^{-1}$ and the length $\ell_D = \ell_0\mathrm{Pe}^{-3/4}$ is the diffusion scale. Substituting Eq. (4.29) into Eq. (4.28) and using Eq. (4.26), we obtain

$$\varepsilon_\theta = D^{(\theta)}\,\frac{\langle\theta^2\rangle}{4\pi}\int_{k_0}^{k_D} k^2\,\tilde{E}_\theta(k)\,dk\int_0^{2\pi}d\varphi\int_0^{\pi}\sin\vartheta\,d\vartheta. \tag{4.30}$$

For the integration in the $\boldsymbol{k}$ space of Eq. (4.30), we use Eq. (4.15) and

$$\int_{k_0}^{k_D} k^2\,\tilde{E}_\theta(k)\,dk = \frac{\mathrm{Pe}}{2\ell_0^2}. \tag{4.31}$$

This yields the dissipation rate of E_θ in a homogeneous isotropic and incompressible turbulence:

$$\varepsilon_\theta = \frac{E_\theta}{\tau_0}. \tag{4.32}$$

4.1.3 Budget Equation for Turbulent Heat Flux

In this section, we derive a budget equation for the turbulent heat flux. We multiply Eq. (4.1) by θ and the equation for temperature fluctuations (4.18) by $\boldsymbol{u}$, and add them. This yields

$$\begin{aligned}\theta\frac{\partial u_i}{\partial t} + u_i\frac{\partial\theta}{\partial t} + \theta\nabla_j u_j u_i + u_i\nabla_j u_j\theta - u_i\langle(\boldsymbol{u}\cdot\nabla)\theta\rangle &= -\frac{(\theta\nabla_i)P}{\rho}\\ + \nu\,\theta\Delta u_i + D^{(\theta)}u_i\Delta\theta + \rho^{-1}\,\theta f_i - u_i u_j\nabla_i\overline{T}.\end{aligned} \tag{4.33}$$

To derive the budget equation for turbulent heat flux, we use the following identities:

$$\theta \frac{\partial u_i}{\partial t} + u_i \frac{\partial \theta}{\partial t} = \frac{\partial}{\partial t}(\theta u_i), \tag{4.34}$$

$$\theta \nabla_j u_j u_i + u_i \nabla_j u_j \theta = \nabla_j \left(u_i u_j \theta\right) = \mathrm{div}\,(\boldsymbol{u}\, u_i \theta)\,, \tag{4.35}$$

$$\begin{aligned} \nu\, \theta \Delta u_i + D^{(\theta)} u_i \Delta \theta &= \nu\, \theta \nabla_j \nabla_j u_i + D^{(\theta)} u_i \nabla_j \nabla_j \theta = \nu\, \nabla_j \left(\theta \nabla_j u_i\right) \\ &\quad - \nu\, (\nabla_j \theta)(\nabla_j u_i) + D^{(\theta)} \nabla_j \left(u_i \nabla_j \theta\right) - D^{(\theta)} (\nabla_j u_i)(\nabla_j \theta) \\ &= \mathrm{div}\left(\nu \theta \boldsymbol{\nabla} u_i + D^{(\theta)} u_i \boldsymbol{\nabla} \theta\right) - \left(\nu + D^{(\theta)}\right) (\nabla_j \theta)(\nabla_j u_i). \end{aligned} \tag{4.36}$$

Using Eqs. (4.34)–(4.36), we rewrite Eq. (4.33) and average it over an ensemble of turbulent velocity field. This yields the budget equation for the turbulent heat flux $\boldsymbol{F}^{(\theta)} = \langle \boldsymbol{u} \theta \rangle$ as

$$\frac{\partial F_i^{(\theta)}}{\partial t} + \mathrm{div}\, \boldsymbol{\Phi}_{ij}^{(F)} = -\varepsilon_i^{(F)} + \Pi_i^{(F)}, \tag{4.37}$$

where

$$\boldsymbol{\Phi}_{ij}^{(F)} = \langle u_i u_j \theta \rangle - \nu \langle \theta \nabla_j u_i \rangle - D^{(\theta)} \langle u_i \nabla_j \theta \rangle \tag{4.38}$$

is the flux of $F_i^{(\theta)}$,

$$\varepsilon_i^{(F)} = \left(\nu + D^{(\theta)}\right) \langle (\nabla_j \theta)(\nabla_j u_i) \rangle \tag{4.39}$$

is the dissipation rate of $F_i^{(\theta)}$, and

$$\Pi_i^{(F)} = -\langle u_i u_j \rangle \nabla_j \overline{T} - \rho^{-1} \langle \theta \nabla_i P \rangle + \rho^{-1} \langle \theta f_i \rangle \tag{4.40}$$

is the production rate of $F_i^{(\theta)}$. Here we take into account that $\langle u_i \, \langle (\boldsymbol{u} \cdot \boldsymbol{\nabla}) \theta \rangle \rangle = \langle u_i \rangle \, \langle (\boldsymbol{u} \cdot \boldsymbol{\nabla}) \theta \rangle = 0$.

To calculate the dissipation rate of the turbulent heat flux in homogeneous isotropic and incompressible turbulence, we determine the following correlation function using the multi-scale approach:

$$\begin{aligned} \langle (\nabla_j \theta)(\nabla_j u_i) \rangle &= \lim_{\boldsymbol{x} \to \boldsymbol{y}} \left\langle \left[\nabla_j u_i(t, \boldsymbol{x})\right] \left[\nabla_j \theta(t, \boldsymbol{y})\right] \right\rangle = \int \langle (\nabla_j u_i)_{\boldsymbol{k}} (\nabla_j \theta)_{-\boldsymbol{k}} \rangle \, d\boldsymbol{k} \\ &= \int \left\langle \left[\mathrm{i}\, k_j u_i(t, \boldsymbol{k})\right] \left[-\mathrm{i}\, k_j \theta(t, -\boldsymbol{k})\right] \right\rangle d\boldsymbol{k} = \int k^2 \langle u_i(t, \boldsymbol{k}) \theta(t, -\boldsymbol{k}) \rangle \, d\boldsymbol{k}. \end{aligned} \tag{4.41}$$

We use the following model for the second moments, $\langle u_i(t, \boldsymbol{k})\, \theta(t, -\boldsymbol{k}) \rangle$, in an isotropic, homogeneous and incompressible turbulence in a Fourier space:

$$\langle u_i(t, \boldsymbol{k})\, \theta(t, -\boldsymbol{k}) \rangle = \frac{\langle u_i \theta \rangle \, E_F(k)}{4\pi k^2}, \tag{4.42}$$

where the spectrum function of $F_i^{(\theta)}$ is $E_F(k) = k_0^{-1}\,(2/3)\,(k/k_0)^{-5/3}$, $k_0 \le k \le k_\nu$, the wave number $k_0 = 1/\ell_0$, the length ℓ_0 is the integral scale, the wave number $k_\nu = \ell_\nu^{-1}$ and the length $\ell_\nu = \ell_0 \mathrm{Re}^{-3/4}$ is the viscous scale. Substituting Eq. (4.42) into Eq. (4.41) and using Eq. (4.39), we arrive at the following equation:

$$\varepsilon_i^{(F)} = \left(\nu + D^{(\theta)}\right) \frac{\langle u_i \theta \rangle}{4\pi} \int_{k_0}^{k_\nu} k^2\, E_F(k)\, dk \int_0^{2\pi} d\varphi \int_0^{\pi} \sin\vartheta\, d\vartheta. \tag{4.43}$$

For the integration in the $\boldsymbol{k}$ space of Eq. (4.43), we use Eq. (4.15) and

$$\int_{k_0}^{k_\nu} k^2\, E_F(k)\, dk = \frac{\mathrm{Re}}{2\ell_0^2}. \tag{4.44}$$

This yields the dissipation rate of the turbulent heat flux in a homogeneous isotropic and incompressible turbulence as

$$\varepsilon_i^{(F)} = \left(1 + \mathrm{Pr}^{-1}\right) \frac{F_i^{(\theta)}}{2\tau_0}. \tag{4.45}$$

4.1.4 Analysis for Non-Stratified Stationary Homogeneous and Isotropic Turbulence

We consider steady-state solutions for the budget equations for the second moments E_K, E_θ and $F_i^{(\theta)}$ for homogeneous turbulence. This implies that the left-hand sides of the budget equations (4.8), (4.24) and (4.37) are assumed to be zero, so that the second moments E_K, E_θ and $F_i^{(\theta)}$ are given by

$$E_K = \tau_0\, \rho\, \langle \boldsymbol{u} \cdot \boldsymbol{f} \rangle, \tag{4.46}$$

$$E_\theta = -\tau_0 \left(\boldsymbol{F}^{(\theta)} \cdot \boldsymbol{\nabla}\right) \overline{T}, \tag{4.47}$$

$$F_i^{(\theta)} = -D_{ij}^{(T)}\, \nabla_j \overline{T}, \tag{4.48}$$

where

$$D_{ij}^{(T)} = \frac{2\tau_0 \langle u_i u_j \rangle}{\left(1 + \mathrm{Pr}^{-1}\right)} \tag{4.49}$$

is the turbulent diffusion tensor. To obtain Eqs. (4.46)–(4.48), we use the dissipation rates of E_K, E_θ and $F_i^{(\theta)}$ given by Eqs. (4.17), (4.32) and (4.45), respectively. We also assume that the first term in the production rate of the turbulent heat flux is much larger than other terms in Eq. (4.40). For isotropic turbulence,

$\langle u_i u_j \rangle = \langle \boldsymbol{u}^2 \rangle \, \delta_{ij}/3$, so that the turbulent diffusion tensor is given by $D_{ij}^{(T)} = D_T \, \delta_{ij}$, where the turbulent diffusion coefficient D_T is

$$D_T = \frac{2\tau_0 \langle \boldsymbol{u}^2 \rangle}{3\left(1 + \mathrm{Pr}^{-1}\right)}. \tag{4.50}$$

The turbulent heat flux $\boldsymbol{F}^{(\theta)}$ and the second moment E_θ of temperature fluctuations for an isotropic turbulence are given by

$$\boldsymbol{F}^{(\theta)} = -D_T \nabla \overline{T}, \tag{4.51}$$

$$E_\theta = \tau_0 D_T \left(\nabla \overline{T}\right)^2 = \frac{C_D}{2} \, \ell_0^2 \left(\nabla \overline{T}\right)^2, \tag{4.52}$$

so that the level of temperature fluctuations is estimated as

$$\langle \theta^2 \rangle = C_D \ell_0^2 \left(\nabla \overline{T}\right)^2, \qquad C_D = \frac{4}{3\left(1 + \mathrm{Pr}^{-1}\right)}. \tag{4.53}$$

Equations (4.50)–(4.53) are in agreement with those derived by the dimensional analysis and the tau approach for large Péclet and Reynolds numbers (see Chapter 1).

4.2 Budget Equations for Stably Stratified Turbulence

In this section, we derive budget equations and perform analysis for stably stratified turbulence using the Boussinesq approximation.

4.2.1 Boussinesq Approximation

In compressible flows, the following four cases can be considered:

- **Compressible flow**. For a compressible fluid flow, the fluid density is determined by the continuity equation:

$$\frac{\partial \rho}{\partial t} + \nabla \cdot (\rho \, \boldsymbol{u}) = 0. \tag{4.54}$$

  The first term on the left-hand side of Eq. (4.54), $|\partial \rho / \partial t| \sim \mathrm{Ma}^2$, where $\mathrm{Ma} = u/c_s$ is the Mach number and c_s is the sound speed. Therefore, Eq. (4.54) should be used when the Mach number is not small.
- **Anelastic approximation**. When the Mach number is small and the fluid density is nonuniform, the anelastic approximation can be used. In the framework of the anelastic approximation, the term $\partial \rho / \partial t$ in Eq. (4.54) is neglected, so that the continuity equation reads

$$\nabla \cdot (\rho \, \boldsymbol{u}) = 0, \tag{4.55}$$

where the fluid density $\rho(\boldsymbol{r})$ depends on the spatial coordinate. This approximation excludes the sound waves from the consideration.

- **Incompressible flow**. When the Mach number is small and the fluid density ρ is independent of the spatial coordinates, the continuity equation is reduced to

$$\nabla \cdot \boldsymbol{u} = 0, \tag{4.56}$$

which corresponds to an incompressible flow.

- **Boussinesq approximation.**[1] The fluid velocity field $\boldsymbol{u}$ in a gravity field is determined by the Navier-Stokes equation:

$$\frac{\partial \boldsymbol{u}}{\partial t} + (\boldsymbol{u} \cdot \nabla)\boldsymbol{u} = -\frac{\nabla P}{\rho} + \nu \, \Delta \boldsymbol{u} + \boldsymbol{g}, \tag{4.57}$$

where $\boldsymbol{g}$ is the acceleration caused by the gravity field. The basic reference state for Eq. (4.57) with $\boldsymbol{u} = 0$ is the hydrostatic equilibrium:

$$\nabla P_0(z) = \rho_0(z) \, \boldsymbol{g}, \tag{4.58}$$

where P_0 and ρ_0 are the fluid pressure and density in the basic reference state, respectively, and z is the vertical coordinate. We consider a small deviation from the basic reference state caused by slow motions $\boldsymbol{u}$, i.e., $P = P_0 + \tilde{P}$, $\rho = \rho_0 + \tilde{\rho}$, where $|\tilde{P}| \ll P_0$ and $|\tilde{\rho}| \ll \rho_0$. We now determine $\nabla P/\rho$:

$$\begin{aligned} \frac{\nabla P}{\rho} &= \frac{\nabla (P_0 + \tilde{P})}{\rho_0 (1 + \tilde{\rho}/\rho_0)} = \frac{1}{\rho_0} \left(1 - \frac{\tilde{\rho}}{\rho_0}\right) \nabla (P_0 + \tilde{P}) \\ &= \frac{\nabla P_0}{\rho_0} + \frac{\nabla \tilde{P}}{\rho_0} - \frac{\nabla P_0}{\rho_0^2} \tilde{\rho} = \boldsymbol{g} + \frac{\nabla \tilde{P}}{\rho_0} - \frac{\boldsymbol{g}}{\rho_0} \tilde{\rho}, \end{aligned} \tag{4.59}$$

where we perform the Taylor expansion of the fluid density, $(1 + \tilde{\rho}/\rho_0)^{-1} = 1 - \tilde{\rho}/\rho_0 + \mathrm{O}[(\tilde{\rho}/\rho_0)^2]$, keeping only linear terms in perturbations. We also assume that there is hydrostatic equilibrium [see Eq. (4.58)] in the basic reference state. For slow motions (i.e., for small Mach numbers), small changes of the fluid temperature $\tilde{T}$ cause mainly small changes of fluid density, i.e.,

$$\tilde{\rho} = \left(\frac{\partial \rho}{\partial T}\right)_P \tilde{T} \equiv -\rho_0 \, \alpha_{TE} \, \tilde{T}, \tag{4.60}$$

where

$$\alpha_{TE} = -\rho_0^{-1} \left(\frac{\partial \rho}{\partial T}\right)_P \tag{4.61}$$

[1] Joseph Valentin Boussinesq (1842–1929) was a French mathematician and physicist who made significant contributions to fluid dynamics and the theory of vibrations.

is the temperature expansion coefficient. Therefore,

$$-\frac{\nabla P}{\rho}+\boldsymbol{g}=-\frac{\nabla \tilde{P}}{\rho_0}-\boldsymbol{g}\,\alpha_{TE}\,\tilde{T}, \tag{4.62}$$

where we use Eqs. (4.59)–(4.61). Substituting Eq. (4.62) into Eq. (4.57), we obtain the Navier-Stokes equation written in the Boussinesq approximation:

$$\frac{\partial \boldsymbol{u}}{\partial t}+(\boldsymbol{u}\cdot\nabla)\boldsymbol{u}=-\frac{\nabla \tilde{P}}{\rho_0}-\boldsymbol{g}\,\alpha_{TE}\,\tilde{T}+\nu\,\Delta\boldsymbol{u}, \tag{4.63}$$

where $-\boldsymbol{g}\,\alpha_{TE}\,\tilde{T}$ is the buoyancy force. Equation (4.63) implies that variations of density are taken into account only in the buoyancy force, but the velocity field satisfies the continuity equation $\operatorname{div}\boldsymbol{u}=0$.

4.2.2 Potential Temperature and Entropy

We introduce potential temperature Θ as

$$\Theta=T\left(\frac{P}{P_0}\right)^{1/\gamma-1}, \tag{4.64}$$

where $\gamma=c_{\rm p}/c_{\rm v}$ is the ratio of the specific heats at constant pressure $c_{\rm p}$ and constant volume $c_{\rm v}$, and P_0 is the pressure in the equilibrium. Deviations of potential temperature $\tilde{\Theta}$ from the equilibrium are given by

$$\frac{\tilde{\Theta}}{\Theta_0}=\frac{\tilde{T}}{T_0}+\left(\frac{1}{\gamma}-1\right)\frac{\tilde{P}}{P_0}, \tag{4.65}$$

where $\Theta_0=T_0$. To derive Eq. (4.65), we calculate $\ln\Theta$ using Eq. (4.64) and determine variations of $\ln\Theta$ relative to the basic reference state (the equilibrium). Now we define entropy as

$$s=c_{\rm v}\ln\left(P\,\rho^{-\gamma}\right). \tag{4.66}$$

Deviations of the entropy from the equilibrium are

$$\tilde{s}=c_{\rm v}\left(\frac{\tilde{P}}{P_0}-\gamma\frac{\tilde{\rho}}{\rho_0}\right), \tag{4.67}$$

and the equation of state for the ideal gas implies that

$$\frac{\tilde{P}}{P_0}=\frac{\tilde{T}}{T_0}+\frac{\tilde{\rho}}{\rho_0}, \tag{4.68}$$

where T_0 and ρ_0 are the fluid temperature and density at the equilibrium. Using Eqs. (4.67) and (4.68), we rewrite variations of the entropy as

$$\tilde{s} = c_{\rm p} \left[\frac{\tilde{T}}{T_0} + \left(\frac{1}{\gamma} - 1 \right) \frac{\tilde{P}}{P_0} \right] \equiv c_{\rm p} \, \frac{\tilde{\Theta}}{T_0}, \tag{4.69}$$

so that the entropy variations are proportional to the potential temperature variations.

4.2.3 Budget Equations

The Navier-Stokes equation for velocity fluctuations with the buoyancy force written in the Boussinesq approximation in terms of the potential temperature reads:

$$\frac{\partial \boldsymbol{u}}{\partial t} + (\boldsymbol{u} \cdot \nabla)\boldsymbol{u} - \langle (\boldsymbol{u} \cdot \nabla)\boldsymbol{u} \rangle = -\frac{\nabla P'}{\rho_0} - \boldsymbol{\beta}\, \Theta' + \nu\, \Delta \boldsymbol{u} + \boldsymbol{f}, \tag{4.70}$$

where $\boldsymbol{\beta} = \boldsymbol{g}/T_0$ and the acceleration due to the gravity $\boldsymbol{g}$ is directed opposite to the vertical z-axis. The equation for potential temperature fluctuations Θ' reads:

$$\frac{\partial \Theta'}{\partial t} + (\boldsymbol{u} \cdot \nabla)\Theta' - \big\langle (\boldsymbol{u} \cdot \nabla)\Theta' \big\rangle = -(\boldsymbol{u} \cdot \nabla)\overline{\Theta} + D^{(\theta)}\, \Delta \Theta', \tag{4.71}$$

where $\overline{\Theta} = \langle \Theta \rangle$ is the mean potential temperature. The budget equation for the density of turbulent kinetic energy $E_K = \rho\, \langle \boldsymbol{u}^2 \rangle / 2$ has an additional term $\boldsymbol{\beta}\, F_z^{(\Theta)}$ caused by the buoyancy force:

$$\frac{\partial E_K}{\partial t} + {\rm div}\, \boldsymbol{\Phi}_K = \Pi_{_K} + \beta\, F_z^{(\Theta)} - \frac{E_K}{\tau_0}, \tag{4.72}$$

where $\boldsymbol{F}^{(\Theta)} = \langle \Theta' \boldsymbol{u} \rangle$ is the turbulent flux of the potential temperature and $\beta = |\boldsymbol{g}|/T_0$ is the buoyancy parameter. Since the vertical turbulent flux $F_z^{(\Theta)}$ of the potential temperature is negative in a stably stratified turbulence (i.e., it is directed downward), the negative buoyancy term $\beta\, F_z^{(\Theta)}$ decreases the turbulent kinetic energy density. For the same reason, the budget equation for the turbulent flux of the potential temperature $\boldsymbol{F}^{(\Theta)}$ has additional terms $2\beta\, E_\Theta\, \delta_{iz}$ and $-(\boldsymbol{F}^{(\Theta)} \cdot \nabla)\overline{U}_i$ caused by the buoyancy

$$\begin{aligned} \frac{\partial F_i^{(\Theta)}}{\partial t} + {\rm div}\, \boldsymbol{\Phi}_{ij}^{(F)} &= -\frac{1}{\rho_0} \langle \Theta' \nabla_i P' \rangle - \langle u_i u_j \rangle \nabla_j \overline{\Theta} - \frac{F_i^{(\Theta)}}{C_F \tau_0} \\ &\quad + 2\beta\, E_\Theta\, \delta_{iz} - \left(\boldsymbol{F}^{(\Theta)} \cdot \nabla \right) \overline{U}_i, \end{aligned} \tag{4.73}$$

where $C_F = 2 \left(1 + {\rm Pr}^{-1}\right)^{-1}$ is the constant. We will show below that

$$\frac{1}{\rho_0} \langle \Theta' \nabla_z P' \rangle \approx \frac{2}{3}\, \beta\, E_\Theta, \tag{4.74}$$

so that

$$2\beta\, E_\Theta - \frac{1}{\rho_0}\, \langle \Theta' \nabla_z P' \rangle \approx \frac{4}{3}\, \beta\, E_\Theta. \tag{4.75}$$

Indeed, taking the divergence of the momentum equation (4.70) written for div $\boldsymbol{u} = 0$, we obtain the following equation:

$$\frac{\Delta P'}{\rho_0} = \beta\, \nabla_z \Theta'. \tag{4.76}$$

Applying the inverse Laplacian to Eq. (4.76), we get

$$\frac{P'}{\rho_0} = \beta\, \Delta^{-1} \nabla_z \Theta'. \tag{4.77}$$

Next we take ∇_z of Eq. (4.77), multiply the obtained equation by Θ' and average it over an ensemble of turbulent velocity field. The result is given by

$$\frac{1}{\rho_0}\, \langle \Theta' \nabla_z P' \rangle = \beta\, \langle \Theta' \Delta^{-1} \nabla_z^2 \Theta' \rangle. \tag{4.78}$$

Now we determine $\langle \Theta' \Delta^{-1} \nabla_z^2 \Theta' \rangle$:

$$\begin{aligned}\langle \Theta'(t, \boldsymbol{x}) \Delta^{-1} \nabla_z^2 \Theta'(t, \boldsymbol{x}) \rangle &= \lim_{\boldsymbol{x} \to \boldsymbol{y}} \langle \Theta'(t, \boldsymbol{x}) \Delta^{-1} \nabla_z^2 \Theta'(t, \boldsymbol{y}) \rangle \\ &= \int \left(\frac{k_z^2}{k^2} \right) \langle \Theta'(\boldsymbol{k}) \Theta'(-\boldsymbol{k}) \rangle \, d\boldsymbol{k} \\ &= \frac{\langle \Theta'^2 \rangle}{4\pi} \int_{k_0}^{k_D} E_\Theta(k)\, dk \int_0^{2\pi} d\varphi \int_0^{\pi} \sin\vartheta\, d\vartheta\, \frac{k_z^2}{k^2}. \end{aligned} \tag{4.79}$$

Here we use the following model for the second moments, $\langle \Theta'(\boldsymbol{k}) \Theta'(-\boldsymbol{k}) \rangle$, of potential temperature fluctuations in a homogeneous and incompressible turbulence in a Fourier space:

$$\langle \Theta'(\boldsymbol{k}) \Theta'(-\boldsymbol{k}) \rangle = \frac{\langle \Theta'^2 \rangle \tilde{E}_\Theta(k)}{4\pi k^2}, \tag{4.80}$$

where the spectrum function is $\tilde{E}_\Theta(k) = k_0^{-1}\, (2/3)\, (k/k_0)^{-5/3}$. Here $k_0 \le k \le k_D$, the wave number $k_0 = 1/\ell_0$, the length ℓ_0 is the integral scale, the wave number $k_D = \ell_D^{-1}$ and $\ell_D = \ell_0 \mathrm{Pe}^{-3/4}$ is the diffusion scale. After integration in the $\boldsymbol{k}$ space we obtain Eq. (4.74), where for the integration over angles in the $\boldsymbol{k}$ space, we use the following integral:

$$\int_0^{2\pi} d\varphi \int_0^{\pi} \sin\vartheta\, d\vartheta\, \frac{k_i\, k_j}{k^2} = \frac{4\pi}{3}\, \delta_{ij}. \tag{4.81}$$

Therefore, the budget equation for the turbulent flux of the potential temperature $\boldsymbol{F}^{(\Theta)}$ reads

$$\frac{\partial F_i^{(\Theta)}}{\partial t} + \mathrm{div}\, \boldsymbol{\Phi}_{ij}^{(F)} = \frac{4}{3}\beta\, E_\Theta \delta_{iz} - \langle u_i u_j \rangle \nabla_j \overline{\Theta} - F_j^{(\Theta)}\, \nabla_j \overline{U}_i - \frac{F_i^{(\Theta)}}{C_F \tau_0}. \quad (4.82)$$

The budget equation for the second moment $E_\Theta = \langle \Theta'^2 \rangle / 2$ of potential temperature fluctuations reads

$$\frac{\partial E_\Theta}{\partial t} + \mathrm{div}\, \boldsymbol{\Phi}_\Theta = -(\boldsymbol{F}^{(\Theta)} \cdot \boldsymbol{\nabla})\overline{\Theta} - \frac{E_\Theta}{\tau_0}. \quad (4.83)$$

In the next section, we use these budget equations for the energetic analysis of a shear-produced stably stratified turbulence.

4.2.4 Energetic Analysis for Shear-Produced Stably Stratified Turbulence

We introduce

$$N^2 = \beta \nabla_z \overline{\Theta}, \quad (4.84)$$

where N is the Brunt Väisälä frequency. If the production of the turbulence is caused by the mean velocity shear, the rate of production of the turbulent kinetic energy density, $\Pi_K \equiv -\langle u_i u_j \rangle \nabla_j \overline{U}_i = \nu_T S^2$, where ν_T is the turbulent viscosity, $S = \partial \overline{U}_x / \partial z$ is the mean velocity shear and $\overline{U}_x(z)$ is the mean velocity that is directed along, e.g., the horizontal x-axis, and it is nonuniform in the vertical direction. The ratio of N^2 to S^2 is the gradient Richardson number,

$$\mathrm{Ri} = \frac{N^2}{S^2}. \quad (4.85)$$

We use the steady-state solution of the budget equation (4.72) for the density of turbulent kinetic energy $E_K = \rho \langle \boldsymbol{u}^2 \rangle / 2$ for a homogeneous turbulence with div $\boldsymbol{\Phi}_K = 0$, so that

$$E_K = \tau_0\, S^2\, D_T\, \left(\mathrm{Pr}_T - \mathrm{Ri}\right), \quad (4.86)$$

where $\mathrm{Pr}_T = \nu_T / D_T$ is the turbulent Prandtl number, and we take into account an estimate $F_z^{(\Theta)} = -D_T \nabla_z \overline{\Theta}$ for the vertical turbulent heat flux obtained by the dimensional analysis or by the tau approximation. Equation (4.86) implies that an increase of the vertical gradient of the mean potential temperature enhances the buoyancy and decreases the turbulent kinetic energy density, so that at some gradient Richardson number, $\mathrm{Ri} = \mathrm{Ri}_{\rm cr}$, the turbulent kinetic energy density can vanish, $E_K \to 0$. This implies that turbulence should decay.

Since Richardson (1920), it was generally believed that in stationary homogeneous flows the velocity shear becomes incapable of maintaining turbulence when gradient Richardson number Ri exceeds some critical value $\mathrm{Ri}_{\mathrm{cr}}$ (with the conventional value of $\mathrm{Ri}_{\mathrm{cr}} = 0.25$–$1$). However, this conclusion is in disagreement with numerous meteorological observations, laboratory experiments and numerical simulations (see more discussions with corresponding references in the next section).

To resolve this paradox, we introduce the density of turbulent potential energy as

$$E_P = \frac{\beta^2}{N^2} E_\Theta, \tag{4.87}$$

so that the budget equation for the density of turbulent potential energy reads

$$\frac{\partial E_P}{\partial t} + \mathrm{div}\, \mathbf{\Phi}_P = -\beta\, F_z^{(\Theta)} - \frac{E_P}{\tau_0}, \tag{4.88}$$

where

$$\mathbf{\Phi}_P = \frac{\beta^2}{N^2} \mathbf{\Phi}_\Theta \tag{4.89}$$

is the flux of E_P. The buoyancy flux $\beta F_z^{(\Theta)}$ appears in Eq. (4.72) for E_K and in Eq. (4.88) for E_P with opposite signs and describes the energy exchange between the turbulent kinetic and turbulent potential energies. Introducing the density of the total turbulent energy density $E_T = E_K + E_P$ and using Eqs. (4.72) and (4.88), we obtain the budget equation for E_T as

$$\frac{\partial E_T}{\partial t} + \mathrm{div}\, \mathbf{\Phi}_T = \Pi_K - \frac{E_T}{\tau_0}. \tag{4.90}$$

Since the buoyancy fluxes $\pm\beta\, F_z^{(\Theta)}$ in the budget equations (4.72) and (4.88) for the turbulent kinetic and potential energies enter with opposite signs, they cancel each other in the budget equation (4.90) for the total turbulent energy density. Therefore, as follows from Eq. (4.90), the total turbulent energy density is independent of the buoyancy. This implies that there are no grounds to consider the buoyancy-flux term in an equation for the turbulent kinetic energy density E_K as an ultimate "killer" of turbulence. When the rates of the production and dissipation of the total turbulent energy are compensated, the total turbulent energy is conserved. Therefore, an increase in the vertical gradient of the mean potential temperature enhances the buoyancy and decreases the density of turbulent kinetic energy. On the other hand, this increases the turbulent potential energy, so that the total turbulent energy is conserved.

What is the main mechanism for the self-regulation of a stably stratified turbulence? For a homogeneous turbulence ($\mathrm{div}\, \mathbf{\Phi}_K = 0$) in a steady state, budget equation (4.82) for the vertical turbulent flux $F_z^{(\Theta)}$ of the potential temperature yields

$$F_z^{(\Theta)} = -C_F \left[D_T \nabla_z \overline{\Theta} - \frac{4}{3} \tau_0 \, \beta \, E_\Theta \right], \tag{4.91}$$

where $D_T = \tau_0 \langle u_z^2 \rangle = \tau_0 \langle \boldsymbol{u}^2 \rangle / 3$ is the turbulent diffusion coefficient. An increase of the vertical gradient of the mean potential temperature increases the turbulent potential energy and E_Θ, but it also decreases the vertical flux of potential temperature [see Eq. (4.91)]. Therefore, this feedback closes a loop, i.e., this effect decreases the buoyancy and maintains the stably stratified turbulence for any Richardson numbers. Therefore, the correct mechanism of self-existence of turbulence includes two steps:

- conversion of turbulent kinetic energy into turbulent potential energy with increasing the vertical gradient of the mean potential temperature; and
- self-control feedback of the negative down-gradient turbulent heat transfer through efficient generation of the counteracting positive non-gradient heat transfer by turbulent potential energy.

Due to this feedback, the stably stratified turbulence is maintained up to strongly supercritical stratifications. This explains the absence of the critical Richardson number as a threshold for the existence of stably stratified turbulence. This explanation and the theory for a stably stratified turbulence based on the budget equations have been formulated by S. Zilitinkevich, T. Elperin, N. Kleeorin and I. Rogachevskii (Zilitinkevich et al., 2007, 2013).

Budget equation (4.83) and Eq. (4.91) in a steady state for a homogeneous turbulence yield the level of potential temperature fluctuations as

$$\left\langle \Theta'^2 \right\rangle = \frac{2\ell_0^2 \left(\nabla \overline{\Theta} \right)^2}{3 + 4C_F (\tau_0 \, N)^2}. \tag{4.92}$$

Equation (4.92) takes into account a two-way coupling between velocity and temperature fluctuations. For a weak stratification, $(\tau_0 N)^2 \ll 1$, Eq. (4.92) is reduced to $\langle \Theta'^2 \rangle \approx \ell_0^2 \, (\nabla \overline{\Theta})^2$. The latter equation coincides with that derived using the dimensional analysis and the spectral tau approach (see Chapter 1). The condition $(\tau_0 N)^2 \ll 1$ allows one to neglect the feedback effect of temperature fluctuations on velocity fluctuations, which implies a one-way coupling between velocity and temperature fluctuations.

4.3 Further Reading

The classical theory of turbulent flows is based on seminal papers by L. Richardson, L. Prandtl, A. Kolmogorov and A. Obukhov; see also various books and reviews (Batchelor, 1953; Hinze, 1959; Monin and Yaglom, 1971, 2013; Panchev, 1971; Bradshaw, 1971; Tennekes and Lumley, 1972; Turner, 1973; Townsend, 1980; McComb, 1990; Tikhomirov, 1991; Chorin, 1994; Frisch, 1995; Pope, 2000; Bernard and Wallace, 2002; Lesieur, 2008; Tsinober, 2009, 2014; Davidson, 2015, 2013; Landau and Lifshits, 1987). This theory implies that any turbulent flow can be considered as a superposition of the fully organized mean flow and the fully chaotic turbulence characterized by the forward energy cascade from the larger eddies to smaller, resulting in viscous energy dissipation at the smallest eddies, with a pronounced inertial interval in the energy spectrum that is fully determined by the energy-dissipation rate. The local characteristics of turbulence, in particular, turbulent fluxes that appear in the mean-flow equations, are generally controlled by the local features of the mean flow. It is also assumed that the turbulent flux of any quantity can be expressed as a product of the mean gradient of the quantity multiplied by a turbulent-exchange coefficient. This concept of down-gradient transport reduces the closure problem to a determination of the turbulent-exchange coefficients, usually taken proportionally with the turbulent kinetic energy and timescale. This has been formulated for an isothermal turbulence, i.e., a neutrally stratified turbulence.

4.3.1 Stably Stratified Turbulence

The majority of efforts in the development of the turbulence closure models for meteorological applications over half a century have been limited to "mechanical closures" based on the sole use of the turbulent kinetic energy equation, disregarding turbulent potential energy, and daring only cautious corrections to the concept of down-gradient transport. The equation for the squared potential temperature fluctuations, although it has been well-known for decades, has been ignored in turbulent closure models. Only rather recently this equation has been considered in terms of the turbulent potential energy (Dalaudier and Sidi, 1987; Ostrovsky and Troitskaya, 1987; Canuto and Minotti, 1993; Kaimal and Finnigan, 1994; Hanazaki and Hunt, 1996).

In stable stratification, such closures have resulted in the erroneous conclusion that shear-generated turbulence inevitably decays and that the flow becomes laminar in "supercritical" stratifications at Richardson numbers exceeding some critical value (Richardson, 1920; Chandrasekhar, 2013; Monin and Yaglom, 1971, 2013). Obvious contradictions of this conclusion via the well-documented universal presence of turbulence in strongly "supercritical" conditions typical of the free atmosphere and the deep ocean were attributed to some unknown mechanisms and, in practical applications, mastered heuristically (Kondo et al., 1978; Bertin et al., 1997; Strang and Fernando, 2001; Ohya, 2001; Rehmann and Koseff, 2004; Mahrt and Vickers, 2005; Grachev et al., 2007; Stretch et al., 2010; Mahrt, 2010, 2014; Cohen et al., 2014; Li et al., 2016). The decay of strongly stably stratified turbulence in direct numerical simulations (DNS) has been explained by the effect of diminishing the effective Reynolds number, which comes into play in the not-high-Reynolds-number flows in DNS, but remains insignificant in the very-high-Reynolds-number atmospheric flows.

The energy- and flux-budget (EFB) theory of turbulence closure based on the basic budget equations for stably stratified dry atmospheric flows has been recently developed (Zilitinkevich et al., 2007, 2008, 2009, 2010, 2013). In accordance with wide experimental evidence, the EFB theory shows that high-Reynolds-number turbulence is maintained by shear in any stratification, and the "critical Richardson number," treated over decades as a threshold between the turbulent and laminar regimes, factually separates two turbulent regimes: the strong turbulence, typical of atmospheric boundary layers, and the weak three-dimensional turbulence, typical of the free atmosphere or deep ocean and characterized by a sharp decrease in heat transfer compared to the momentum transport. The principal aspects of the EFB theory have been verified against scarce data from the atmospheric experiments, direct numerical simulations, large-eddy simulations (LES) and laboratory experiments relevant to the steady-state turbulence regime. Following the EFB theory (Zilitinkevich et al., 2007), other recently published turbulent closure models (Mauritsen et al., 2007; Galperin et al., 2007; Canuto et al., 2008; L'vov et al., 2008; Sukoriansky and Galperin, 2008) also do not imply critical Richardson numbers.

Key ingredients of stably stratified turbulence are internal gravity waves (Beer, 1974; Miropol'sky, 2001; Nappo, 2013; Sutherland, 2010). In the atmosphere, internal gravity waves exist at scales ranging from meters to kilometers, and are measured by direct probing or remote sensing using radars and lidars. The sources of internal gravity waves can be flows over complex terrain, strong wind shears, convective and other local-scale motions underlying the stably stratified layer, and wave-wave interactions. There are many interesting studies related to wave turbulence (Zakharov et al., 2012; Nazarenko, 2011).

In stably stratified turbulence, large-scale internal gravity waves (IGW) cause additional vertical turbulent flux of momentum and additional productions of turbulent kinetic energy (TKE), turbulent potential energy (TPE) and turbulent flux of potential temperature (Zilitinkevich et al., 2009; Kleeorin et al., 2019). For the stationary, homogeneous regime, the EFB theory in the absence of the large-scale IGW (Zilitinkevich et al., 2007, 2013) yields universal dependencies of the basic dimensionless parameters on the gradient Richardson number (i.e., the ratio of TKE to TPE, the normalized vertical turbulent fluxes of momentum and heat, the flux Richardson number and the turbulent Prandtl number are the universal dependencies on the gradient Richardson number), where the flux Richardson number is the ratio of the decay rate of TKE due to the buoyancy to the rate of turbulence production due to the shear and the turbulent Prandtl number is the ratio of turbulent viscosity to turbulent diffusivity. Due to the large-scale IGW, these dependencies lose their universality. The maximum value of the flux Richardson number (universal constant 0.2–0.25 in the no-IGW regime) becomes strongly variable in the turbulence with large-scale IGW. In the vertically homogeneous stratification, the flux Richardson number increases with increasing wave energy and can even exceed 1. The large-scale internal gravity waves also reduce the anisotropy of turbulence. These waves increase the share of TPE in the turbulent total energy, which is the sum of the turbulent kinetic and potential energies. A well-known effect of IGW is their direct contribution to the vertical transport of momentum. Depending on the direction (downward or upward), IGW either strengthen or weaken the total vertical flux of momentum. Predictions from the EFB theory which take into account the large-scale internal gravity waves (Zilitinkevich et al., 2009; Kleeorin et al., 2019) are consistent with available data from atmospheric and laboratory experiments, DNS and LES (Kondo et al., 1978; Bertin et al., 1997; Strang and Fernando, 2001; Ohya, 2001; Rehmann and Koseff, 2004; Mahrt and Vickers, 2005; Grachev et al., 2007; Stretch et al., 2010; Chung and Matheou, 2012; Cohen et al., 2014; Li et al., 2016).

Numerous high-resolution DNS of stably stratified turbulence with IGW have been also performed (Lindborg, 2006; Clark di Leoni and Mininni, 2015; Pouquet et al., 2018; Druzhinin et al., 2013; Druzhinin and Ostrovsky, 2015). The role of IGW and their interaction with the large-scale flow of vertically sheared horizontal winds was studied by Clark di Leoni and Mininni (2015) and showed that most of the energy is concentrated along a dispersion relation that is Doppler shifted by the horizontal winds. When uniform winds are formed in each horizontal layer of the flow, waves whose phase velocity is equal to the horizontal wind speed, have negligible energy, which indicates a nonlocal transfer of their energy to the mean flow. Scaling laws for mixing and dissipation in unforced stratified turbulence have been found by Pouquet et al. (2018). Three regimes characterized by the Froude number, namely (i) dominant waves, (ii) eddy-wave interactions and (iii) strong turbulence, have been observed by Pouquet et al. (2018). An interaction between large-scale IGW and the turbulent layer above the pycnocline (where the density gradient is largest) has been studied using DNS (Druzhinin et al., 2013; Druzhinin and Ostrovsky, 2015). These studies have demonstrated that in the absence of IGW, the turbulence decays and most of the turbulent energy is concentrated at the pycnocline center. The turbulent eddies are collapsed in the vertical direction and acquire the "pancake" shape. The internal gravity waves significantly increase turbulent energy (Druzhinin et al., 2013; Druzhinin and Ostrovsky, 2015).

The theoretical predictions based on the budget equations for turbulent kinetic and potential energies, and turbulent heat flux, are also in agreement with the experimental results obtained for both stably and unstably stratified forced turbulent flows (Bukai et al., 2011; Eidelman et al., 2013). In these experiments, turbulence has been produced by two oscillating grids located near the side walls of the chamber.

4.3.2 Turbulent Convection

Another example of a temperature stratified turbulence is a turbulent convection in a system with a heated bottom boundary (Turner, 1973; Zilitinkevich, 1991; Ching, 2014). Buoyancy-driven structures, such as plumes, jets and large-scale circulation patterns, are observed in numerous laboratory experiments. The large-scale circulations caused by convection in the Rayleigh-Bénard apparatus are often called the "mean wind" (Krishnamurti and Howard, 1981). There are several open questions concerning these flows, e.g., how they arise and what their characteristics and dynamics are.

The large-scale circulations (or semi-organized coherent structures) that are usually observed in laboratory experiments (Fitzjarrald, 1976; Krishnamurti and Howard, 1981; Niemela et al., 2001; Eidelman et al., 2006b; Ahlers et al., 2009; Bukai et al., 2009; Lohse and Xia, 2010) and the atmospheric convective boundary layers (Etling and Brown, 1993; Atkinson and Wu Zhang, 1996; Zilitinkevich, 1991) essentially differ from turbulence. First of all, they exist and retain their specific forms during much longer periods than the largest time scales of turbulence. In the atmospheric shear-free convection, the structures represent large three-dimensional long-lived Bénard-type cells composed of narrow uprising plumes and wide downdraughts. They usually embrace the entire convective boundary layer (of the order of 1–3 km in height) and include pronounced convergence flow patterns close to the surface. In the sheared convective flows, the structures represent convective–boundary–layer–scale rolls stretched along the mean wind. The lifetimes and spatial scales of coherent semi-organized structures are much larger compared to the largest turbulent time scales. Their spectral properties differ from those of small-scale turbulence. They are characterized by narrow spectra and do not exhibit direct energy-cascade behavior (from larger to smaller scales). As a result, the turbulence and the structures interact in practically the same way as the turbulence and the mean flow. In other words, the structures show more similarity in their behavior with mean regular flows than with turbulence.

In spite of a number of theoretical and numerical studies (Busse and Whitehead, 1971, 1974; Busse, 1983; Hartlep et al., 2003; Parodi et al., 2004; Chillà and Schumacher, 2012; Van Reeuwijk et al., 2005; Grossmann and Lohse, 2011; Ching, 2014), the nature of large-scale coherent structures in turbulent convection is a topic of discussion. Hartlep et al. (2003) noted that there are two points of view on the origin of large-scale circulation in turbulent convection. "According to one point of view, the rolls which develop at low Rayleigh numbers near the onset of convection continually increase their size as the Rayleigh numbers is increased and continue to exist in an average sense at even the highest Rayleigh numbers reached in the experiments. Another hypothesis holds that the large-scale circulation is a genuine high Rayleigh number effect" (Krishnamurti and Howard, 1981).

A mean-field theory of turbulent convection has been recently developed (Elperin et al., 2002b, 2006a,b). In the shear-free regime, this theory predicts convective wind instability, which causes the formation of large-scale semi-organized motions in the form of cells. In the presence of wind shear, there is another type of instability, which results in the formation of large-scale rolls and the generation of convective-shear waves propagating perpendicular to the convective rolls. The spatial characteristics of these structures, such as the minimum size of the growing perturbations and the size of perturbations with the maximum growth rate, are determined. The increase of shear promotes excitation of the convective shear instability. The predictions of this theory are in good agreement with observations of the semi-organized structures in the atmospheric convective boundary layer (Elperin et al., 2006b).

In various flows, a complex terrain (i.e., canopy and various topography) strongly affects the stratified turbulence (Baines, 1997). It changes local temperature gradients, heat and mass fluxes, and affects a local structure of fluid flows. Therefore, one of the important questions is raised: What is the effect of complex terrain on convective and stably stratified turbulence? To model the effect of complex terrain, the Rayleigh-Bénard convection (RBC) with modulated boundaries has been investigated in various studies (Weiss et al., 2014; Toppaladoddi et al., 2017; Zhu et al., 2017; Barel et al., 2020).

The roughness effect on the heat transport in RBC has been investigated in two-dimensional numerical simulations by varying the height and wavelength of the roughness elements where the sinusoidal roughness profile has been chosen (Toppaladoddi et al., 2017; Zhu et al., 2017). The ultimate regime of thermal convection (when the boundary layers undergo a transition leading to the generation of smaller scales near the boundaries that increase the system's efficiency in transporting the heat) has recently been found (Toppaladoddi et al., 2017). This regime, in which the heat flux becomes independent of the molecular properties of the fluid, has been predicted (Kraichnan, 1962; Grossmann and Lohse, 2011). One of the key roles of the roughness elements is the production and release of the plumes from the roughness elements, resulting in the formation of larger plumes. This can cause an increase in the efficiency of the heat transfer (Toppaladoddi et al., 2017). The existence of two universal regimes in RBC, namely the ultimate regime and the classical boundary-layer-controlled regime with an increased Rayleigh number, has been demonstrated (Zhu et al., 2017). The transition from the first to the second regime is determined by the competition between bulk and boundary layer flow. The bulk-dominated regime corresponds to the ultimate regime.

Convection over a wavy heated bottom wall in the air flow has been studied in experiments with the Rayleigh number $\sim 10^8$ (Barel et al., 2020). It is shown that the mean temperature gradient in the flow core inside a large-scale circulation is directed upward, which corresponds to the stably stratified flow. In experiments with a wavy heated bottom wall, large-scale standing internal gravity waves excited in the regions with the stably stratified flow have been detected. The wavelength and the period of these waves are much larger than the turbulent spatial and time scales, respectively. In particular, the frequencies of the observed large-scale waves vary from 0.006 Hz to 0.07 Hz, while the turbulent time in the integral scale is about 0.5 s. The measured spectra of these waves contains several localized maxima, that implies an existence of waveguide resonators for the large-scale standing internal gravity waves. For comparison, experiments with convection over a smooth plane bottom wall at the same mean temperature difference between the bottom and upper walls have been conducted as well. In these experiments, various locations with a stably stratified flow are also found, and the large-scale standing internal gravity waves are observed in these regions (Barel et al., 2020).

4.3.3 Rotating Turbulence

The study of rotating turbulence is of interest for a wide range of problems, ranging from geophysics (oceans, the atmosphere of the Earth, gaseous planets) to astrophysics (solar physics, galactic and accretion discs), as well as engineering (Pedlosky, 1987; Greenspan, 1990; Salmon, 1998; Davidson, 2013). Inertial and Rossby waves arise in rotating flows and are observed in the atmosphere of the Earth (Pedlosky, 1987; Zeitlin, 2007; Connaughton et al., 2015; Gill, 2016; Elperin et al., 2017a). These waves can also exist in the solar convective zone. Rossby waves are caused by the combined effect of rotation and curvature of the surface, while inertial waves are only caused by rotation.

In turbulent rotating flows, inertial waves are damped due to high turbulent viscosity. However, large-scale inertial waves are observed in turbulent rotating flows. A mechanism of excitation of large-scale inertial waves in a rotating inhomogeneous turbulence due to an excitation of large-scale instability has been suggested by Elperin et al. (2005). This instability is caused by a combined effect of the inhomogeneity of the turbulence and the uniform mean rotation. The source of the large-scale instability is the energy of the small-scale turbulence. The range of parameters at which the large-scale instability occurs, the growth rate of the instability and the frequency of the generated large-scale inertial waves have been found by Elperin et al. (2005).

Density-stratified or inhomogeneous fast-rotating turbulent convection can form large-scale vortical structures. Recent DNS have shown that large-scale vortices in rapidly rotating turbulent convection can be formed in compressible (Käpylä et al., 2011) or Boussinesq fluids (Guervilly et al., 2014). An instability resulting in the generation of large-scale vorticity in a fast-rotating turbulent convection with inhomogeneous fluid density along the rotational axis has been recently found in theoretical study (Rogachevskii and Kleeorin, 2019). The large-scale instability causes the excitation of modes with dominant vertical vorticity. For this mode, the mean entropy is depleted inside the cyclonic vortices, while it is enhanced inside the anti-cyclonic vortices. The developed theory (Rogachevskii and Kleeorin, 2019) may be relevant for explaining the origin of large spots observed as immense storms in great planets, e.g., the Great Red Spot in Jupiter and large spots in Saturn (Marcus, 1993). It may be also useful for explanation of an origin of high-latitude spots in rapidly rotating late-type stars (Käpylä et al., 2011).

5

Path-Integral Approach

In this chapter, we study the turbulent transport of particles, temperature and magnetic fields in different types of random flows with a very short correlated time using the path-integral approach. Such velocity is modeled by a random field with a constant renewal time. We derive mean-field equations for particle number density, temperature and magnetic fields and obtain expressions for the turbulent flux of particles, turbulent heat flux, turbulent electromotive force, turbulent diffusion tensor and the α tensor for an anisotropic turbulence.

5.1 Turbulent Transport of Temperature Field

In this section, we study turbulent transport of temperature field in isotropic inhomogeneous and incompressible turbulence using the path-integral approach.

5.1.1 Solution for the Equation for Temperature Field

The equation for the evolution of instantaneous fluid temperature field $T(t, \boldsymbol{x})$ in incompressible fluid velocity field $\boldsymbol{U}(t, \boldsymbol{x})$ reads:

$$\frac{\partial T}{\partial t} + (\boldsymbol{U} \cdot \nabla) T = D^{(\theta)} \, \Delta T, \tag{5.1}$$

where $D^{(\theta)}$ is the coefficient of the molecular diffusion of the temperature field. The fluid velocity field satisfies to the continuity equation $\nabla \cdot \boldsymbol{U} = 0$. The solution of Eq. (5.1) with the initial condition $T(t = t_0, \boldsymbol{x}) = T_0(\boldsymbol{x})$ is given by

$$\boxed{T(t, \boldsymbol{x}) = \langle T_0[\boldsymbol{\xi}(t, \boldsymbol{x}|t_0)] \rangle_{\mathbf{w}} ,} \tag{5.2}$$

where $\boldsymbol{\xi}(t, \boldsymbol{x}|t_0)$ is the Wiener[1] trajectory defined as follows:

$$\boxed{\boldsymbol{\xi}(t, \boldsymbol{x}|t_0) = \boldsymbol{x} - \int_{t_0}^{t} \boldsymbol{U}[\tau, \boldsymbol{\xi}(t, \boldsymbol{x}|\tau)]\, d\tau + \sqrt{2D^{(\theta)}}\, \mathbf{w}(t - t_0),} \tag{5.3}$$

where $\mathbf{w}(t)$ is the Wiener random process that describes the molecular diffusion and has the following properties,

$$\langle \mathbf{w}(t) \rangle_{\mathbf{W}} = 0, \quad \left\langle \mathbf{w}_i(t + \tau)\mathbf{w}_j(t) \right\rangle_{\mathbf{W}} = \tau \delta_{ij}, \tag{5.4}$$

and $\langle \ldots \rangle_{\mathbf{W}}$ denotes the averaging over the statistics of the Wiener process. The velocity $\boldsymbol{U}[\tau, \boldsymbol{\xi}(t, \boldsymbol{x}|\tau)]$ describes the Eulerian velocity calculated at the Wiener trajectory. Equation (5.3) describes a set of random trajectories which pass through the point $\boldsymbol{x}$ at time t. Equation (5.2) is the Feynman[2]-Kac formula.

First, we show that Eq. (5.2) is a solution of Eq. (5.1). The field $T(t, \boldsymbol{x})$ in Eq. (5.2) is specified at time t. Let us determine the field $T(t + \Delta t, \boldsymbol{x})$ specified at time $t + \Delta t$ by means of the replacements $t \to t + \Delta t$ and $t_0 \to t$ in Eq. (5.2):

$$T(t + \Delta t, \boldsymbol{x}) = \langle T[t, \boldsymbol{\xi}(t + \Delta t, \boldsymbol{x}|t)] \rangle_{\mathbf{W}}, \tag{5.5}$$

where

$$\boldsymbol{\xi}_{\Delta t} \equiv \boldsymbol{\xi}(t + \Delta t, \boldsymbol{x}|t) = \boldsymbol{x} - \int_{t}^{t+\Delta t} \boldsymbol{U}[\tau, \boldsymbol{\xi}(t + \Delta t, \boldsymbol{x}|\tau)]\, d\tau + \sqrt{2D^{(\theta)}}\, \mathbf{w}(\Delta t), \tag{5.6}$$

We expand the function $T[t, \boldsymbol{\xi}_{\Delta t}] \equiv T[t, \boldsymbol{\xi}(t + \Delta t, \boldsymbol{x}|t)]$ in Eq. (5.5) in a Taylor series in the vicinity of the point $\boldsymbol{x}$ as

$$T\left(t, \boldsymbol{\xi}_{\Delta t}\right) = T(t, \boldsymbol{x}) + \frac{\partial T(t, \boldsymbol{x})}{\partial x_m} \left(\boldsymbol{\xi}_{\Delta t} - \boldsymbol{x}\right)_m + \frac{1}{2} \frac{\partial^2 T(t, \boldsymbol{x})}{\partial x_m \partial x_n} \left(\boldsymbol{\xi}_{\Delta t} - \boldsymbol{x}\right)_m \left(\boldsymbol{\xi}_{\Delta t} - \boldsymbol{x}\right)_n + \ldots, \tag{5.7}$$

and we rewrite the equation for the Wiener trajectory as

$$\left(\boldsymbol{\xi}_{t_2 - t_1} - \boldsymbol{x}\right)_m = -\int_{0}^{t_2 - t_1} U_m(t_s, \boldsymbol{\xi}_s)\, ds + \sqrt{2D^{(\theta)}}\, \mathrm{w}_m(t_2 - t_1), \tag{5.8}$$

[1] Norbert Wiener (1894–1964) was an American mathematician well-known for his works in stochastic processes, electronic engineering, electronic communication, control systems, computer science and artificial intelligence, biology and neuroscience. Wiener is considered the originator of cybernetics.

[2] Richard Feynman (1918–1988) was an American theoretical physicist well-known for his works in quantum physics (the path integral formulation of quantum mechanics, the theory of quantum electrodynamics, the physics of the superfluidity of supercooled liquid helium), as well as in particle physics (he proposed the parton model). Feynman received the Nobel Prize in Physics in 1965 jointly with Julian Schwinger and Shinichiro Tomonaga.

where $t_s = t_2 - s$, $\boldsymbol{\xi}(t_2, t_2 - s) \equiv \boldsymbol{\xi}_s$. We evaluate the integral in Eq. (5.8) using the "mean value" theorem as

$$\left(\boldsymbol{\xi}_{\Delta t} - \mathbf{x}\right)_m = -U_m(\tilde{t}, \mathbf{x})\Delta t + \sqrt{2D^{(\theta)}}\mathrm{w}_m + \mathrm{O}[(\Delta t)^2], \tag{5.9}$$

where $\tilde{t}$ is within the interval $(t, t + \Delta t)$. Substituting Eq. (5.9) into Eq. (5.7), we obtain an expression for the field $T(t + \Delta t, \boldsymbol{\xi}_{\Delta t})$:

$$T(t + \Delta t, \boldsymbol{\xi}_{\Delta t}) = T(t, \mathbf{x}) + \frac{\partial T(t, \mathbf{x})}{\partial x_m}\left[-U_m(\tilde{t}, \mathbf{x})\Delta t + \sqrt{2D^{(\theta)}}\,\mathrm{w}_m\right]$$
$$+\frac{1}{2}\frac{\partial^2 T(t, \mathbf{x})}{\partial x_m \partial x_n}\left[-U_m(\tilde{t}, \mathbf{x})\Delta t + \sqrt{2D^{(\theta)}}\,\mathrm{w}_m\right]\left[-U_n(\tilde{t}, \mathbf{x})\Delta t + \sqrt{2D^{(\theta)}}\,\mathrm{w}_n\right]$$
$$+\mathrm{O}[(\Delta t)^2]. \tag{5.10}$$

Averaging Eq. (5.10) over statistics of Wiener trajectories and taking into account the properties of the Wiener random process given by Eq. (5.4), we obtain an equation for $T(t + \Delta t, \mathbf{x})$:

$$T(t + \Delta t, \mathbf{x}) = \left\langle T(t + \Delta t, \boldsymbol{\xi}_{\Delta t})\right\rangle_{\mathbf{w}} = T(t, \mathbf{x}) + \frac{\partial T(t, \mathbf{x})}{\partial x_m}\left[-U_m(\tilde{t}, \mathbf{x})\Delta t\right]$$
$$+\frac{\partial^2 T(t, \mathbf{x})}{\partial x_m \partial x_m} D^{(\theta)}\,\Delta t + \mathrm{O}[(\Delta t)^2]. \tag{5.11}$$

Using Eq. (5.11), we calculate the limit:

$$\frac{\partial T}{\partial t} \equiv \left(\frac{T(t + \Delta t, \mathbf{x}) - T(t, \mathbf{x})}{\Delta t}\right)_{\Delta t \to 0} = -(\boldsymbol{U} \cdot \boldsymbol{\nabla})T + D^{(\theta)}\,\Delta T. \tag{5.12}$$

Equation (5.12) coincides with Eq. (5.1), which demonstrates that Eq. (5.2) is a solution of Eq. (5.1).

5.1.2 Mean-Field Temperature Equation: Turbulent Diffusion

In this section, we derive the mean-field equation using solution (5.2) of Eq. (5.1) for the fluid temperature. Let us determine the field $T(t + \Delta t, \boldsymbol{x})$ specified at time $t + \Delta t$ by means of the replacements $t \to t + \Delta t$ and $t_0 \to t$ in Eq. (5.2):

$$T(t + \Delta t, \boldsymbol{x}) = \langle T[t, \boldsymbol{\xi}(t + \Delta t, \boldsymbol{x}|t)]\rangle_{\mathbf{w}}, \tag{5.13}$$

where $\boldsymbol{\xi}(t + \Delta t, \boldsymbol{x}|t)$ is given by Eq. (5.6). We expand the function $T[t, \boldsymbol{\xi}_{\Delta t}] \equiv T[t, \boldsymbol{\xi}(t + \Delta t, \boldsymbol{x}|t)]$ in Eq. (5.13) in a Taylor series in the vicinity of the point $\boldsymbol{x}$:

$$T\left(t, \boldsymbol{\xi}_{\Delta t}\right) = T(t, \boldsymbol{x}) + \frac{\partial T(t, \boldsymbol{x})}{\partial x_m}\left(\boldsymbol{\xi}_{\Delta t} - \boldsymbol{x}\right)_m$$
$$+\frac{1}{2}\frac{\partial^2 T(t, \boldsymbol{x})}{\partial x_m \partial x_l}\left(\boldsymbol{\xi}_{\Delta t} - \boldsymbol{x}\right)_m\left(\boldsymbol{\xi}_{\Delta t} - \boldsymbol{x}\right)_n + \dots. \tag{5.14}$$

We rewrite Eq. (5.6) for the Wiener trajectory as follows:

$$\left(\boldsymbol{\xi}_{t_2-t_1} - \boldsymbol{x}\right)_n = -\int_0^{t_2-t_1} U_n(t_\sigma, \boldsymbol{\xi}_\sigma)\, d\sigma + \sqrt{2D^{(\theta)}}\, \mathrm{w}_n(t_2 - t_1), \tag{5.15}$$

where $\boldsymbol{\xi}_\sigma \equiv \boldsymbol{\xi}(t_2, t_2 - \sigma)$. We expand the velocity $U_m(t_s, \boldsymbol{\xi}_s)$ in Eq. (5.8) in a Taylor series in the vicinity of the point $\boldsymbol{x}$ and use Eq. (5.15):

$$\begin{aligned} U_m(t_s, \boldsymbol{\xi}_s) &= U_m(t, \boldsymbol{x}) + \frac{\partial U_m(t, \boldsymbol{x})}{\partial x_n} \left(\boldsymbol{\xi}_s - \boldsymbol{x}\right)_n + \ldots = U_m(t, \boldsymbol{x}) \\ &+ \frac{\partial U_m(t, \boldsymbol{x})}{\partial x_n} \left[-\int_0^s U_n(t_\sigma, \boldsymbol{\xi}_\sigma)\, d\sigma + \sqrt{2D^{(\theta)}}\, \mathrm{w}_n(s) \right] + \ldots, \end{aligned} \tag{5.16}$$

and $t_\sigma = t + \Delta t - \sigma$, and $\boldsymbol{\xi}(t_2, t_1) \equiv \boldsymbol{\xi}_{t_2-t_1}$, i.e., $\boldsymbol{\xi}_\sigma = \boldsymbol{\xi}(t + \Delta t, t_\sigma)$. Substituting Eq. (5.16) into Eq. (5.8) and calculating the integrals, we obtain:

$$\begin{aligned} \left(\boldsymbol{\xi}_{t_2-t_1} - \boldsymbol{x}\right)_m &= -\int_0^{t_2-t_1} \left[U_m(t, \boldsymbol{x}) + \frac{\partial U_m(t, \boldsymbol{x})}{\partial x_n} \Big(-U_n(t, \boldsymbol{x}) s \right. \\ &\left. + \sqrt{2D^{(\theta)}}\, \mathrm{w}_n(s) \Big) \right] ds + \sqrt{2D^{(\theta)}}\; \mathrm{w}_m(t_2 - t_1) = -U_m\, (t_2 - t_1) \\ &+ \frac{1}{2} U_n \frac{\partial U_m}{\partial x_n} (t_2 - t_1)^2 - \sqrt{2D^{(\theta)}}\, \frac{\partial U_m(t, \boldsymbol{x})}{\partial x_n} \int_0^{t_2-t_1} \mathrm{w}_n(s)\, ds \\ &+ \sqrt{2D^{(\theta)}}\, \mathrm{w}_m(t_2 - t_1) + O[(t_2 - t_1)^3]. \end{aligned} \tag{5.17}$$

Substituting Eq. (5.17) into Eq. (5.14), averaging the obtained equation over statistics of Wiener trajectories and taking into account the properties of the Wiener random process given by Eq. (5.4), we obtain an equation for $T(t + \Delta t, \boldsymbol{x})$:

$$\begin{aligned} T(t + \Delta t, \boldsymbol{x}) &= \left\langle T[t + \Delta t, \boldsymbol{\xi}_{\Delta t}] \right\rangle_{\mathbf{w}} = T(t, \boldsymbol{x}) + \frac{\partial T(t, \boldsymbol{x})}{\partial x_m} \left[-U_m\, \Delta t \right. \\ &\left. + \frac{1}{2} U_n \frac{\partial U_m}{\partial x_n} (\Delta t)^2 \right] + \frac{1}{2} \frac{\partial^2 T(t, \boldsymbol{x})}{\partial x_m \partial x_l} \left[U_m U_n\, (\Delta t)^2 + 2D^{(\theta)} \delta_{mn} \right] + O[(\Delta t)^3]. \end{aligned} \tag{5.18}$$

Now we use a model of a short-correlated in time random velocity field, $\boldsymbol{u} = \boldsymbol{U} - \overline{\boldsymbol{U}}$, where $\overline{\boldsymbol{U}}$ is the mean velocity field. The velocity field is Gaussian, delta-correlated in time random process (Kazantsev, 1968; Kraichnan, 1968):

$$\left\langle u_i(t, \boldsymbol{x})\, u_j(t', \boldsymbol{y}) \right\rangle_{\boldsymbol{u}} = 2\delta(t - t')\, f_{ij}(t, \boldsymbol{x}, \boldsymbol{y}), \tag{5.19}$$

where $\langle \ldots \rangle_{\boldsymbol{u}}$ is averaging over a statistics of random velocity field, $\delta(t - t')$ is the Dirac delta function, and $f_{ij}(t, \boldsymbol{x}, \boldsymbol{y}) = \left\langle u_i(t, \boldsymbol{x}) u_j(t, \boldsymbol{y}) \right\rangle_{\boldsymbol{u}}$. The concept of a short-correlated in time random velocity field can be introduced as follows (Zeldovich et al., 1988, 1990):

- We assume that the velocity $\boldsymbol{u}$ is constant (time-independent) at small time intervals $(0, \Delta t)$; $(\Delta t, 2\Delta t)$; $\cdots$, and changes every small time Δt. This implies that we consider a random velocity field with a constant renewal time.
- The velocity is statistically independent at different time intervals (the renewal velocity field). This velocity field cannot be considered as a stationary velocity field for small times $\sim \Delta t$; however, it behaves like a stationary field for $t \gg \Delta t$.
- The random velocity and the random Wiener process are independent random processes.

Averaging Eq. (5.18) over the ensemble of the turbulent velocity field and using a model (5.19) of a short-correlated in time random velocity field, we obtain the mean temperature field $\overline{T}(t+\Delta t, \boldsymbol{x}) = \langle T(t+\Delta t, \boldsymbol{x})\rangle_u$:

$$\frac{\overline{T}(t+\Delta t, \boldsymbol{x}) - \overline{T}(t, \boldsymbol{x})}{\Delta t} = \frac{\partial \overline{T}(t, \boldsymbol{x})}{\partial x_m}\left[-\overline{U}_m + \frac{1}{2}\nabla_n \langle \Delta t\, U_m\, U_n\rangle_u\right] + \frac{\partial^2 \overline{T}(t, \boldsymbol{x})}{\partial x_m \partial x_n}\left[\frac{1}{2}\langle \Delta t\, U_m\, U_n\rangle_{\boldsymbol{u}} + D^{(\theta)}\delta_{mn}\right] + O[(\Delta t)^2]. \tag{5.20}$$

Using Eq. (5.20) and calculating the limit

$$\frac{\partial \overline{T}}{\partial t} \equiv \left(\frac{\overline{T}(t+\Delta t, \boldsymbol{x}) - \overline{T}(t, \boldsymbol{x})}{\Delta t}\right)_{\Delta t \to 0}, \tag{5.21}$$

we arrive at the mean-field equation for the temperature field for an anisotropic inhomogeneous and incompressible turbulence as

$$\boxed{\frac{\partial \overline{T}}{\partial t} + \left(\overline{\boldsymbol{U}}\cdot\nabla\right)\overline{T} = \nabla_m \nabla_n \left[\left(D^{(\theta)}\delta_{mn} + D_{mn}^{(T)}\right)\overline{T}\right].} \tag{5.22}$$

Here the tensor of turbulent diffusion of temperature field is given by

$$\boxed{D_{mn}^{(T)} = \langle \tau u_m\, u_n\rangle = \frac{1}{2}\lim_{\Delta t \to 0}\langle \Delta t\, U_m\, U_n\rangle_{\boldsymbol{u}},} \tag{5.23}$$

where we neglect a weak dependence of τ on the coordinates. For an isotropic turbulence, $D_{mn}^{(T)} = D_T \delta_{mn}$, so that the equation for the mean temperature for isotropic inhomogeneous and incompressible turbulence is given by

$$\boxed{\frac{\partial \overline{T}}{\partial t} + \left(\overline{\boldsymbol{U}}\cdot\nabla\right)\overline{T} = \Delta\left[\left(D^{(\theta)} + D_T\right)\overline{T}\right].} \tag{5.24}$$

For large Péclet numbers, the time τ is determined by the correlation time of turbulent flow, $\tau = \tau_0$, and Eq. (5.23) coincides with Eq. (1.63); see Section 1.3.5. On the other hand, for small Péclet numbers, the time τ is determined by the molecular diffusion time, $\tau = \ell_0^2/D^{(\theta)}$, and Eq. (5.23) coincides with Eq. (1.67).

5.2 Turbulent Transport of Particles

In this section, we study turbulent transport of particles in inhomogeneous and density-stratified turbulence using the path-integral approach.

5.2.1 Solution of Convective-Diffusion Equation for Particles

The equation for the evolution of instantaneous particle number density $n(t, \boldsymbol{x})$ reads:

$$\frac{\partial n}{\partial t} + \nabla \cdot (\boldsymbol{U} n) = D^{(n)} \Delta n, \tag{5.25}$$

where $D^{(n)}$ is the Brownian diffusion coefficient. The solution of Eq. (5.25) with the initial condition $n(t = t_0, \boldsymbol{x}) = n_0(\boldsymbol{x})$ is given by the Feynman-Kac formula

$$\boxed{n(t, \boldsymbol{x}) = \langle G(t, t_0)\, n_0[\boldsymbol{\xi}(t, \boldsymbol{x}|t_0)]\rangle_{\mathbf{w}}\,,} \tag{5.26}$$

where $\boldsymbol{\xi}(t, \boldsymbol{x}|t_0)$ is the Wiener trajectory defined as follows:

$$\boxed{\boldsymbol{\xi}(t, \boldsymbol{x}|t_0) = \boldsymbol{x} - \int_{t_0}^{t} \boldsymbol{U}[\tau, \boldsymbol{\xi}(t, \boldsymbol{x}|\tau)]\, d\tau + \sqrt{2D^{(n)}}\, \mathbf{w}(t - t_0),} \tag{5.27}$$

where $\mathbf{w}(t)$ is the Wiener random process that describes the molecular diffusion and has the following properties

$$\langle \mathbf{w}(t)\rangle_{\mathbf{w}} = 0, \quad \left\langle \mathbf{w}_i(t+\tau)\mathbf{w}_j(t)\right\rangle_{\mathbf{w}} = \tau\,\delta_{ij}, \tag{5.28}$$

and $\langle \ldots \rangle_{\mathbf{w}}$ denotes the averaging over the statistics of the Wiener process. The Green function $G(t, t_0)$ is determined by

$$\boxed{G(t, t_0) = \exp\left\{-\int_{t_0}^{t} \mathrm{Co}[\sigma, \boldsymbol{\xi}(t, \boldsymbol{x}|\sigma)]\, d\sigma\right\},} \tag{5.29}$$

where $\mathrm{Co} = \nabla \cdot \boldsymbol{U}$. The velocity $U_i[\tau, \boldsymbol{\xi}(t, \boldsymbol{x}|\tau)]$ describes the Eulerian velocity determined on the Wiener trajectory. Equation (5.27) describes a set of random trajectories that pass through the point $\boldsymbol{x}$ at time t.

First, we show that Eq. (5.26) is a solution of Eq. (5.25). The field $n(t, \boldsymbol{x})$ in Eq. (5.26) is specified at time t. Let us determine the field $n(t + \Delta t, \boldsymbol{x})$ specified at time $t + \Delta t$ by means of the replacements $t \to t + \Delta t$ and $t_0 \to t$ in Eq. (5.26):

$$n(t + \Delta t, \boldsymbol{x}) = \langle G(t + \Delta t, t)\, n[t, \boldsymbol{\xi}(t + \Delta t, \boldsymbol{x}|t)]\rangle_{\mathbf{w}}\,, \tag{5.30}$$

where

$$\boldsymbol{\xi}_{\Delta t} \equiv \boldsymbol{\xi}(t+\Delta t, \boldsymbol{x}|t) = \boldsymbol{x} - \int_t^{t+\Delta t} \boldsymbol{U}[\sigma, \boldsymbol{\xi}(t+\Delta t, \boldsymbol{x}|\sigma)]\, d\sigma + \sqrt{2D^{(n)}}\, \mathbf{w}(\Delta t), \tag{5.31}$$

and

$$G(t+\Delta t, t) = \exp\left\{- \int_t^{t+\Delta t} \mathrm{Co}[\sigma, \boldsymbol{\xi}(t+\Delta t, \boldsymbol{x}|\sigma)]\, d\sigma\right\}. \tag{5.32}$$

We expand the function $n[t, \boldsymbol{\xi}_{\Delta t}]$ in Eq. (5.30) in a Taylor series in the vicinity of the point $\boldsymbol{x}$:

$$\begin{aligned} n\left(t, \boldsymbol{\xi}_{\Delta t}\right) =& n(t, \boldsymbol{x}) + \frac{\partial n(t, \boldsymbol{x})}{\partial x_m}\left(\boldsymbol{\xi}_{\Delta t} - \boldsymbol{x}\right)_m + \frac{1}{2}\frac{\partial^2 n(t, \boldsymbol{x})}{\partial x_m \partial x_\ell}\left(\boldsymbol{\xi}_{\Delta t} - \boldsymbol{x}\right)_m\left(\boldsymbol{\xi}_{\Delta t} - \boldsymbol{x}\right)_\ell \\ &+ \dots . \end{aligned} \tag{5.33}$$

We rewrite the equation for the Wiener trajectory as follows:

$$\left(\boldsymbol{\xi}_{t_2 - t_1} - \boldsymbol{x}\right)_m = - \int_0^{t_2 - t_1} U_m(t_s, \boldsymbol{\xi}_s)\, ds + \sqrt{2D^{(n)}}\, \mathrm{w}_m(t_2 - t_1), \tag{5.34}$$

where $t_s = t_2 - s$, $\boldsymbol{\xi}(t_2, t_2 - s) \equiv \boldsymbol{\xi}_s$. We evaluate the integral in Eq. (5.34) using the "mean value" theorem as

$$\left(\boldsymbol{\xi}_{\Delta t} - \mathbf{x}\right)_m = -U_m(\tilde{t}, \mathbf{x})\Delta t + \sqrt{2D^{(n)}}\mathrm{w}_m + \mathrm{O}[(\Delta t)^2], \tag{5.35}$$

where $\tilde{t}$ is within the interval $(t, t + \Delta t)$. Substituting Eq. (5.35) into Eq. (5.33), we obtain an expression for the particle number density $n(t + \Delta t, \boldsymbol{\xi}_{\Delta t})$ as

$$\begin{aligned} n(t+\Delta t, \boldsymbol{\xi}_{\Delta t}) =& n(t, \mathbf{x}) + \frac{\partial n(t, \mathbf{x})}{\partial x_m}\left[-U_m(\tilde{t}, \mathbf{x})\Delta t + \sqrt{2D^{(n)}}\, \mathrm{w}_m\right] \\ &+ \frac{1}{2}\frac{\partial^2 n(t, \mathbf{x})}{\partial x_m \partial x_\ell}\left[-U_m(\tilde{t}, \mathbf{x})\Delta t + \sqrt{2D^{(n)}}\, \mathrm{w}_m\right]\Big[- U_\ell(\tilde{t}, \mathbf{x})\Delta t \\ &+ \sqrt{2D^{(n)}}\, \mathrm{w}_\ell\Big] + \mathrm{O}[(\Delta t)^2]. \end{aligned} \tag{5.36}$$

Expanding the function $G(t + \Delta t, t)$ in Eq. (5.32) in a Taylor series for small Δt and evaluating the integral, $\int_t^{t+\Delta t} \mathrm{Co}\left(\sigma, \boldsymbol{\xi}_\sigma\right) d\sigma$, using the "mean value" theorem, we determine the function $G(t + \Delta t, t)$ as

$$G(t+\Delta t, t) = 1 - \int_t^{t+\Delta t} \mathrm{Co}\left(\sigma, \boldsymbol{\xi}_\sigma\right) \, d\sigma + \mathrm{O}[(\Delta t)^2]$$
$$= 1 - \mathrm{Co}\left(\tilde{t}, \mathbf{x}\right) \Delta t + \mathrm{O}[(\Delta t)^2], \tag{5.37}$$

where $\tilde{t}$ is within the interval $(t, t + \Delta t)$. Substituting Eqs. (5.36) and (5.37) into Eq. (5.30), and averaging obtained equation over statistics of Wiener trajectories, we obtain an equation for the field $n(t + \Delta t, \mathbf{x})$:

$$n(t+\Delta t, \boldsymbol{x}) = n(t, \mathbf{x}) \left[1 - \mathrm{Co}(\tilde{t}, \mathbf{x}) \, \Delta t\right] + \frac{\partial n(t, \mathbf{x})}{\partial x_m} \left[-U_m(\tilde{t}, \mathbf{x}) \Delta t\right]$$
$$+ \frac{\partial^2 n(t, \mathbf{x})}{\partial x_m \partial x_\ell} D^{(n)} \, \delta_{m\ell} \, \Delta t + \mathrm{O}[(\Delta t)^2]. \tag{5.38}$$

Using Eq. (5.38), we determine the limit

$$\frac{\partial n}{\partial t} \equiv \left(\frac{n(t+\Delta t, \mathbf{x}) - n(t, \mathbf{x})}{\Delta t}\right)_{\Delta t \to 0}, \tag{5.39}$$

which yields the equation for the particle number density:

$$\frac{\partial n}{\partial t} = -n(\boldsymbol{\nabla} \cdot \boldsymbol{U}) - (\boldsymbol{U} \cdot \boldsymbol{\nabla}) n + D^{(n)} \Delta n. \tag{5.40}$$

Equation (5.40) coincides with Eq. (5.25), so that Eq. (5.26) is a solution of Eq. (5.25).

5.2.2 Mean-Field Equation for Particle Number Density

In this section, we derive the mean-field equation using solution (5.26) of Eq. (5.25) for particle number density. Let us determine the field $n(t+\Delta t, \boldsymbol{x})$ specified at time $t + \Delta t$ by means of the replacements $t \to t + \Delta t$ and $t_0 \to t$ in Eq. (5.26):

$$n(t+\Delta t, \boldsymbol{x}) = \langle G(t+\Delta t, t) \, n[t, \boldsymbol{\xi}(t+\Delta t, \boldsymbol{x}|t)] \rangle_{\mathbf{w}}, \tag{5.41}$$

where $\boldsymbol{\xi}(t + \Delta t, \boldsymbol{x}|t)$ is given by Eq. (5.31). We expand the function $n\left(t, \boldsymbol{\xi}_{\Delta t}\right)$ in Eq. (5.41) in a Taylor series in the vicinity of the point $\boldsymbol{x}$:

$$n\left(t, \boldsymbol{\xi}_{\Delta t}\right) = n(t, \boldsymbol{x}) + \frac{\partial n(t, \boldsymbol{x})}{\partial x_m} \left(\boldsymbol{\xi}_{\Delta t} - \boldsymbol{x}\right)_m + \frac{1}{2} \frac{\partial^2 n(t, \boldsymbol{x})}{\partial x_m \partial x_\ell} \left(\boldsymbol{\xi}_{\Delta t} - \boldsymbol{x}\right)_m \left(\boldsymbol{\xi}_{\Delta t} - \boldsymbol{x}\right)_\ell$$
$$+ \dots . \tag{5.42}$$

We expand the velocity $U_m(t_s, \boldsymbol{\xi}_s)$ in Eq. (5.31) in a Taylor series in the vicinity of the point $\boldsymbol{x}$,

$$U_m(t_s, \boldsymbol{\xi}_s) = U_m(t, \boldsymbol{x}) + \frac{\partial U_m(t, \boldsymbol{x})}{\partial x_\ell} \left(\boldsymbol{\xi}_s - \boldsymbol{x}\right)_\ell + \ldots = U_m(t, \boldsymbol{x})$$

$$
+\frac{\partial U_m(t,\boldsymbol{x})}{\partial x_\ell}\left[-\int_0^s U_\ell(t_\sigma,\boldsymbol{\xi}_\sigma)\,d\sigma+\sqrt{2D^{(n)}}\,\mathrm{w}_\ell(s)\right]+\ldots, \tag{5.43}
$$

where $t_\sigma = t + \Delta t - \sigma$ and $\boldsymbol{\xi}(t_2, t_1) \equiv \boldsymbol{\xi}_{t_2-t_1}$, i.e., $\boldsymbol{\xi}_\sigma = \boldsymbol{\xi}(t + \Delta t, t_\sigma)$. Substituting Eq. (5.43) into Eq. (5.31) and calculating the integrals, we obtain

$$
\begin{aligned}
\left(\boldsymbol{\xi}_{t_2-t_1}-\boldsymbol{x}\right)_m &= -\int_0^{t_2-t_1}\left[U_m(t,\boldsymbol{x})+\frac{\partial U_m(t,\boldsymbol{x})}{\partial x_\ell}\left(-U_\ell(t,\boldsymbol{x})s\right.\right.\\
&\quad\left.\left.+\sqrt{2D^{(n)}}\,\mathrm{w}_\ell(s)\right)\right]ds+\sqrt{2D^{(n)}}\,\mathrm{w}_m(t_2-t_1) = -U_m\,(t_2-t_1)\\
&\quad+\frac{1}{2}U_\ell\,\frac{\partial U_m}{\partial x_\ell}\,(t_2-t_1)^2-\sqrt{2D^{(n)}}\,\frac{\partial U_m(t,\boldsymbol{x})}{\partial x_\ell}\int_0^{t_2-t_1}\mathrm{w}_\ell(s)\,ds\\
&\quad+\sqrt{2D^{(n)}}\,\mathrm{w}_m(t_2-t_1)+O\left[(t_2-t_1)^3\right].
\end{aligned} \tag{5.44}
$$

Substituting Eq. (5.44) into Eq. (5.42), we obtain an equation for $n\left(t+\Delta t, \boldsymbol{\xi}_{\Delta t}\right)$:

$$
\begin{aligned}
n\left(t+\Delta t,\boldsymbol{\xi}_{\Delta t}\right) &= n(t,\boldsymbol{x})+\frac{\partial n(t,\boldsymbol{x})}{\partial x_m}\left[-U_m\,\Delta t+\frac{1}{2}U_\ell\,\frac{\partial U_m}{\partial x_\ell}\,(\Delta t)^2\right]\\
&\quad+\frac{1}{2}\frac{\partial^2 n(t,\boldsymbol{x})}{\partial x_m\partial x_\ell}\left[U_m U_\ell\,(\Delta t)^2+2D^{(n)}\mathrm{w}_m\mathrm{w}_\ell\right]+O[(\Delta t)^3].
\end{aligned} \tag{5.45}
$$

Next, we expand the function $\mathrm{Co}(\sigma, \boldsymbol{\xi}_\sigma)$ in $G(t + \Delta t, t)$ in a Taylor series in the vicinity of the point $\mathbf{x}$:

$$
\begin{aligned}
\mathrm{Co}(\sigma,\boldsymbol{\xi}_\sigma) &= \mathrm{Co}(t,\boldsymbol{x})+\frac{\partial \mathrm{Co}(t,\boldsymbol{x})}{\partial x_m}\left(\boldsymbol{\xi}_\sigma-\boldsymbol{x}\right)_m+\ldots\\
&= \mathrm{Co}(t,\boldsymbol{x})+\frac{\partial \mathrm{Co}(t,\boldsymbol{x})}{\partial x_m}\left[-U_m\sigma+\sqrt{2D^{(n)}}\mathrm{w}_m(\sigma)\right]+\ldots.
\end{aligned} \tag{5.46}
$$

We calculate the integral $\int_t^{t+\Delta t}\mathrm{Co}(\sigma, \boldsymbol{\xi}_\sigma)\,d\sigma$:

$$
\begin{aligned}
\int_t^{t+\Delta t}\mathrm{Co}(\sigma,\boldsymbol{\xi}_\sigma)\,d\sigma &= \mathrm{Co}(t,\mathbf{x})\Delta t-\frac{1}{2}U_m\frac{\partial \mathrm{Co}}{\partial x_m}(\Delta t)^2\\
&\quad+\sqrt{2D^{(n)}}\frac{\partial \mathrm{Co}}{\partial x_m}\int_t^{t+\Delta t}\mathrm{w}_m\,d\sigma+O[(\Delta t)^3].
\end{aligned} \tag{5.47}
$$

Using Eq. (5.47), we calculate the function $G(t + \Delta t, t)$:

$$
\begin{aligned}
G(t+\Delta t,t) &= 1-\mathrm{Co}\,\Delta t+\frac{1}{2}U_m\frac{\partial \mathrm{Co}}{\partial x_m}(\Delta t)^2+\frac{1}{2}\mathrm{Co}^2(\Delta t)^2\\
&\quad-\sqrt{2D^{(n)}}\frac{\partial \mathrm{Co}}{\partial x_m}\int_t^{t+\Delta t}\mathrm{w}_m\,d\sigma+O[(\Delta t)^3].
\end{aligned} \tag{5.48}
$$

Substituting Eqs. (5.45) and (5.48) into Eq. (5.41), averaging obtained equation over statistics of Wiener trajectories and using Eq. (5.28), we obtain an equation for $n(t+\Delta t, \boldsymbol{x}) = \langle n[t+\Delta t, \boldsymbol{\xi}_{\Delta t}]\rangle_{\mathbf{w}}$:

$$
\begin{aligned}
n(t+\Delta t, \boldsymbol{x}) &= n(t, \boldsymbol{x}) + n(t, \boldsymbol{x}) \left[-\mathrm{Co} + \frac{1}{2} U_m \frac{\partial \mathrm{Co}}{\partial x_m} \Delta t + \frac{1}{2} \mathrm{Co}^2 \, \Delta t \right] \Delta t \\
&+ \frac{\partial n(t, \boldsymbol{x})}{\partial x_m} \left[-U_m + U_m \, \mathrm{Co} \, \Delta t + \frac{1}{2} U_\ell \frac{\partial U_m}{\partial x_\ell} \Delta t \right] \Delta t \\
&+ \frac{\partial^2 n(t, \boldsymbol{x})}{\partial x_m \partial x_\ell} \left[D^{(n)} \delta_{m\ell} + \frac{1}{2} U_m U_\ell \, \Delta t \right] \Delta t. \qquad (5.49)
\end{aligned}
$$

The velocity field is Gaussian, delta-correlated in a time random process: $\langle u_i(t, \boldsymbol{x}) u_j(t', \boldsymbol{y})\rangle_{\boldsymbol{u}} = 2\delta(t-t') f_{ij}(t, \boldsymbol{x}, \boldsymbol{y})$. This velocity field is modeled by a renovating flow, in which

- we assume that the velocity $\boldsymbol{u}$ is constant (time-independent) at small time intervals $(0, \Delta t)$; $(\Delta t, 2\Delta t)$; $\cdots$, and changes every small time Δt, i.e., we consider a random velocity field with a constant renewal time;
- the velocity is statistically independent at different time intervals, and this velocity field cannot be considered as a stationary velocity field for small times $\sim \Delta t$, but it behaves like a stationary field for $t \gg \Delta t$; and
- the random velocity and the random Wiener process are independent random processes.

Here $\langle \ldots \rangle_{\boldsymbol{u}}$ is averaging over statistics of random velocity field, $\delta(t)$ is the Dirac delta function and $f_{ij}(t, \boldsymbol{x}, \boldsymbol{y}) = \langle u_i(t, \boldsymbol{x}) u_j(t, \boldsymbol{y})\rangle_{\boldsymbol{u}}$.

Averaging Eq. (5.49) over an ensemble of the turbulent velocity field, we obtain the mean field $\overline{n}(t+\Delta t, \boldsymbol{x}) = \langle n(t+\Delta t, \boldsymbol{x})\rangle_{\boldsymbol{u}}$:

$$
\begin{aligned}
\frac{\overline{n}(t+\Delta t, \boldsymbol{x}) - \overline{n}(t, \boldsymbol{x})}{\Delta t} &= -\nabla_m \left[\overline{n} \left(\overline{U}_m - \frac{1}{2} \langle \Delta t \, U_m \mathrm{Co}\rangle_{\boldsymbol{u}} \right) \right] \\
&+ \nabla_m \nabla_\ell \left[\overline{n} \left(D^{(n)} \delta_{m\ell} + \frac{1}{2} \langle \Delta t \, U_m \, U_\ell \rangle_{\boldsymbol{u}} \right) \right] + O[(\Delta t)^2], \qquad (5.50)
\end{aligned}
$$

where we take into account that $U_m \nabla_m \mathrm{Co} = \nabla_m (U_m \mathrm{Co}) - \mathrm{Co}^2$, $U_\ell \nabla_\ell U_m = \nabla_\ell (U_m U_\ell) - U_m \mathrm{Co}$ and $\overline{\mathrm{Co}} = \nabla_m \overline{U}_m$. In Eq. (5.50), we calculate the limit at $\Delta t \to 0$, i.e.,

$$
\frac{\partial \overline{n}}{\partial t} \equiv \left(\frac{\overline{n}(t+\Delta t, \boldsymbol{x}) - \overline{n}(t, \boldsymbol{x})}{\Delta t} \right)_{\Delta t \to 0}. \qquad (5.51)
$$

5.2.3 Turbulent Diffusion Tensor and Effective Pumping Velocity for Particles

The procedure described in Section 5.2.2 yields the mean-field equation for the particle number density in inhomogeneous and density-stratified turbulence as

$$\boxed{\frac{\partial \overline{n}}{\partial t} + \nabla \cdot \left[\left(\overline{U} + V^{\rm eff}\right) \overline{n}\right] = \nabla_m \nabla_\ell \left[\left(D^{(n)} \delta_{m\ell} + D^{(T)}_{m\ell}\right) \overline{n}\right],} \tag{5.52}$$

where

$$\boxed{D^{(T)}_{m\ell} = \langle \tau u_m \, u_\ell \rangle = \frac{1}{2} \lim_{\Delta t \to 0} \langle \Delta t \, U_m \, U_\ell \rangle_{\boldsymbol{u}}} \tag{5.53}$$

is the turbulent diffusion tensor and

$$\boxed{V^{\rm eff} = - \langle \tau \boldsymbol{u} \, (\nabla \cdot \boldsymbol{u}) \rangle = -\frac{1}{2} \lim_{\Delta t \to 0} \langle \Delta t \, \boldsymbol{U} \, Co \rangle_{\boldsymbol{u}}} \tag{5.54}$$

is the effective pumping velocity. For large Péclet numbers, the time τ is determined by the correlation time of turbulent flow, $\tau = \tau_0$, and Eq. (5.54) coincides with Eq. (2.17) (see Section 2.1.6), while for small Péclet numbers, the time τ is determined by the molecular diffusion time, $\tau = \ell_0^2/D^{(n)}$, and Eq. (5.54) coincides with Eq. (2.28). The effective pumping velocity of particles describes the phenomenon of turbulent thermal diffusion of particles and causes the formation of inhomogeneous large-scale particle spatial distributions in small-scale stratified turbulence.

5.3 Turbulent Transport of Magnetic Field

In this section, we study turbulent transport of magnetic field in incompressible inhomogeneous and helical turbulence using the path-integral approach.

5.3.1 Solution for the Induction Equation

The equation for the evolution of the instantaneous magnetic field $B_i(t, \boldsymbol{x})$ reads:

$$\frac{\partial B_i}{\partial t} + (\boldsymbol{U} \cdot \nabla) B_i = (\boldsymbol{B} \cdot \nabla) U_i + \eta \, \Delta B_i, \tag{5.55}$$

where η is the coefficient of the magnetic diffusion caused by the electrical conductivity of the fluid. The solution of Eq. (5.55) with the initial condition $B_i(t = t_0, \boldsymbol{x}) = B_{i0}(\boldsymbol{x})$ is given by the Feynman-Kac formula:

$$\boxed{B_i(t, \boldsymbol{x}) = \left\langle G_{ij}(t, t_0) \, B_{j0}[\boldsymbol{\xi}(t, \boldsymbol{x}|t_0)] \right\rangle_{\mathbf{w}},} \tag{5.56}$$

where $\boldsymbol{\xi}(t, \boldsymbol{x}|t_0)$ is the Wiener trajectory defined as

$$\boxed{\boldsymbol{\xi}(t, \boldsymbol{x}|t_0) = \boldsymbol{x} - \int_{t_0}^{t} U[\tau, \boldsymbol{\xi}(t, \boldsymbol{x}|\tau)]\, d\tau + \sqrt{2\eta}\; \mathbf{w}(t - t_0),} \tag{5.57}$$

and $\mathbf{w}(t)$ is a Wiener process that describes the magnetic diffusion caused by the electrical conductivity of the fluid and the velocity $U_i[\tau, \boldsymbol{\xi}(t, \boldsymbol{x}|\tau)]$ is the Eulerian velocity calculated at the Wiener trajectory. Equation (5.57) describes a set of random trajectories that pass through the point $\boldsymbol{x}$ at time t. The Wiener random process has the following properties:

$$\langle \mathbf{w}(t) \rangle_{\mathbf{w}} = 0, \quad \big\langle \mathbf{w}_i(t+\tau)\mathbf{w}_j(t) \big\rangle_{\mathbf{w}} = \tau \delta_{ij}, \tag{5.58}$$

and $\langle \dots \rangle_{\mathbf{w}}$ denotes the averaging over the statistics of the Wiener process. The Green function G_{ij} is determined by the equation

$$\boxed{\frac{d}{ds} G_{ij}(t_s, t_0) = N_{ik} G_{kj},} \tag{5.59}$$

with the initial condition $G_{ij} = \delta_{ij}$ for $t_s = t_0$ and $N_{ik} = \nabla_k U_i$.

First, we show that Eq. (5.56) is a solution of Eq. (5.55). The field $B_i(t, \boldsymbol{x})$ in Eq. (5.56) is specified at time t. Let us determine the field $B_i(t + \Delta t, \boldsymbol{x})$ specified at time $t + \Delta t$ by means of the replacements $t \to t + \Delta t$ and $t_0 \to t$ in Eq. (5.56):

$$B_i(t + \Delta t, \boldsymbol{x}) = \big\langle G_{ij}(t + \Delta t, t)\, B_j[t, \boldsymbol{\xi}(t + \Delta t, \boldsymbol{x}|t)] \big\rangle_{\mathbf{w}}, \tag{5.60}$$

where

$$\boldsymbol{\xi}_{\Delta t} \equiv \boldsymbol{\xi}(t + \Delta t, \boldsymbol{x}|t) = \boldsymbol{x} - \int_{t}^{t+\Delta t} U[\tau, \boldsymbol{\xi}(t + \Delta t, \boldsymbol{x}|\tau)]\, d\tau + \sqrt{2\eta}\; \mathbf{w}(\Delta t). \tag{5.61}$$

To find the function $G_{ij}(t + \Delta t, t)$ we solve equation (5.59) by iterations, so that

$$G_{ij}(t + \Delta t, t) = \delta_{ij} + \int_0^{\Delta t} N_{ij}(t_\sigma, \boldsymbol{\xi}_\sigma)\, d\sigma + O[(\Delta t)^2]. \tag{5.62}$$

We expand the function $B_j(t, \boldsymbol{\xi}_{\Delta t}) \equiv B_j[t, \boldsymbol{\xi}(t + \Delta t, \boldsymbol{x}|t)]$ in Eq. (5.60) in a Taylor series in the vicinity of the point $\boldsymbol{x}$:

$$\begin{aligned} B_j\left(t, \boldsymbol{\xi}_{\Delta t}\right) &= B_j(t, \boldsymbol{x}) + \frac{\partial B_j(t, \boldsymbol{x})}{\partial x_m} \left(\boldsymbol{\xi}_{\Delta t} - \boldsymbol{x}\right)_m \\ &\quad + \frac{1}{2} \frac{\partial^2 B_j(t, \boldsymbol{x})}{\partial x_m \partial x_\ell} \left(\boldsymbol{\xi}_{\Delta t} - \boldsymbol{x}\right)_m \left(\boldsymbol{\xi}_{\Delta t} - \boldsymbol{x}\right)_\ell + \dots . \end{aligned} \tag{5.63}$$

We rewrite the equation for the Wiener trajectory as

$$\left(\boldsymbol{\xi}_{t_2-t_1} - \boldsymbol{x}\right)_m = -\int_0^{t_2-t_1} U_m(t_s, \boldsymbol{\xi}_s)\, ds + \sqrt{2\eta}\; \mathrm{w}_m(t_2 - t_1), \tag{5.64}$$

where $t_s = t_2 - s$, $\boldsymbol{\xi}(t_2, t_2 - s) \equiv \boldsymbol{\xi}_s$. The integral $\int_0^{t_2-t_1} U_m(t_s, \boldsymbol{\xi}_s)\, ds$ in Eq. (5.64) is evaluated using the "mean value" theorem, which yields

$$(\boldsymbol{\xi}_{\Delta t} - \mathbf{x})_m = -U_m(\tilde{t}, \mathbf{x})\Delta t + \sqrt{2\eta}\; \mathrm{w}_m + \mathrm{O}[(\Delta t)^2], \tag{5.65}$$

where $\tilde{t}$ is within the interval $(t, t + \Delta t)$. Substituting Eq. (5.65) into Eq. (5.63), we obtain expression for the field $B_i(t + \Delta t, \boldsymbol{\xi}_{\Delta t})$:

$$\begin{aligned} B_j(t + \Delta t, \boldsymbol{\xi}_{\Delta t}) &= B_j(t, \mathbf{x}) + \frac{\partial B_j(t, \mathbf{x})}{\partial x_m} \left[-U_m(\tilde{t}, \mathbf{x})\Delta t + \sqrt{2\eta}\; \mathrm{w}_m\right] \\ &+ \frac{1}{2} \frac{\partial^2 B_j(t, \mathbf{x})}{\partial x_m \partial x_\ell} \left[-U_m(\tilde{t}, \mathbf{x})\Delta t + \sqrt{2\eta}\; \mathrm{w}_m\right]\left[-U_\ell(\tilde{t}, \mathbf{x})\Delta t + \sqrt{2\eta}\; \mathrm{w}_\ell\right] \\ &+ \mathrm{O}[(\Delta t)^2]. \end{aligned} \tag{5.66}$$

To determine the function $G_{ij}(t + \Delta t, t)$ given by Eq. (5.62), we evaluate the integral, $\int_0^{\Delta t} N_{ij}(t_\sigma, \boldsymbol{\xi}_\sigma)\, d\sigma$, using the "mean value" theorem, so that the function $G_{ij}(t + \Delta t, t)$ is given by

$$G_{ij}(t + \Delta t, t) = \delta_{ij} + \nabla_j U_i(\tilde{t}, \boldsymbol{x})\Delta t + \mathrm{O}[(\Delta t)^2], \tag{5.67}$$

where $\tilde{t}$ is within the interval $(t, t + \Delta t)$. Substituting Eqs. (5.66) and (5.67) into Eq. (5.60) and averaging over the statistics of Wiener trajectories, we obtain an equation for the field $B_i(t + \Delta t, \mathbf{x})$:

$$\begin{aligned} B_i(t + \Delta t, \boldsymbol{x}) &= \left[B_i(t, \mathbf{x}) + B_j(t, \mathbf{x}) \frac{\partial U_i(\tilde{t}, \mathbf{x})}{\partial x_j} - U_m(\tilde{t}, \mathbf{x}) \frac{\partial B_i(t, \mathbf{x})}{\partial x_m}\right] \Delta t \\ &+ \eta \frac{\partial^2 B_i(t, \mathbf{x})}{\partial x_m \partial x_m} \Delta t + \mathrm{O}[(\Delta t)^2]. \end{aligned} \tag{5.68}$$

Using Eq. (5.68), we calculate the limit

$$\frac{\partial B_i}{\partial t} \equiv \left(\frac{B_i(t + \Delta t, \mathbf{x}) - B_i(t, \mathbf{x})}{\Delta t}\right)_{\Delta t \to 0}, \tag{5.69}$$

which yields the equation for the magnetic field:

$$\frac{\partial B_i}{\partial t} = (\boldsymbol{B} \cdot \nabla)U_i - (\boldsymbol{U} \cdot \nabla)B_i + \eta \Delta B_i. \tag{5.70}$$

Equation (5.70) coincides with Eq. (5.55), so that Eq. (5.56) is a solution of Eq. (5.55).

5.3.2 Mean-Field Equation for Magnetic Field

In this section, we derive the mean-field induction equation using solution (5.56) of the induction equation (5.55). Let us determine the field $B_i(t+\Delta t, \boldsymbol{x})$ specified at time $t+\Delta t$ by means of the replacements $t \to t+\Delta t$ and $t_0 \to t$ in Eq. (5.56):

$$B_i(t+\Delta t, \boldsymbol{x}) = \big\langle G_{ij}(t+\Delta t, t)\, B_j[t, \boldsymbol{\xi}(t+\Delta t, \boldsymbol{x}|t)]\big\rangle_{\mathbf{w}}, \tag{5.71}$$

where $\boldsymbol{\xi}(t+\Delta t, \boldsymbol{x}|t)$ is given by Eq. (5.61) and the function $G_{ij}(t+\Delta t, t)$ can be determined using Eq. (5.59). We expand the function $B_j(t, \boldsymbol{\xi}_{\Delta t})$ in Eq. (5.71) in a Taylor series in the vicinity of the point $\boldsymbol{x}$:

$$\begin{aligned} B_j\left(t, \boldsymbol{\xi}_{\Delta t}\right) &= B_j(t, \boldsymbol{x}) + \frac{\partial B_j(t, \boldsymbol{x})}{\partial x_m} \left(\boldsymbol{\xi}_{\Delta t} - \boldsymbol{x}\right)_m \\ &\quad + \frac{1}{2} \frac{\partial^2 B_j(t, \boldsymbol{x})}{\partial x_m \partial x_\ell} \left(\boldsymbol{\xi}_{\Delta t} - \boldsymbol{x}\right)_m \left(\boldsymbol{\xi}_{\Delta t} - \boldsymbol{x}\right)_\ell + \ldots . \end{aligned} \tag{5.72}$$

We rewrite Eq. (5.61) for the Wiener trajectory as

$$\left(\boldsymbol{\xi}_{t_2-t_1} - \boldsymbol{x}\right)_m = - \int_0^{t_2-t_1} U_m(t_s, \boldsymbol{\xi}_s)\, ds + \sqrt{2\eta}\; \mathrm{w}_m(t_2 - t_1), \tag{5.73}$$

where $t_s = t + \Delta t - s$, and $\boldsymbol{\xi}(t_2, t_1) \equiv \boldsymbol{\xi}_{t_2-t_1}$, i. e., $\boldsymbol{\xi}_s = \boldsymbol{\xi}(t+\Delta t, t_s)$. To find the function $G_{ij}(t+\Delta t, t)$, we solve equation (5.59) by iterations:

$$\begin{aligned} G_{ij}(t+\Delta t, t) &= \delta_{ij} + \int_0^{\Delta t} N_{ij}(\sigma, \boldsymbol{\xi}_\sigma)\, d\sigma + \int_0^{\Delta t} N_{ik}(s, \boldsymbol{\xi}_s)\, ds \int_0^{s} N_{kj}(\sigma, \boldsymbol{\xi}_\sigma)\, d\sigma \\ &\quad + O\left[(\Delta t)^3\right]. \end{aligned} \tag{5.74}$$

We also expand the velocity $U_m(t_s, \boldsymbol{\xi}_s)$ in Eq. (5.73) in a Taylor series in the vicinity of the point $\boldsymbol{x}$:

$$\begin{aligned} U_m(t_s, \boldsymbol{\xi}_s) &= U_m(t, \boldsymbol{x}) + \frac{\partial U_m(t, \boldsymbol{x})}{\partial x_\ell} \left(\boldsymbol{\xi}_s - \boldsymbol{x}\right)_\ell + \ldots \\ &= U_m(t, \boldsymbol{x}) + \frac{\partial U_m(t, \boldsymbol{x})}{\partial x_\ell} \left[- \int_0^{s} U_\ell(t_\sigma, \boldsymbol{\xi}_\sigma)\, d\sigma + \sqrt{2\eta}\; \mathrm{w}_\ell(s)\right] + \ldots , \end{aligned} \tag{5.75}$$

where $t_\sigma = t + \Delta t - \sigma$ and $\boldsymbol{\xi}_\sigma = \boldsymbol{\xi}(t+\Delta t, t_\sigma)$. Substituting Eq. (5.75) into Eq. (5.73) and calculating the integrals in Eq. (5.73), we obtain

$$\begin{aligned} \left(\boldsymbol{\xi}_{\Delta t} - \boldsymbol{x}\right)_m &= - \int_0^{\Delta t} \left[U_m(t, \boldsymbol{x}) + \frac{\partial U_m(t, \boldsymbol{x})}{\partial x_\ell} \left(-U_\ell(t, \boldsymbol{x}) s + \sqrt{2\eta}\; \mathrm{w}_\ell(s)\right)\right] ds \\ &\quad + \sqrt{2\eta}\; \mathrm{w}_m(\Delta t) = -U_m\, \Delta t + \frac{1}{2} U_\ell \frac{\partial U_m}{\partial x_\ell} (\Delta t)^2 + \sqrt{2\eta}\, \mathrm{w}_m(\Delta t) \\ &\quad - \sqrt{2\eta}\; \frac{\partial U_m(t, \boldsymbol{x})}{\partial x_\ell} \int_0^{\Delta t} \mathrm{w}_\ell(s)\, ds + O[(\Delta t)^3]. \end{aligned} \tag{5.76}$$

Substituting Eq. (5.76) into Eq. (5.72), we obtain expression for the magnetic field $B_j(t + \Delta t, \boldsymbol{\xi}_{\Delta t})$:

$$B_j\left(t + \Delta t, \boldsymbol{\xi}_{\Delta t}\right) = B_j(t, \boldsymbol{x}) + \frac{\partial B_j(t, \boldsymbol{x})}{\partial x_m}\left[-U_m\, \Delta t + \frac{1}{2} U_\ell\, \frac{\partial U_m}{\partial x_\ell}\, (\Delta t)^2\right] + \frac{\partial^2 B_j(t, \boldsymbol{x})}{\partial x_m \partial x_\ell}\left[\frac{1}{2} U_m U_\ell\, (\Delta t)^2 + \eta\, \mathrm{w}_m \mathrm{w}_\ell\right] + O\left[(\Delta t)^3\right]. \tag{5.77}$$

Similar calculations for the function $N_{ij}(\sigma, \boldsymbol{\xi}_\sigma)$ yield

$$N_{ij}(\sigma, \boldsymbol{\xi}_\sigma) = N_{ij}(t, \mathbf{x}) + \frac{\partial N_{ij}(t, \mathbf{x})}{\partial x_m}\left(-U_m\, \sigma + \sqrt{2\eta}\; \mathrm{w}_m\right) + \eta\, \frac{\partial^2 N_{ij}(t, \mathbf{x})}{\partial x_m \partial x_\ell}\, \mathrm{w}_m \mathrm{w}_\ell + O\left(\sigma^2\right). \tag{5.78}$$

We determine also the integral $\int_0^s N_{ij}(\sigma, \boldsymbol{\xi}_\sigma)\, d\sigma$ as

$$\int_0^s N_{ij}(\sigma, \boldsymbol{\xi}_\sigma)\, d\sigma = N_{ij}(t, \mathbf{x}) s - \frac{1}{2} U_m \frac{\partial N_{ij}}{\partial x_m} s^2 + \sqrt{2\eta}\, \frac{\partial N_{ij}}{\partial x_m} \int_o^s \mathrm{w}_m(\sigma)\, d\sigma + \eta\, \frac{\partial^2 N_{ij}}{\partial x_m \partial x_\ell} \int_0^s \mathrm{w}_m \mathrm{w}_\ell\, d\sigma + O\left(s^3\right). \tag{5.79}$$

Using Eqs. (5.74) and (5.79), we calculate the function $G_{ij}(t + \Delta t, t)$:

$$G_{ij}(t + \Delta t, t) = \delta_{ij} + N_{ij}(t, \mathbf{x})\Delta t + \frac{1}{2}\left(-U_k \frac{\partial N_{ij}(t, \mathbf{x})}{\partial x_k} + N_{ik} N_{kj}\right)(\Delta t)^2 + \sqrt{2\eta}\, \frac{\partial N_{ij}}{\partial x_q} \int_0^{\Delta t} \mathrm{w}_q(\sigma)\, d\sigma + \eta\, \frac{\partial^2 N_{ij}}{\partial x_m \partial x_\ell} \int_0^{\Delta t} \mathrm{w}_m \mathrm{w}_\ell\, ds + O\left[(\Delta t)^3\right]. \tag{5.80}$$

Taking into account that the velocity field is incompressible, we obtain

$$\frac{\partial}{\partial x_k}\left(U_i \frac{\partial U_k}{\partial x_j}\right) = \frac{\partial U_i}{\partial x_k} \frac{\partial U_k}{\partial x_j}. \tag{5.81}$$

This allows us to rewrite Eq. (5.80) as

$$G_{ij}(t + \Delta t, t) = \delta_{ij} + N_{ij}(t, \boldsymbol{x})\Delta t + \frac{1}{2}\left[\frac{\partial}{\partial x_k}\left(U_i \frac{\partial U_k}{\partial x_j} - U_k \frac{\partial U_i}{\partial x_j}\right)\right](\Delta t)^2 + \sqrt{2\eta} \frac{\partial N_{ij}}{\partial x_q} \int_0^{\Delta t} \mathrm{w}_q\, d\sigma + \eta\, \frac{\partial^2 N_{ij}}{\partial x_m \partial x_\ell} \int_0^{\Delta t} \mathrm{w}_m \mathrm{w}_\ell\, ds + O[(\Delta t)^3]. \tag{5.82}$$

Substituting Eqs. (5.77) and (5.82) into Eq. (5.71), averaging over statistics of Wiener trajectories and using Eq. (5.58), we determine $B_i(t + \Delta t, \boldsymbol{x})$:

$$B_i(t+\Delta t,\boldsymbol{x}) = B_i(t,\boldsymbol{x}) + Q_i(t,\boldsymbol{x})\Delta t + P_i(t,\boldsymbol{x})(\Delta t)^2 + O\left[(\Delta t)^3\right], \tag{5.83}$$

where

$$Q_i = B_m \frac{\partial U_i}{\partial x_m} - U_m \frac{\partial B_i}{\partial x_m} + \eta \delta_{mn} \frac{\partial^2 B_i}{\partial x_m \partial x_n}, \tag{5.84}$$

$$P_i = \frac{1}{2} B_n \left[\frac{\partial}{\partial x_m} \left(U_i \frac{\partial U_m}{\partial x_n} - U_m \frac{\partial U_i}{\partial x_n} \right) \right] - \frac{\partial B_n}{\partial x_m} \left(U_m \frac{\partial U_i}{\partial x_n} \right) + \frac{1}{2} \frac{\partial B_i}{\partial x_m} \left(U_n \frac{\partial U_m}{\partial x_n} \right) + \frac{1}{2} U_m U_n \frac{\partial^2 B_i}{\partial x_m \partial x_n}. \tag{5.85}$$

We consider the following model of a short-correlated in time random velocity field. The velocity field is Gaussian, delta-correlated in time random process: $\langle u_i(t,\boldsymbol{x}) u_j(t',\boldsymbol{y})\rangle_{\boldsymbol{u}} = 2\delta(t-t') f_{ij}(\boldsymbol{x},\boldsymbol{y})$, where the angular brackets $\langle \ldots \rangle_{\boldsymbol{u}}$ denote averaging over a statistics of random velocity field, $\delta(t)$ is the Dirac delta function and $f_{ij}(t,\boldsymbol{x},\boldsymbol{y}) = \langle u_i(t,\boldsymbol{x}) u_j(t,\boldsymbol{y})\rangle_{\boldsymbol{u}}$. We assume that (i) the velocity is constant (time-independent) at small time intervals $(0, \Delta t)$; $(\Delta t, 2\Delta t)$; $\cdots$, and changes every small time Δt. This implies that we consider a random velocity field with a constant renewal time. (ii) The velocity is statistically independent at different time intervals (the renewal velocity field). This velocity field cannot be considered as a stationary velocity field for small times $\sim \Delta t$; however, it behaves like a stationary field for $t \gg \Delta t$. (iii) The random velocity and the random Wiener process are independent random processes.

5.3.3 Turbulent Diffusion Tensor, α Effect and Effective Pumping Velocity for Magnetic Field

Averaging Eq. (5.83) over the ensemble of the turbulent velocity field, we obtain an equation for the mean field $\overline{\boldsymbol{B}}(t+\Delta t,\boldsymbol{x}) = \langle \boldsymbol{B}(t+\Delta t,\boldsymbol{x})\rangle_u$. Calculating the limit,

$$\frac{\partial \overline{\boldsymbol{B}}}{\partial t} \equiv \left(\frac{\overline{\boldsymbol{B}}(t+\Delta t,\boldsymbol{x}) - \overline{\boldsymbol{B}}(t,\boldsymbol{x})}{\Delta t} \right)_{\Delta t \to 0}, \tag{5.86}$$

at $\Delta t \to 0$ in Eq. (5.83), we obtain an equation for the mean magnetic field:

$$\frac{\partial \overline{B}_i}{\partial t} = \overline{B}_m \nabla_m \overline{U}_i - \overline{U}_m \nabla_m \overline{B}_i + \eta \Delta \overline{B}_i + \overline{B}_n \nabla_m \Big(\langle \tau u_i \nabla_n u_m \rangle - \langle \tau u_m \nabla_n u_i \rangle \Big) - 2(\nabla_m \overline{B}_n)\, \langle \tau u_m \nabla_n u_i \rangle + \nabla_n \left(\eta_{nm}^{(T)} \nabla_m \overline{B}_i \right), \tag{5.87}$$

where

$$\boxed{\eta_{ij}^{(T)} = \left\langle \tau u_i\, u_j \right\rangle = \frac{1}{2} \lim_{\Delta t \to 0} \left\langle \Delta t\, U_i\, U_j \right\rangle_{\boldsymbol{u}}} \tag{5.88}$$

is the turbulent magnetic diffusion tensor, and

$$\langle \tau u_i\, \nabla_n u_l \rangle = \frac{1}{2} \lim_{\Delta t \to 0} \left\langle \Delta t\, U_i\, \nabla_n U_j \right\rangle_{\boldsymbol{u}} . \tag{5.89}$$

Here τ is the scale-dependent momentum relaxation time and $\overline{\boldsymbol{U}} = \langle \boldsymbol{U} \rangle_{\boldsymbol{u}}$ is the mean velocity field. To simplify the induction equation for the mean magnetic field, we use an identity

$$\left\langle \tau u_i \nabla_n u_j \right\rangle = \frac{1}{2} \left(\varepsilon_{ijm} \alpha_{mn} + \delta_{jn} V_i^{\rm eff} - \delta_{in} V_j^{\rm eff} + \nabla_n \eta_{ij}^{(T)} \right), \tag{5.90}$$

where δ_{ij} is the Kronecker tensor and ε_{ijm} is the fully antisymmetric Levi-Civita tensor. In addition,

$$\boxed{\hat{\boldsymbol{\alpha}} \equiv \alpha_{mn} = \frac{1}{2} \left(\varepsilon_{mij} \left\langle \tau u_i \nabla_n u_j \right\rangle + \varepsilon_{nij} \left\langle \tau u_i \nabla_m u_j \right\rangle \right)} \tag{5.91}$$

is the $\hat{\boldsymbol{\alpha}}$ tensor for an anisotropic turbulence, and

$$\boxed{V_m^{\rm eff} = -\frac{1}{2} \nabla_n \eta_{nm}^{(T)}} \tag{5.92}$$

is the effective pumping velocity of the mean magnetic field.

Let us show that Eq. (5.91) in the case of an isotropic turbulence is reduced to the well-known expression for the α effect. We rewrite the tensor $\varepsilon_{mij} \left\langle \tau u_i \nabla_n u_j \right\rangle$ using the identity: $\nabla_n u_j = (\partial u)_{nj} + \varepsilon_{njp} \boldsymbol{\omega}/2$, as

$$\varepsilon_{mij} \left\langle \tau u_i \nabla_n u_j \right\rangle = \frac{1}{2} \Big(\langle \tau u_n \omega_m \rangle - \langle \tau \boldsymbol{u} \cdot \boldsymbol{\omega} \rangle\, \delta_{mn} \Big) + \varepsilon_{mij} \left\langle \tau u_i\, (\partial u)_{nj} \right\rangle, \tag{5.93}$$

where $\boldsymbol{\omega} = \boldsymbol{\nabla} \times \boldsymbol{u}$ and $(\partial u)_{ij} = (\nabla_i u_j + \nabla_j u_i)/2$. Substituting Eq. (5.93) into Eq. (5.91), we rewrite the tensor α_{mn} as

$$\begin{aligned} \alpha_{mn} = \frac{1}{2} \Bigg[& \frac{1}{2} \Big(\langle \tau u_n\, \omega_m \rangle + \langle \tau u_m\, \omega_n \rangle \Big) - \langle \tau \boldsymbol{u} \cdot \boldsymbol{\omega} \rangle\, \delta_{mn} \\ & + \varepsilon_{mij} \left\langle \tau u_i\, (\partial u)_{nj} \right\rangle + \varepsilon_{nij} \left\langle \tau u_i\, (\partial u)_{mj} \right\rangle \Bigg]. \end{aligned} \tag{5.94}$$

For an isotropic turbulence, we obtain that

$$\langle \tau u_i\, \omega_j \rangle = \frac{1}{3} \delta_{ij}\, \langle \tau \boldsymbol{u} \cdot \boldsymbol{\omega} \rangle, \tag{5.95}$$

and

$$\langle \tau u_i \, (\partial u)_{mj} \rangle = \frac{1}{3} \, \delta_{ij} \, \langle \tau u_n \, (\partial u)_{mn} \rangle. \tag{5.96}$$

This yields the $\hat{\boldsymbol{\alpha}}$ tensor for an isotropic turbulence $\alpha_{ij} = \alpha \, \delta_{ij}$, where the α effect is given by

$$\boxed{\alpha = -\frac{1}{3} \, \langle \tau \boldsymbol{u} \cdot (\boldsymbol{\nabla} \times \boldsymbol{u}) \rangle \, .} \tag{5.97}$$

Equation (5.92) for the velocity $\boldsymbol{V}^{\rm eff}$ in the case of isotropic turbulence coincides with the well-known formula for the effective pumping velocity of the mean magnetic field which describes turbulent diamagnetism.

Finally, using Eqs. (5.91)–(5.92) and identity $\nabla_i \overline{B}_j = (\partial \overline{B})_{ij} + \varepsilon_{ijp} (\boldsymbol{\nabla} \times \overline{\boldsymbol{B}})_p / 2$, we rewrite Eq. (5.87) for the mean magnetic field for an isotropic turbulence as

$$\boxed{\frac{\partial \overline{\boldsymbol{B}}}{\partial t} = \boldsymbol{\nabla} \times \left(\overline{\boldsymbol{U}} \times \overline{\boldsymbol{B}} + \hat{\boldsymbol{\alpha}} \overline{\boldsymbol{B}} + \boldsymbol{V}^{\rm eff} \times \overline{\boldsymbol{B}} - \hat{\eta} \boldsymbol{\nabla} \times \overline{\boldsymbol{B}} \right),} \tag{5.98}$$

where the tensor

$$\boxed{\hat{\eta} = (\eta + \eta_{_T}) \, \delta_{ij}} \tag{5.99}$$

is the total magnetic diffusion tensor, which includes the magnetic diffusion η caused by the electrical conductivity of the fluid, $\eta_{_T} = \ell_0 \, u_0 / 3$ is the turbulent magnetic diffusion, $(\partial \overline{B})_{ij} = (\nabla_i \overline{B}_j + \nabla_j \overline{B}_i)/2$, the tensor $\hat{\boldsymbol{\alpha}} \equiv \alpha_{ij} = \alpha \, \delta_{ij}$, the α effect is given by Eq. (5.97), and the effective pumping velocity of the mean magnetic field is given by Eq. (5.92). Thus we recover the standard form of the mean induction equation derived in Chapter 3.

5.4 Further Reading

The path integral approach has been developed by R. Feynman in quantum mechanics and M. Kac in statistical physics (Feynman and Hibbs, 1965; Kac, 1959). This approach has been also adopted for investigating the turbulent transport of magnetic fields and passive scalars in a short-correlated or finite-in-time random incompressible renewal flow by Ya. Zeldovich, S. Molchanov, A. Ruzmaikin and D. Sokoloff (Zeldovich et al., 1988, 1990; Piterbarg and Ostrovskii, 2013). Later, the path integral approach has been applied for compressible random renewal flow with short-correlated or finite correlation time to study turbulent transport of particles and magnetic field (Elperin et al., 1995, 1996c,b, 1997c, 2000b, 2001a, 2002a; Rogachevskii and Kleeorin, 1997; Kleeorin et al., 2002a).

The path integral approach has been also used in various applied fields including financial mathematics and technologies (Lipton, 2001).

Turbulent transport of passive scalar and vector fields has been studied intensively during the last decades. The main progress has been achieved in the understanding of the properties of the high-order moments of a passive scalar, e.g., non-Gaussian statistics and anomalous scalings (Zeldovich et al., 1988, 1990; Kraichnan, 1994; Frisch, 1995; Gawedzki and Kupiainen, 1995; Shraiman and Siggia, 1995, 2000; Falkovich et al., 2001; Cardy et al., 2008). While the analysis of high-order moments of a passive scalar has been a subject of numerous studies, far fewer investigations discuss the dynamics of the mean passive scalar field (Avellaneda and Majda, 1990; Elperin et al., 1995, 1996c, 1997c; Majda and Kramer, 1999). However, the mean-field theory for turbulent transport of particles and gases is of great importance in view of numerous applications. In particular, this theory is applied for the analysis of transport of aerosols, pollutants and cloud droplets in the atmospheric turbulence of the Earth and other planets (Twomey, 1977; Csanady, 2012; Pasquill and Smith, 1983; Seinfeld and Pandis, 2016), dust transport in interstellar turbulence and turbulent transport of particles and gases in industrial flows (Bałdyga and Bourne, 1999).

Mean-field equations for the particle number density, temperature and magnetic fields discussed in this chapter are derived for the δ correlated in time random velocity field. When a random velocity field has small yet finite correlation time, the mean-field equation is much more complicated. As was shown by Elperin et al. (2000b), where a model of a random compressible velocity field with a constant renewal time has been used, the mean-field equation for the number density of particles contains high-order spatial derivatives. In the study by Elperin et al. (2000b), three models of a random velocity field have been considered: (i) a velocity field with a small renewal time; (ii) the Gaussian approximation for Lagrangian trajectories; and (iii) a small inhomogeneity of the velocity and mean passive scalar fields. For a small renewal time, the mean-field equation is similar to that for the δ-correlated in time random velocity field. The finite renewal time and compressibility of the velocity field cause a depletion of turbulent diffusion and a modification of an effective drift velocity. For a compressible velocity field, the form of the mean-field equation for a passive scalar depends on the details of the velocity field, i.e., the universality is lost (Elperin et al., 2000b), while for an incompressible velocity field, the universality exists even for finite renewal time.

In spite of the fact that the simplified model of a random compressible velocity field with a constant renewal time considered by Elperin et al. (2000b) allows one to derive the mean-field equation, this model is limited by the assumption of a constant renewal time and does not allow one to describe a real turbulent velocity field. A turbulent velocity field with a random renewal time is more realistic than that with a constant renewal time. In the turbulent model with a random renewal time, there are two characteristic time scales: the correlation time of a random velocity field τ_c and the mean renewal time τ. The mean-field equation for a passive scalar has been derived by Elperin et al. (2001a) for different ratios of τ/τ_c, where the turbulent diffusion coefficient is determined by the minimum of the times τ_c and τ. When the mean renewal time is very large in comparison with the correlation time of velocity field, the results obtained for a turbulent flow without renewal are recovered. When $\tau_c \ll \tau \ll \tau_n$, the form of the mean-field equation for a passive scalar is independent of the statistics of the velocity field (Elperin et al., 2001a), where τ_n is the characteristic time of variations of a mean passive scalar field. In a general case for arbitrary ratio τ/τ_c, the mean-field equation is an integral equation. However, in various applications, the second-order differential equations (in spatial variables) for the mean passive scalar field are usually used. Elperin et al. (2001a) have found conditions when the mean-field equation can be reduced to a second-order differential equation that depend on a chosen model of a turbulent velocity field. The model of the turbulent velocity field with a random renewal time reproduces important features of some real turbulent flows. For example, the interstellar turbulence is driven by supernovae explosions, and it loses memory at the instants of explosions, which can be described by the Poisson process.

6

Practice Problems and Solutions

In this chapter, we formulate different problems on turbulent transport of particles, temperature and magnetic fields and give solutions for these problems. These problems are strongly related to the content of this book and can be used for exercises and examinations.

6.1 Practice Problems on Turbulent Transport

6.1.1 Practice Problems for Chapter 1

1. To find spectrum for velocity fluctuations and scale-dependence of the characteristic turbulent time $\tau(k)$ in the inertial range, $k_0 \le k \le k_\nu$, for fully developed homogeneous and isotropic hydrodynamic turbulence applying the dimensional analysis. To determine the Kolmogorov viscous length-scale $\ell_\nu = k_\nu^{-1}$ and Kolmogorov time-scale τ_ν versus the Reynolds number, $\mathrm{Re} = u_0 \, \ell_0/\nu \gg 1$, applying the dimensional analysis, where the length $\ell_0 = k_0^{-1}$ is the integral scale of turbulence, u_0 is the characteristic turbulent velocity in the scale ℓ_0 and ν is the kinematic viscosity.

2. To determine the molecular diffusion length-scale ℓ_D and time-scale τ_D of temperature field versus Péclet number, $\mathrm{Pe} = u_0 \, \ell_0/D^{(\theta)} \gg 1$, applying the dimensional analysis, where the length ℓ_0 is the integral scale of turbulence, u_0 is the characteristic turbulent velocity in the scale ℓ_0 and $D^{(\theta)}$ is the molecular diffusion coefficient of the temperature field.

3. To determine the spectrum of isotropic temperature fluctuations produced in the inertial range $\ell_\nu \le \ell_D < \ell < \ell_0$ of the incompressible turbulent fluid flow for Prandtl number $\mathrm{Pr} \equiv \nu/D^{(\theta)} < 1$ applying the dimensional analysis, where $\ell_\nu = \ell_0 \, \mathrm{Re}^{-3/4}$ is the viscous Kolmogorov scale, $\ell_D = \ell_0 \, \mathrm{Pe}^{-3/4}$ is the characteristic scale of the molecular diffusion of temperature field, $\mathrm{Re} \equiv u_0\ell_0/\nu$ is the Reynolds number, ν is the kinematic viscosity, $\mathrm{Pe} \equiv u_0\ell_0/D^{(\theta)}$ is the Péclet number and $D^{(\theta)}$ is the molecular diffusion coefficient of the temperature field.

4. To determine the spectrum of isotropic temperature fluctuations produced in the viscous range $\ell_D < \ell < \ell_\nu$ of the incompressible turbulent fluid flow for Prandtl number $\mathrm{Pr} \equiv \nu/D^{(\theta)} > 1$ applying the dimensional analysis, where $\ell_\nu = \ell_0\,\mathrm{Re}^{-3/4}$ is the viscous Kolmogorov scale, $\ell_D = \ell_0\,\mathrm{Pe}^{-3/4}$ is the characteristic scale of the molecular diffusion of temperature field, $\mathrm{Re} \equiv u_0\ell_0/\nu$ is the Reynolds number and $\mathrm{Pe} \equiv u_0\ell_0/D^{(\theta)}$ is the Péclet number. Here the length ℓ_0 is the integral scale of turbulence, u_0 is the characteristic turbulent velocity in the scale ℓ_0 and $D^{(\theta)}$ is the molecular diffusion coefficient of the temperature field.

5. To determine the spectrum of anisotropic temperature fluctuations produced by the tangling of the gradient of the mean temperature by velocity fluctuations in the inertial range $\ell_\nu < \ell \ll \ell_D \ll \ell_0$ of the incompressible turbulent fluid flow for small Prandtl numbers $\mathrm{Pr} \equiv \nu/D^{(\theta)} \ll 1$ applying the dimensional analysis, where $\ell_\nu = \ell_0\,\mathrm{Re}^{-3/4}$ is the viscous Kolmogorov scale, $\ell_D = \ell_0\,\mathrm{Pe}^{-3/4}$ is the characteristic scale of the molecular diffusion of temperature field, $\mathrm{Re} \equiv u_0\ell_0/\nu$ is the Reynolds number and $\mathrm{Pe} \equiv u_0\ell_0/D^{(\theta)}$ is the Péclet number. Here the length ℓ_0 is the integral scale of turbulence, u_0 is the characteristic turbulent velocity in the scale ℓ_0 and $D^{(\theta)}$ is the molecular diffusion coefficient of the temperature field.

6. To determine the spectrum of anisotropic temperature fluctuations produced by the tangling of the gradient of the mean temperature by velocity fluctuations in the inertial range $\ell_\nu \ll \ell_D < \ell \ll \ell_0$ of the incompressible turbulent fluid flow for small Prandtl numbers $\mathrm{Pr} \equiv \nu/D^{(\theta)} \ll 1$ using the dimensional analysis, where $\ell_\nu = \ell_0\,\mathrm{Re}^{-3/4}$ is the viscous Kolmogorov scale, $\ell_D = \ell_0\,\mathrm{Pe}^{-3/4}$ is the characteristic scale of the molecular diffusion of temperature field, $\mathrm{Re} \equiv u_0\ell_0/\nu$ is the Reynolds number and $\mathrm{Pe} \equiv u_0\ell_0/D^{(\theta)}$ is the Péclet number. Here the length ℓ_0 is the integral scale of turbulence, u_0 is the characteristic turbulent velocity in the scale ℓ_0 and $D^{(\theta)}$ is the molecular diffusion coefficient of the temperature field.

7. To determine the turbulent diffusion coefficient D_T of the mean temperature field and the turbulent heat flux $\langle \boldsymbol{u}\,\theta \rangle$ applying the dimensional analysis for small Péclet numbers $\mathrm{Pe} \equiv u_0\ell_0/D^{(\theta)} \ll 1$, where θ and $\boldsymbol{u}$ are temperature and velocity fluctuations, the length ℓ_0 is the maximum scale of random motions, u_0 is the characteristic turbulent velocity in the scale ℓ_0, $D^{(\theta)}$ is the coefficient of the molecular diffusion of temperature field in a random velocity field in incompressible, homogeneous and isotropic fluid flow:

$$\langle u_i\, u_j \rangle = \frac{1}{3} \langle \boldsymbol{u}^2 \rangle\, \delta_{ij}.$$

Here δ_{ij} is the Kronecker unit tensor.

8. To determine the turbulent diffusion coefficient D_T of the mean temperature field and the turbulent heat flux $\langle \boldsymbol{u}\, \theta \rangle$ applying the dimensional analysis for large Péclet numbers $\mathrm{Pe} \equiv u_0 \ell_0 / D^{(\theta)} \gg 1$, where θ and $\boldsymbol{u}$ are temperature and velocity fluctuations, the length ℓ_0 is the maximum scale of random motions, u_0 is the characteristic turbulent velocity in the scale ℓ_0, $D^{(\theta)}$ is the coefficient of the molecular diffusion of temperature field in a turbulent velocity field in incompressible, homogeneous and isotropic fluid flow:

$$\langle u_i \, u_j \rangle = \frac{1}{3} \langle \boldsymbol{u}^2 \rangle \, \delta_{ij} .$$

Here δ_{ij} is the Kronecker unit tensor.

9. To determine the level $\langle \theta^2 \rangle$ of temperature fluctuations θ in a random incompressible fluid flow with an imposed mean temperature gradient, $\boldsymbol{\nabla} T$, for small Péclet numbers $\mathrm{Pe} = u_0 \, \ell_0 / D^{(\theta)} \ll 1$ applying the dimensional analysis, where the length ℓ_0 is the integral scale of turbulence, u_0 is the characteristic turbulent velocity in the scale ℓ_0 and $D^{(\theta)}$ is the molecular diffusion coefficient of the temperature field. To assume that the random fluid flow is nearly isotropic,

$$\langle u_i \, u_j \rangle = \frac{1}{3} \langle \boldsymbol{u}^2 \rangle \, \delta_{ij} ,$$

for a small mean temperature gradient, where δ_{ij} is the Kronecker unit tensor.

10. To determine the level $\langle \theta^2 \rangle$ of temperature fluctuations θ in a turbulent incompressible flow with an imposed mean temperature gradient $\boldsymbol{\nabla} T$ for large Péclet numbers $\mathrm{Pe} = u_0 \, \ell_0 / D^{(\theta)} \gg 1$, applying the dimensional analysis, where the length ℓ_0 is the integral scale of turbulence, u_0 is the characteristic turbulent velocity in the scale ℓ_0 and $D^{(\theta)}$ is the molecular diffusion coefficient of the temperature field. To assume that the turbulence is nearly isotropic,

$$\langle u_i \, u_j \rangle = \frac{1}{3} \langle \boldsymbol{u}^2 \rangle \, \delta_{ij} ,$$

for a small mean temperature gradient, where δ_{ij} is the Kronecker unit tensor.

11. To find the dependence of the turbulent heat flux $\langle \boldsymbol{u}\, \theta \rangle$ on the mean temperature gradient, $\boldsymbol{\nabla} T$, for small Péclet numbers $\mathrm{Pe} \ll 1$ applying the quasi-linear approach in a random fluid flow that is determined by the following correlation function written in the Fourier space:

$$\langle u_i(\omega, \boldsymbol{k}) \, u_j(-\omega, -\boldsymbol{k}) \rangle = \langle \boldsymbol{u}^2 \rangle \, \frac{E(k)\, \delta(\omega)}{8\pi \, k^2} \left[\delta_{ij} - \frac{k_i \, k_j}{k^2} \right] ,$$

where θ and $\boldsymbol{u}$ are temperature and velocity fluctuations, δ_{ij} is the Kronecker tensor, the energy spectrum function is $E(k) = (2/3)\, k_0^{-1}\, (k/k_0)^{-5/3}$ for $k_0 < k < k_0\, \mathrm{Re}^{3/4}$, the wave number $k_0 = 1/\ell_0$, the length ℓ_0 is the maximum scale of random motions, u_0 is the characteristic turbulent velocity in the scale ℓ_0 and $\mathrm{Re} = u_0\, \ell_0/\nu \gg 1$. The function $\delta(\omega)$ is the Dirac's delta-function which has the following property: $\int_a^b f(y)\, \delta(y-c)\, dy = f(c)$, where $a < c < b$. How does the turbulent diffusion coefficient of the mean temperature field depend on the Péclet number?

12. To find the dependence of the second moment $\langle \theta^2 \rangle$ of temperature fluctuations θ on the mean temperature gradient, $\boldsymbol{\nabla} T$, for small Péclet numbers $\mathrm{Pe} = u_0\, \ell_0/D^{(\theta)} \ll 1$ applying the quasi-linear approach in a random fluid flow that is determined by the following correlation function written in the Fourier space:

$$\langle u_i(\omega, \mathbf{k})\, u_j(-\omega, -\mathbf{k}) \rangle = \left\langle \boldsymbol{u}^2 \right\rangle \frac{E(k)\, \delta(\omega)}{8\pi\, k^2} \left[\delta_{ij} - \frac{k_i\, k_j}{k^2} \right],$$

where $D^{(\theta)}$ is the molecular diffusion coefficient, δ_{ij} is the Kronecker tensor, the energy spectrum function is $E(k) = (2/3)\, k_0^{-1}\, (k/k_0)^{-5/3}$ for $k_0 < k < k_0\, \mathrm{Re}^{3/4}$, the wave number $k_0 = 1/\ell_0$, the length ℓ_0 is the maximum scale of random motions, u_0 is the characteristic turbulent velocity in the scale ℓ_0, $\mathrm{Re} = u_0\, \ell_0/\nu \gg 1$, ν is the kinematic viscosity, The function $\delta(y)$ is the Dirac's delta-function which has the following property: $\int_a^b f(y)\, \delta(y-c)\, dy = f(c)$.

13. To find the dependence of the turbulent heat flux $\langle \boldsymbol{u}\, \theta \rangle$ on the mean temperature gradient $\boldsymbol{\nabla} T$ for small Péclet numbers $\mathrm{Pe} \ll 1$ applying the quasi-linear approach in a random fluid flow that is determined by the following correlation function written in the Fourier space:

$$\langle u_i(\omega, \boldsymbol{k})\, u_j(-\omega, -\boldsymbol{k}) \rangle = \left\langle \boldsymbol{u}^2 \right\rangle \frac{E(k)\, \delta(\omega)}{8\pi\, k^2} \left[\delta_{ij} - \frac{k_i\, k_j}{k^2} \right],$$

where θ and $\boldsymbol{u}$ are temperature and velocity fluctuations, δ_{ij} is the Kronecker tensor, the energy spectrum function is $E(k) = \delta(k - k_0)$, the wave number $k_0 = 1/\ell_0$, the length ℓ_0 is the integral scale of random motions, u_0 is the characteristic turbulent velocity in the scale ℓ_0 and $\mathrm{Re} = u_0\, \ell_0/\nu \gg 1$. The function $\delta(y)$ is the Dirac's delta-function which has the following property: $\int_a^b f(y)\, \delta(y-c)\, dy = f(c)$, where $a < c < b$.

14. To find the dependence of the turbulent heat flux $\langle \boldsymbol{u}\, \theta \rangle$ on the mean temperature gradient $\boldsymbol{\nabla} T$ for large Péclet numbers $\mathrm{Pe} \gg 1$ applying the spectral tau (relaxation) approach in a turbulent fluid flow that is determined by the following correlation function written in the Fourier space:

$$\langle u_i(\boldsymbol{k})\, u_j(-\boldsymbol{k})\rangle = \langle \boldsymbol{u}^2\rangle \,\frac{E(k)}{8\pi\, k^2}\left[\delta_{ij} - \frac{k_i\, k_j}{k^2}\right],$$

where θ and $\boldsymbol{u}$ are temperature and velocity fluctuations, the energy spectrum function is $E(k) = (2/3)\, k_0^{-1}\, (k/k_0)^{-5/3}$ and the turbulent correlation time is $\tau(k) = 2\,\tau_0\, (k/k_0)^{-2/3}$ for $k_0 < k < k_0\, \mathrm{Re}^{3/4}$, the wave number $k_0 = 1/\ell_0$, the length ℓ_0 is the maximum scale of random motions, u_0 is the characteristic turbulent velocity in the scale ℓ_0 and $\mathrm{Re} = u_0\, \ell_0/\nu \gg 1$. Does the turbulent diffusion coefficient of the mean temperature field depend on the Péclet number?

15. To find the dependence of the second moment $\langle\theta^2\rangle$ of temperature fluctuations θ on the mean temperature gradient $\boldsymbol{\nabla} T$ for large Péclet numbers $\mathrm{Pe} = u_0\, \ell_0/D^{(\theta)} \gg 1$ applying the spectral τ approach in a turbulent fluid flow that is determined by the following correlation function written in the Fourier space:

$$\langle u_i(\mathbf{k})\, u_j(-\mathbf{k})\rangle = \langle \boldsymbol{u}^2\rangle \,\frac{E(k)}{8\pi\, k^2}\left[\delta_{ij} - \frac{k_i\, k_j}{k^2}\right],$$

where $D^{(\theta)}$ is the molecular diffusion coefficient of the temperature field, δ_{ij} is the Kronecker tensor, the energy spectrum function is $E(k) = (2/3)\, k_0^{-1}\, (k/k_0)^{-5/3}$ for $k_0 < k < k_0\, \mathrm{Re}^{3/4}$ and the turbulent correlation time is $\tau(k) = 2\,\tau_0\, (k/k_0)^{-2/3}$ for $k_0 < k < k_0\, \mathrm{Re}^{3/4}$, the wave number $k_0 = 1/\ell_0$, the length ℓ_0 is the integral scale of turbulence, u_0 is the characteristic turbulent velocity in the scale ℓ_0, $\mathrm{Re} = u_0\, \ell_0/\nu \gg 1$ and ν is the kinematic viscosity.

16. To find the dependence of the turbulent heat flux $\langle \boldsymbol{u}\, \theta\rangle$ on the mean temperature gradient $\boldsymbol{\nabla} T$ for large Péclet numbers $\mathrm{Pe} \gg 1$ applying the spectral tau (relaxation) approach in a turbulent fluid flow that is determined by the following correlation function written in the Fourier space:

$$\langle u_i(\mathbf{k})\, u_j(-\mathbf{k})\rangle = \langle \boldsymbol{u}^2\rangle \,\frac{E(k)}{8\pi\, k^2}\left[\delta_{ij} - \frac{k_i\, k_j}{k^2}\right],$$

where θ and $\boldsymbol{u}$ are temperature and velocity fluctuations, the energy spectrum function is $E(k) = \delta(k - k_0)$, the wave number $k_0 = 1/\ell_0$, the length ℓ_0 is the integral scale of random motions, u_0 is the characteristic turbulent velocity in the scale ℓ_0 and $\mathrm{Re} = u_0\, \ell_0/\nu \gg 1$. The function $\delta(y)$ is the Dirac's delta-function which has the following property: $\int_a^b f(y)\, \delta(y - c)\, dy = f(c)$, where $a < c < b$.

17. To find the dependence of the second moment $\langle\theta^2\rangle$ of temperature fluctuations θ on the mean temperature gradient $\boldsymbol{\nabla} T$ for large Péclet numbers $\mathrm{Pe} = u_0\, \ell_0/D^{(\theta)} \gg 1$ applying the spectral tau (relaxation) approach in a turbulent fluid flow that is determined by the following correlation function written in the Fourier space:

$$\langle u_i(\boldsymbol{k})\, u_j(-\boldsymbol{k})\rangle = \langle \boldsymbol{u}^2\rangle \, \frac{E(k)}{8\pi\, k^2} \left[\delta_{ij} - \frac{k_i\, k_j}{k^2}\right],$$

where θ and $\boldsymbol{u}$ are temperature and velocity fluctuations, the energy spectrum function is $E(k) = \delta(k - k_0)$, the wave number $k_0 = 1/\ell_0$, the length ℓ_0 is the integral scale of random motions, u_0 is the characteristic turbulent velocity in the scale ℓ_0 and $\mathrm{Re} = u_0\, \ell_0/\nu \gg 1$. The function $\delta(y)$ is the Dirac's delta-function which has the following property: $\int_a^b f(y)\, \delta(y - c)\, dy = f(c)$, where $a < c < b$.

6.1.2 Practice Problems for Chapter 2

18. To determine the molecular diffusion length-scale $\ell_D^{(n)}$ and time-scale $\tau_D^{(n)}$ of the particle number density versus Péclet number, $\mathrm{Pe}^{(n)} = u_0\, \ell_0/D^{(n)} \gg 1$, applying the dimensional analysis, where the length ℓ_0 is the integral scale of turbulence, u_0 is the characteristic turbulent velocity in the scale ℓ_0 and $D^{(n)}$ is the molecular (Brownian) diffusion coefficient of the particle number density.

19. To determine the turbulent diffusion D_T coefficient and the effective particle velocity $\boldsymbol{V}^{\mathrm{eff}}$ of the mean particle number density, and the turbulent flux of particles $\langle \boldsymbol{u}\, n'\rangle$, applying the dimensional analysis for small Péclet numbers $\mathrm{Pe}^{(n)} \equiv u_0\ell_0/D^{(n)} \ll 1$, where n' and $\boldsymbol{u}$ are fluctuations of particle number density and velocity, $D^{(n)}$ is the coefficient of the molecular diffusion of small particles in a random velocity field in homogeneous and isotropic fluid flow in anelastic approximation:

$$\langle u_i\, u_j\rangle = \frac{1}{3} \langle \boldsymbol{u}^2\rangle\, \delta_{ij}.$$

Here δ_{ij} is the Kronecker unit tensor.

20. To determine the turbulent diffusion D_T coefficient and the effective particle velocity $\boldsymbol{V}^{\mathrm{eff}}$ of the mean particle number density, and the turbulent flux of particles, $\langle \boldsymbol{u}\, n'\rangle$, applying the dimensional analysis for large Péclet numbers $\mathrm{Pe}^{(n)} \equiv u_0\ell_0/D^{(n)} \gg 1$, where n' and $\boldsymbol{u}$ are fluctuations of particle number density and velocity, $D^{(n)}$ is the coefficient of the molecular diffusion of small particles in a random velocity field in homogeneous and isotropic fluid flow in anelastic approximation:

$$\langle u_i\, u_j\rangle = \frac{1}{3} \langle \boldsymbol{u}^2\rangle\, \delta_{ij}.$$

Here δ_{ij} is the Kronecker unit tensor.

21. To determine the level $\langle n'^2 \rangle$ of particle number density fluctuations n' in a random density-stratified fluid flow with an imposed mean particle number density gradient $\nabla \overline{n}$ for small Péclet numbers $\mathrm{Pe}^{(n)} = u_0 \, \ell_0 / D^{(n)} \ll 1$ applying the dimensional analysis, where the length ℓ_0 is the integral scale of turbulence, u_0 is the characteristic turbulent velocity in the scale ℓ_0 and $D^{(n)}$ is the molecular diffusion coefficient of the particle number density. To assume that the random fluid flow is nearly isotropic,

$$\langle u_i \, u_j \rangle = \frac{1}{3} \left\langle \boldsymbol{u}^2 \right\rangle \delta_{ij},$$

for a small mean fluid density gradient, where δ_{ij} is the Kronecker unit tensor.

22. To determine the level $\langle n'^2 \rangle$ of particle number density fluctuations n' in a turbulent density-stratified fluid flow with an imposed mean particle number density gradient $\nabla \overline{n}$ for large Péclet numbers $\mathrm{Pe}^{(n)} = u_0 \, \ell_0 / D^{(n)} \gg 1$ applying the dimensional analysis, where the length ℓ_0 is the integral scale of turbulence, u_0 is the characteristic turbulent velocity in the scale ℓ_0 and $D^{(n)}$ is the molecular diffusion coefficient of the particle number density. To assume that the turbulent fluid flow is nearly isotropic,

$$\langle u_i \, u_j \rangle = \frac{1}{3} \left\langle \boldsymbol{u}^2 \right\rangle \delta_{ij},$$

for a small mean fluid density gradient, where δ_{ij} is the Kronecker unit tensor.

23. To determine the turbulent diffusion D_T coefficient and the effective particle velocity $\boldsymbol{V}^{\mathrm{eff}}$ of the mean particle number density, and the turbulent flux of particles $\langle \boldsymbol{u} \, n' \rangle$, applying the quasi-linear approach for small Péclet numbers $\mathrm{Pe}^{(n)} \equiv u_0 \ell_0 / D^{(n)} \ll 1$, where $D^{(n)}$ is the coefficient of the molecular diffusion of small particles in a random velocity field in homogeneous fluid flow in anelastic approximation:

$$\langle u_i(\omega, \boldsymbol{k}) \, u_j(-\omega, -\boldsymbol{k}) \rangle = \left\langle \boldsymbol{u}^2 \right\rangle \frac{E(k)\delta(\omega)}{8\pi \, k^2} \left[\delta_{ij} - \frac{k_i \, k_j}{k^2} + \frac{\mathrm{i}}{k^2} \left(\lambda_i \, k_j - \lambda_j \, k_i \right) \right].$$

Here n' and $\boldsymbol{u}$ are fluctuations of particle number density and velocity, $\lambda_i = -\nabla_i \rho / \rho$, ρ is the fluid density, δ_{ij} is the Kronecker tensor, the energy spectrum function is $E(k) = (2/3) \, k_0^{-1} \, (k/k_0)^{-5/3}$ for $k_0 < k < k_0 \, \mathrm{Re}^{3/4}$, the wave number $k_0 = 1/\ell_0$, the length ℓ_0 is the maximum scale of random motions and $\mathrm{Re} \gg 1$. The function $\delta(\omega)$ is the Dirac's delta-function which has the following property: $\int_a^b f(y) \, \delta(y - c) \, dy = f(c)$, where $a < c < b$. To determine the spatial distribution of the mean number density of non-inertial particles in a steady state when the mean fluid density is a function of the vertical coordinate z.

24. To determine the turbulent diffusion D_T coefficient and the effective particle velocity $\boldsymbol{V}^{\rm eff}$ of the mean particle number density, and the turbulent flux of particles $\langle \boldsymbol{u}\, n' \rangle$, applying the quasi-linear approach for small Péclet numbers $\mathrm{Pe}^{(n)} \equiv u_0 \ell_0 / D^{(n)} \ll 1$, where $D^{(n)}$ is the coefficient of the molecular diffusion of small particles in a random velocity field in homogeneous fluid flow in anelastic approximation:

$$\langle u_i(\omega, \boldsymbol{k})\, u_j(-\omega, -\boldsymbol{k}) \rangle = \langle \boldsymbol{u}^2 \rangle \, \frac{E(k)\delta(\omega)}{8\pi\, k^2} \left[\delta_{ij} - \frac{k_i\, k_j}{k^2} + \frac{\mathrm{i}}{k^2} \left(\lambda_i\, k_j - \lambda_j\, k_i \right) \right].$$

Here n' and $\boldsymbol{u}$ are fluctuations of particle number density and velocity, $\lambda_i = -\nabla_i \rho / \rho$, ρ is the fluid density, δ_{ij} is the Kronecker tensor, the energy spectrum function is $E(k) = \delta(k - k_0)$, the wave number $k_0 = 1/\ell_0$, the length ℓ_0 is the integral scale of random motions, u_0 is the characteristic turbulent velocity in the scale ℓ_0 and $\mathrm{Re} = u_0\, \ell_0 / \nu \gg 1$. The function $\delta(y)$ is the Dirac's delta-function which has the following property: $\int_a^b f(y)\, \delta(y - c)\, dy = f(c)$, where $a < c < b$.

25. To determine the level of particle number density fluctuations $\langle n'^2 \rangle$ for small Péclet numbers $\mathrm{Pe}^{(n)} \equiv u_0 \ell_0 / D^{(n)} \ll 1$ applying the quasi-linear approach, where $D^{(n)}$ is the coefficient of the molecular diffusion of small particles in a random velocity field in homogeneous fluid flow in anelastic approximation:

$$\langle u_i(\omega, \mathbf{k})\, u_j(-\omega, -\mathbf{k}) \rangle = \langle \boldsymbol{u}^2 \rangle \, \frac{E(k)\delta(\omega)}{8\pi\, k^2} \left[\delta_{ij} - \frac{k_i\, k_j}{k^2} + \frac{\mathrm{i}}{k^2} \left(\lambda_i\, k_j - \lambda_j\, k_i \right) \right].$$

Here $\lambda_i = -\nabla_i \rho / \rho$, ρ is the fluid density, δ_{ij} is the Kronecker tensor, the energy spectrum function is $E(k) = (2/3)\, k_0^{-1}\, (k/k_0)^{-5/3}$ for $k_0 < k < k_0\, \mathrm{Re}^{3/4}$, the wave number $k_0 = 1/\ell_0$, the length ℓ_0 is the maximum scale of random motions and $\mathrm{Re} \gg 1$. The function $\delta(y)$ is the Dirac's delta-function which has the following property: $\int_a^b f(y)\, \delta(y - c)\, dy = f(c)$.

26. To determine the turbulent diffusion D_T coefficient and the effective particle velocity $\boldsymbol{V}^{\rm eff}$ of the mean particle number density, and the turbulent flux of particles $\langle \boldsymbol{u}\, n' \rangle$, applying the spectral tau (relaxation) approach for large Péclet numbers $\mathrm{Pe}^{(n)} \equiv u_0 \ell_0 / D^{(n)} \gg 1$, where n' and $\boldsymbol{u}$ are fluctuations of particle number density and velocity, $D^{(n)}$ is the coefficient of the molecular diffusion of small particles in a random velocity field in homogeneous fluid flow in anelastic approximation:

$$\langle u_i(\boldsymbol{k})\, u_j(-\boldsymbol{k}) \rangle = \langle \boldsymbol{u}^2 \rangle \, \frac{E(k)}{8\pi\, k^2} \left[\delta_{ij} - \frac{k_i\, k_j}{k^2} + \frac{\mathrm{i}}{k^2} \left(\lambda_i\, k_j - \lambda_j\, k_i \right) \right].$$

Here $\lambda_i = -\nabla_i \rho / \rho$, ρ is the fluid density, δ_{ij} is the Kronecker tensor, the energy spectrum function is $E(k) = (2/3)\, k_0^{-1}\, (k/k_0)^{-5/3}$ and the turbulent correlation time is $\tau(k) = 2\, \tau_0\, (k/k_0)^{-2/3}$ for $k_0 < k < k_0\, \mathrm{Re}^{3/4}$, the wave number $k_0 = 1/\ell_0$, the length ℓ_0 is the maximum scale of random motions and $\mathrm{Re} \gg 1$. To determine the spatial distribution of the mean number density of non-inertial particles in a steady state when the mean fluid density is a function of the vertical coordinate z.

27. To determine the turbulent diffusion D_T coefficient and the effective particle velocity $\boldsymbol{V}^{\rm eff}$ of the mean particle number density, and the turbulent flux of particles $\langle \boldsymbol{u}\, n' \rangle$, applying the spectral tau (relaxation) approach for large Péclet numbers $\mathrm{Pe}^{(n)} \equiv u_0 \ell_0 / D^{(n)} \gg 1$, where n' and $\boldsymbol{u}$ are fluctuations of particle number density and velocity, $D^{(n)}$ is the coefficient of the molecular diffusion of small particles in a random velocity field in homogeneous fluid flow in anelastic approximation:

$$\langle u_i(\boldsymbol{k})\, u_j(-\boldsymbol{k}) \rangle = \langle \boldsymbol{u}^2 \rangle \, \frac{E(k)}{8\pi\, k^2} \left[\delta_{ij} - \frac{k_i\, k_j}{k^2} + \frac{\mathrm{i}}{k^2} \left(\lambda_i\, k_j - \lambda_j\, k_i \right) \right].$$

Here $\lambda_i = -\nabla_i \rho / \rho$, ρ is the fluid density, δ_{ij} is the Kronecker tensor, the energy spectrum function is $E(k) = \delta(k - k_0)$, the wave number $k_0 = 1/\ell_0$, the length ℓ_0 is the integral scale of random motions, u_0 is the characteristic turbulent velocity in the scale ℓ_0 and $\mathrm{Re} = u_0\, \ell_0 / \nu \gg 1$. The function $\delta(y)$ is the Dirac's delta-function which has the following property: $\int_a^b f(y)\, \delta(y - c)\, dy = f(c)$, where $a < c < b$.

28. To determine the level of particle number density fluctuations $\langle n'^2 \rangle$ for large Péclet numbers $\mathrm{Pe}^{(n)} \equiv u_0 \ell_0 / D^{(n)} \gg 1$ applying the spectral τ approach, where $D^{(n)}$ is the coefficient of the molecular diffusion of very small particles in random velocity field in homogeneous fluid flow in anelastic approximation:

$$\langle u_i(\mathbf{k})\, u_j(-\mathbf{k}) \rangle = \langle \boldsymbol{u}^2 \rangle \, \frac{E(k)}{8\pi\, k^2} \left[\delta_{ij} - \frac{k_i\, k_j}{k^2} + \frac{\mathrm{i}}{k^2} \left(\lambda_i\, k_j - \lambda_j\, k_i \right) \right].$$

Here $\lambda_i = -\nabla_i \rho / \rho$, ρ is the fluid density, δ_{ij} is the Kronecker tensor, the energy spectrum function is $E(k) = (2/3)\, k_0^{-1}\, (k/k_0)^{-5/3}$ for $k_0 < k < k_0\, \mathrm{Re}^{3/4}$ and the turbulent correlation time is $\tau(k) = 2\, \tau_0\, (k/k_0)^{-2/3}$ for $k_0 < k < k_0\, \mathrm{Re}^{3/4}$, the wave number $k_0 = 1/\ell_0$, the length ℓ_0 is the integral scale of random motions and $\mathrm{Re} \gg 1$.

29. To determine the level of particle number density fluctuations $\langle n'^2 \rangle$ for large Péclet numbers $\mathrm{Pe}^{(n)} \equiv u_0 \ell_0 / D^{(n)} \gg 1$ applying the spectral τ approach, where $D^{(n)}$ is the coefficient of the molecular diffusion of small particles in a random velocity field in homogeneous fluid flow in anelastic approximation:

$$\langle u_i(\mathbf{k})\, u_j(-\mathbf{k}) \rangle = \langle \boldsymbol{u}^2 \rangle \, \frac{E(k)}{8\pi\, k^2} \left[\delta_{ij} - \frac{k_i\, k_j}{k^2} + \frac{\mathrm{i}}{k^2} \left(\lambda_i\, k_j - \lambda_j\, k_i \right) \right].$$

Here $\lambda_i = -\nabla_i \rho / \rho$, ρ is the fluid density, δ_{ij} is the Kronecker tensor, the energy spectrum function is $E(k) = \delta(k - k_0)$, the wave number $k_0 = 1/\ell_0$, the length ℓ_0 is the integral scale of random motions, u_0 is the characteristic turbulent velocity in the scale ℓ_0 and $\mathrm{Re} = u_0\, \ell_0 / \nu \gg 1$. The function $\delta(y)$ is the Dirac's delta-function which has the following property: $\int_a^b f(y)\, \delta(y - c)\, dy = f(c)$, where $a < c < b$.

30. To determine the effective particle velocity $\boldsymbol{V}^{\rm eff}$ of the inertial particles for large Péclet numbers $\mathrm{Pe}^{(n)} \equiv u_0 \ell_0 / D^{(n)} \gg 1$, where $D^{(n)}$ is the coefficient of the molecular diffusion of particles in a random velocity field in homogeneous fluid flow in anelastic approximation. To determine the spatial distribution of the mean number density of inertial particles in a steady state satisfying the boundary conditions with a zero total particle flux at the boundaries. To take into account the terminal fall velocity of inertial particles caused by the gravity field. The mean temperature is a function of the vertical coordinate z.

31. To determine the spectrum of isotropic fluctuations of particle number density produced in the inertial range $\ell_\nu \le \ell_D^{(n)} < \ell < \ell_0$ of the incompressible turbulent fluid flow for Schmidt number $\mathrm{Sc} \equiv \nu / D^{(n)} < 1$ applying the dimensional analysis, where $\ell_\nu = \ell_0 / \mathrm{Re}^{3/4}$ is the viscous Kolmogorov scale, $\ell_D^{(n)} = \ell_0 / \left(\mathrm{Pe}^{(n)}\right)^{3/4}$ is the characteristic scale of the molecular diffusion of the particle number density, $\mathrm{Re} \equiv u_0 \ell_0 / \nu$ is the Reynolds number, ν is the kinematic viscosity, $\mathrm{Pe}^{(n)} \equiv u_0 \ell_0 / D^{(n)}$ is the Péclet number and $D^{(n)}$ is the molecular (Brownian) diffusion coefficient of the particle number density.

32. To determine the spectrum of isotropic fluctuations of particle number density produced in the viscous range $\ell_D^{(n)} < \ell < \ell_\nu$ of the incompressible turbulent fluid flow for Schmidt number $\mathrm{Sc} \equiv \nu / D^{(n)} > 1$ applying the dimensional analysis, where $\ell_\nu = \ell_0 / \mathrm{Re}^{3/4}$ is the viscous Kolmogorov scale, $\ell_D^{(n)} = \ell_0 / \left(\mathrm{Pe}^{(n)}\right)^{3/4}$ is the characteristic scale of the molecular diffusion of the particle number density, $\mathrm{Re} \equiv u_0 \ell_0 / \nu$ is the Reynolds number and $\mathrm{Pe}^{(n)} \equiv u_0 \ell_0 / D^{(n)}$ is the Péclet number. Here the length ℓ_0 is the integral scale of turbulence, u_0 is the characteristic turbulent velocity in the scale ℓ_0 and $D^{(n)}$ is the molecular (Brownian) diffusion coefficient of the particle number density.

33. To determine the spectrum of anisotropic fluctuations of particle number density produced by the tangling of the gradient of the mean particle number density by velocity fluctuations in the inertial range $\ell_\nu < \ell \ll \ell_D^{(n)} \ll \ell_0$ of the incompressible turbulent fluid flow for small Schmidt numbers $\mathrm{Sc} \equiv \nu / D^{(n)} \ll 1$ applying the dimensional analysis, where $\ell_\nu = \ell_0 / \mathrm{Re}^{3/4}$ is the viscous Kolmogorov scale, $\ell_D^{(n)} = \ell_0 / \left(\mathrm{Pe}^{(n)}\right)^{3/4}$ is the characteristic scale of the molecular diffusion of the particle number density, $\mathrm{Re} \equiv u_0 \ell_0 / \nu$ is the Reynolds number and $\mathrm{Pe}^{(n)} \equiv u_0 \ell_0 / D^{(n)}$ is the Péclet number. Here the length ℓ_0 is the integral scale of turbulence, u_0 is the characteristic turbulent velocity in the scale ℓ_0 and $D^{(n)}$ is the molecular (Brownian) diffusion coefficient of the particle number density.

34. To determine the spectrum of anisotropic fluctuations of particle number density produced by the tangling of the gradient of the mean particle number density by velocity fluctuations in the inertial range $\ell_\nu \ll \ell_D^{(n)} < \ell \ll \ell_0$ of the incompressible turbulent fluid flow for small Schmidt numbers $\mathrm{Sc} \equiv \nu/D^{(n)} \ll 1$ using the dimensional analysis, where $\ell_\nu = \ell_0/\mathrm{Re}^{3/4}$ is the viscous Kolmogorov scale, $\ell_D^{(n)} = \ell_0/\left(\mathrm{Pe}^{(n)}\right)^{3/4}$ is the characteristic scale of the molecular diffusion of the particle number density, $\mathrm{Re} \equiv u_0\ell_0/\nu$ is the Reynolds number and $\mathrm{Pe}^{(n)} \equiv u_0\ell_0/D^{(n)}$ is the Péclet number. Here the length ℓ_0 is the integral scale of turbulence, u_0 is the characteristic turbulent velocity in the scale ℓ_0 and $D^{(n)}$ is the molecular (Brownian) diffusion coefficient of the particle number density.

6.1.3 Practice Problems for Chapter 3

35. To determine the molecular magnetic diffusion length-scale ℓ_η (the resistive scale) and time-scale τ_η of magnetic fluctuations versus magnetic Reynolds number $\mathrm{Rm} = u_0\,\ell_0/\eta \gg 1$, applying the dimensional analysis, where the length ℓ_0 is the integral scale of turbulence, u_0 is the characteristic turbulent velocity in the scale ℓ_0 and η is the magnetic diffusion coefficient of the magnetic field caused by the electrical conductivity of the fluid.

36. To determine the spectrum of isotropic magnetic fluctuations produced in the inertial range $\ell_m \ll \ell_\nu < \ell < \ell_0$ of an incompressible turbulent fluid flow for magnetic Prandtl number $\mathrm{Pr}_m \equiv \nu/\eta > 1$ applying the dimensional analysis, where $\ell_\nu = \ell_0/\mathrm{Re}^{3/4}$ is the viscous Kolmogorov scale, $\ell_m = \ell_0/\mathrm{Rm}^{3/4}$ is the characteristic scale of the magnetic diffusion, $\mathrm{Re} \equiv u_0\ell_0/\nu$ is the Reynolds number and $\mathrm{Rm} \equiv u_0\ell_0/\eta$ is the magnetic Reynolds number. Here the length ℓ_0 is the integral scale of turbulence, u_0 is the characteristic turbulent velocity in the scale ℓ_0 and η is the magnetic diffusion coefficient of the magnetic field caused by the electrical conductivity of the fluid.

37. To determine the spectrum of isotropic magnetic fluctuations produced in the viscous range $\ell_m \ll \ell < \ell_\nu \ll \ell_0$ of an incompressible turbulent fluid flow for magnetic Prandtl number $\mathrm{Pr}_m \equiv \nu/\eta > 1$ applying the dimensional analysis, where $\ell_\nu = \ell_0/\mathrm{Re}^{3/4}$ is the viscous Kolmogorov scale, $\ell_m = \ell_0/\mathrm{Rm}^{3/4}$ is the characteristic scale of the magnetic diffusion, $\mathrm{Re} \equiv u_0\ell_0/\nu$ is the Reynolds number and $\mathrm{Rm} \equiv u_0\ell_0/\eta$ is the magnetic Reynolds number. Here the length ℓ_0 is the integral scale of turbulence, u_0 is the characteristic turbulent velocity in the scale ℓ_0 and η is the magnetic diffusion coefficient of the magnetic field caused by the electrical conductivity of the fluid.

38. To determine the spectrum of anisotropic magnetic fluctuations produced by tangling of the mean magnetic field by velocity fluctuations in the inertial range of scales $\ell_\nu \ll \ell \le \ell_m \ll \ell_0$ of an incompressible turbulent fluid flow for small magnetic Prandtl numbers $\mathrm{Pm} \equiv \nu/\eta \ll 1$ applying the dimensional analysis, where $\ell_m = \ell_0/\mathrm{Rm}^{3/4}$ is the characteristic resistive scale, $\ell_\nu = \ell_0/\mathrm{Re}^{3/4}$ is the viscous Kolmogorov scale, $\mathrm{Re} \equiv u_0\ell_0/\nu$ is the Reynolds number and $\mathrm{Rm} \equiv u_0\ell_0/\eta$. Here the length ℓ_0 is the integral scale of turbulence, u_0 is the characteristic turbulent velocity in the scale ℓ_0 and η is the magnetic diffusion coefficient of the magnetic field caused by the electrical conductivity of the fluid.

39. To determine the spectrum of anisotropic magnetic fluctuations produced by tangling of the mean magnetic field by velocity fluctuations in the inertial range of scales $\ell_\nu \ll \ell_m \ll \ell \ll \ell_0$ of an incompressible turbulent fluid flow for small magnetic Prandtl numbers $\mathrm{Pm} \equiv \nu/\eta \ll 1$ applying the dimensional analysis, where $\ell_m = \ell_0/\mathrm{Rm}^{3/4}$ is the characteristic resistive scale, $\ell_\nu = \ell_0/\mathrm{Re}^{3/4}$ is the viscous Kolmogorov scale, $\mathrm{Re} \equiv u_0\ell_0/\nu$ is the Reynolds number and $\mathrm{Rm} \equiv u_0\ell_0/\eta$. Here the length ℓ_0 is the integral scale of turbulence, u_0 is the characteristic turbulent velocity in the scale ℓ_0 and η is the magnetic diffusion coefficient of the magnetic field caused by the electrical conductivity of the fluid.

40. To determine the effective diamagnetic velocity $\boldsymbol{V}^{\rm eff}$ of the mean magnetic field, the turbulent magnetic diffusion coefficient $\eta_{_T}$, the α effect, the turbulent electromotive force (EMF) $\boldsymbol{\mathcal{E}}$, for small magnetic Reynolds numbers $\mathrm{Rm} = u_0\ell_0/\eta \ll 1$ applying the dimensional analysis in random velocity field in homogeneous and isotropic fluid flow:

$$\langle u_i\, u_j\rangle = \frac{1}{3}\left\langle \boldsymbol{u}^2\right\rangle \delta_{ij}.$$

Here δ_{ij} is the Kronecker unit tensor, the length ℓ_0 is the integral scale of turbulence, u_0 is the characteristic turbulent velocity in the scale ℓ_0, and η is the magnetic diffusion coefficient of the magnetic field caused by the electrical conductivity of the fluid.

41. To determine the effective diamagnetic velocity $\boldsymbol{V}^{\rm eff}$ of the mean magnetic field, the turbulent magnetic diffusion coefficient $\eta_{_T}$, the α effect, the turbulent electromotive force (EMF) $\boldsymbol{\mathcal{E}}$, for large magnetic Reynolds numbers $\mathrm{Rm} = u_0\ell_0/\eta \gg 1$ applying the dimensional analysis in random velocity field in homogeneous and isotropic fluid flow:

$$\langle u_i\, u_j\rangle = \frac{1}{3}\left\langle \boldsymbol{u}^2\right\rangle \delta_{ij}.$$

Here δ_{ij} is the Kronecker tensor, the length ℓ_0 is the integral scale of turbulence, u_0 is the characteristic turbulent velocity in the scale ℓ_0, ν is the kinematic viscosity of the fluid and η is the magnetic diffusion coefficient of the magnetic field caused by the electrical conductivity of the fluid.

42. To determine the level $\langle \boldsymbol{b}^2 \rangle$ of magnetic fluctuations $\boldsymbol{b}$ for small magnetic Reynolds numbers $\mathrm{Rm} = u_0 \ell_0/\eta \ll 1$ applying the dimensional analysis, where the length ℓ_0 is the integral scale of turbulence, u_0 is the characteristic turbulent velocity in the scale ℓ_0 and η is the magnetic diffusivity due to the electrical conductivity of the fluid. To assume that the turbulent fluid flow is isotropic,

$$\langle u_i \, u_j \rangle = \frac{1}{3} \left\langle \boldsymbol{u}^2 \right\rangle \delta_{ij},$$

where δ_{ij} is the Kronecker unit tensor.

43. To determine the level $\langle \boldsymbol{b}^2 \rangle$ of magnetic fluctuations $\boldsymbol{b}$ for large magnetic Reynolds numbers $\mathrm{Rm} = u_0 \ell_0/\eta \gg 1$ applying the dimensional analysis, where the length ℓ_0 is the integral scale of turbulence, u_0 is the characteristic turbulent velocity in the scale ℓ_0 and η is the magnetic diffusivity due to the electrical conductivity of the fluid. To assume that the turbulent fluid flow is isotropic,

$$\langle u_i \, u_j \rangle = \frac{1}{3} \left\langle \boldsymbol{u}^2 \right\rangle \delta_{ij},$$

where δ_{ij} is the Kronecker unit tensor.

44. To determine the effective diamagnetic velocity $\boldsymbol{V}^{\rm eff}$ of the mean magnetic field, the turbulent magnetic diffusion coefficient $\eta_{_T}$, the α effect, the turbulent electromotive force (EMF) $\boldsymbol{\mathcal{E}}$, for small magnetic Reynolds numbers $\mathrm{Rm} = u_0 \ell_0/\eta \ll 1$ applying the quasi-linear approach. The turbulent velocity field is determined by the following correlation function written in the Fourier space:

$$\langle u_i(\omega, \boldsymbol{k}) \, u_j(-\omega, -\boldsymbol{k}) \rangle = \frac{E(k)\delta(\omega)}{8\pi \, k^2} \Bigg\{ \left[\delta_{ij} - \frac{k_i \, k_j}{k^2} + \frac{\mathrm{i}}{2k^2} \left(k_i \nabla_j - k_j \nabla_i \right) \right] \left\langle \boldsymbol{u}^2 \right\rangle - \frac{\mathrm{i}}{k^2} \, \varepsilon_{ijp} \, k_p \, \langle \boldsymbol{u} \cdot (\boldsymbol{\nabla} \times \boldsymbol{u}) \rangle \Bigg\},$$

where the energy spectrum function is $E(k) = (2/3) \, k_0^{-1} \, (k/k_0)^{-5/3}$ for $k_0 < k < k_0 \, \mathrm{Re}^{3/4}$, the wave number $k_0 = 1/\ell_0$, the length ℓ_0 is the maximum scale of random motions and $\mathrm{Re} \gg 1$, ν is the kinematic viscosity of the fluid and η is the magnetic diffusivity due to the electrical conductivity of the fluid. The function $\delta(\omega)$ is the Dirac's delta-function which has the following property: $\int_a^b f(y) \, \delta(y - c) \, dy = f(c)$, where $a < c < b$.

45. To determine the effective diamagnetic velocity $\boldsymbol{V}^{\rm eff}$ of the mean magnetic field, the turbulent magnetic diffusion coefficient $\eta_{_T}$, the α effect, the turbulent electromotive force (EMF) $\boldsymbol{\mathcal{E}}$, for small magnetic Reynolds numbers $\mathrm{Rm} = u_0 \ell_0/\eta \ll 1$ applying the quasi-linear approach. The turbulent velocity field is determined by the following correlation function written in the Fourier space:

$$\langle u_i(\omega, \boldsymbol{k})\, u_j(-\omega, -\boldsymbol{k})\rangle = \frac{E(k)\,\delta(\omega)}{8\pi\, k^2}\left\{\left[\delta_{ij} - \frac{k_i\, k_j}{k^2} + \frac{\mathrm{i}}{2k^2}\left(k_i \nabla_j - k_j \nabla_i\right)\right]\langle \boldsymbol{u}^2\rangle - \frac{\mathrm{i}}{k^2}\,\varepsilon_{ijp}\, k_p\, \langle \boldsymbol{u} \cdot (\boldsymbol{\nabla} \times \boldsymbol{u})\rangle\right\},$$

where the energy spectrum function is $E(k) = \delta(k - k_0)$, the wave number $k_0 = 1/\ell_0$, the length ℓ_0 is the integral scale of random motions, u_0 is the characteristic turbulent velocity in the scale ℓ_0 and $\mathrm{Re} = u_0\, \ell_0/\nu \gg 1$. The function $\delta(y)$ is the Dirac's delta-function which has the following property: $\int_a^b f(y)\, \delta(y - c)\, dy = f(c)$, where $a < c < b$. Here ν is the kinematic viscosity of the fluid and η is the magnetic diffusivity due to the electrical conductivity of the fluid.

46. To determine the level $\langle \boldsymbol{b}^2\rangle$ of magnetic fluctuations $\boldsymbol{b}$ for small magnetic Reynolds numbers $\mathrm{Rm} = u_0\ell_0/\eta \ll 1$ applying the quasi-linear approach. The turbulent velocity field is determined by the following correlation function written in the Fourier space:

$$\langle u_i(\omega, \mathbf{k})\, u_j(-\omega, -\mathbf{k})\rangle = \frac{E(k)\,\delta(\omega)}{8\pi\, k^2}\left\{\left[\delta_{ij} - \frac{k_i\, k_j}{k^2} + \frac{\mathrm{i}}{2k^2}\left(k_i \nabla_j - k_j \nabla_i\right)\right]\langle \boldsymbol{u}^2\rangle - \frac{\mathrm{i}}{k^2}\,\varepsilon_{ijp}\, k_p\, \langle \mathbf{u} \cdot (\boldsymbol{\nabla} \times \mathbf{u})\rangle\right\},$$

where the energy spectrum function is $E(k) = (2/3)\, k_0^{-1}\, (k/k_0)^{-5/3}$ for $k_0 < k < k_0\, \mathrm{Re}^{3/4}$, the wave number $k_0 = 1/\ell_0$, the length ℓ_0 is the maximum scale of random motions, the fluid Reynolds number $\mathrm{Re} = u_0\ell_0/\nu \gg 1$, the length ℓ_0 is the integral scale of turbulence, u_0 is the characteristic turbulent velocity in the scale ℓ_0 and η is the magnetic diffusivity due to the electrical conductivity of the fluid.

47. To determine the level $\langle \boldsymbol{b}^2\rangle$ of magnetic fluctuations $\boldsymbol{b}$ for small magnetic Reynolds numbers $\mathrm{Rm} = u_0\ell_0/\eta \ll 1$ applying the quasi-linear approach. The turbulent velocity field is determined by the following correlation function written in the Fourier space:

$$\langle u_i(\omega, \mathbf{k})\, u_j(-\omega, -\mathbf{k})\rangle = \frac{E(k)\,\delta(\omega)}{8\pi\, k^2}\left\{\left[\delta_{ij} - \frac{k_i\, k_j}{k^2} + \frac{\mathrm{i}}{2k^2}\left(k_i \nabla_j - k_j \nabla_i\right)\right]\langle \boldsymbol{u}^2\rangle - \frac{\mathrm{i}}{k^2}\,\varepsilon_{ijp}\, k_p\, \langle \mathbf{u} \cdot (\boldsymbol{\nabla} \times \mathbf{u})\rangle\right\},$$

where the energy spectrum function is $E(k) = \delta(k - k_0)$, the wave number $k_0 = 1/\ell_0$, the length ℓ_0 is the integral scale of random motions, u_0 is the characteristic turbulent velocity in the scale ℓ_0 and $\mathrm{Re} = u_0\, \ell_0/\nu \gg 1$. The function $\delta(y)$ is the Dirac's delta-function which has the following property: $\int_a^b f(y)\, \delta(y - c)$

$dy = f(c)$, where $a < c < b$. Here ν is the kinematic viscosity of the fluid and η is the magnetic diffusivity due to the electrical conductivity of the fluid.

48. To determine the effective diamagnetic velocity $\boldsymbol{V}^{\rm eff}$ of the mean magnetic field, the turbulent magnetic diffusion coefficient $\eta_{_T}$, the α effect, the turbulent electromotive force (EMF) $\boldsymbol{\mathcal{E}}$, for large magnetic Reynolds numbers $\mathrm{Rm} = u_0\ell_0/\eta \gg 1$ applying the spectral tau (relaxation) approach. The turbulent velocity field is determined by the following correlation function written in the Fourier space:

$$\langle u_i(\boldsymbol{k})\, u_j(-\boldsymbol{k})\rangle = \frac{E(k)}{8\pi\, k^2}\biggl\{\biggl[\delta_{ij} - \frac{k_i\, k_j}{k^2} + \frac{\mathrm{i}}{2k^2}\left(k_i \nabla_j - k_j \nabla_i\right)\biggr]\langle \boldsymbol{u}^2\rangle - \frac{\mathrm{i}}{k^2}\,\varepsilon_{ijp}\, k_p\, \langle \boldsymbol{u} \cdot (\nabla \times \boldsymbol{u})\rangle\biggr\},$$

where the energy spectrum function is $E(k) = (2/3)\, k_0^{-1}\, (k/k_0)^{-5/3}$ for $k_0 < k < k_0\,\mathrm{Re}^{3/4}$ and the turbulent correlation time is $\tau(k) = 2\,\tau_0\,(k/k_0)^{-2/3}$ for $k_0 < k < k_0\,\mathrm{Re}^{3/4}$, the wave number $k_0 = 1/\ell_0$, the length ℓ_0 is the maximum scale of random motions and the fluid Reynolds number $\mathrm{Re} \equiv u_0\ell_0/\nu \gg 1$. Here the length ℓ_0 is the integral scale of turbulence, u_0 is the characteristic turbulent velocity in the scale ℓ_0, ν is the kinematic viscosity of the fluid and η is the magnetic diffusion coefficient of the magnetic field caused by the electrical conductivity of the fluid.

49. To determine the effective diamagnetic velocity $\boldsymbol{V}^{\rm eff}$ of the turbulent magnetic field, the turbulent magnetic diffusion coefficient $\eta_{_T}$, the α effect, the mean electromotive force (EMF) $\boldsymbol{\mathcal{E}}$, for large magnetic Reynolds numbers $\mathrm{Rm} = u_0\ell_0/\eta \gg 1$ applying the spectral tau (relaxation) approach. The turbulent velocity field is determined by the following correlation function written in the Fourier space:

$$\langle u_i(\boldsymbol{k})\, u_j(-\boldsymbol{k})\rangle = \frac{E(k)}{8\pi\, k^2}\biggl\{\biggl[\delta_{ij} - \frac{k_i\, k_j}{k^2} + \frac{\mathrm{i}}{2k^2}\left(k_i \nabla_j - k_j \nabla_i\right)\biggr]\langle \boldsymbol{u}^2\rangle - \frac{\mathrm{i}}{k^2}\,\varepsilon_{ijp}\, k_p\, \langle \boldsymbol{u} \cdot (\nabla \times \boldsymbol{u})\rangle\biggr\},$$

where the energy spectrum function is $E(k) = \delta(k - k_0)$, the wave number $k_0 = 1/\ell_0$, the length ℓ_0 is the integral scale of random motions, u_0 is the characteristic turbulent velocity in the scale ℓ_0 and $\mathrm{Re} = u_0\,\ell_0/\nu \gg 1$. The function $\delta(y)$ is the Dirac's delta-function which has the following property: $\int_a^b f(y)\,\delta(y - c)\,dy = f(c)$, where $a < c < b$. Here ν is the kinematic viscosity of the fluid and η is the magnetic diffusion coefficient of the magnetic field caused by the electrical conductivity of the fluid.

50. To determine the level $\langle \boldsymbol{b}^2\rangle$ of magnetic fluctuations $\boldsymbol{b}$ for large magnetic Reynolds numbers $\mathrm{Rm} = u_0\ell_0/\eta \gg 1$ applying the spectral τ approach. The turbulent velocity field is determined by the following correlation function written in the Fourier space:

$$\langle u_i(\omega, \mathbf{k})\, u_j(-\omega, -\mathbf{k})\rangle = \frac{E(k)}{8\pi\, k^2}\left\{\left[\delta_{ij} - \frac{k_i\, k_j}{k^2} + \frac{\mathrm{i}}{2k^2}\left(k_i \nabla_j - k_j \nabla_i\right)\right]\langle \boldsymbol{u}^2\rangle - \frac{\mathrm{i}}{k^2}\,\varepsilon_{ijp}\, k_p\, \langle \mathbf{u} \cdot (\nabla \times \mathbf{u})\rangle\right\},$$

where the energy spectrum function is $E(k) = (2/3)\, k_0^{-1}\, (k/k_0)^{-5/3}$ for $k_0 < k < k_0\, \mathrm{Re}^{3/4}$ and the turbulent correlation time is $\tau(k) = 2\,\tau_0\,(k/k_0)^{-2/3}$ for $k_0 < k < k_0\, \mathrm{Re}^{3/4}$, the wave number $k_0 = 1/\ell_0$, the length ℓ_0 is the maximum scale of random motions, the fluid Reynolds number $\mathrm{Re} = u_0 \ell_0/\nu \gg 1$, the length ℓ_0 is the integral scale of turbulence, u_0 is the characteristic turbulent velocity in the scale ℓ_0 and η is the magnetic diffusivity due to the electrical conductivity of the fluid.

51. To determine the level $\langle \boldsymbol{b}^2\rangle$ of magnetic fluctuations $\boldsymbol{b}$ for large magnetic Reynolds numbers $\mathrm{Rm} = u_0 \ell_0/\eta \gg 1$ applying the spectral τ approach. The turbulent velocity field is determined by the following correlation function written in the Fourier space:

$$\langle u_i(\omega, \mathbf{k})\, u_j(-\omega, -\mathbf{k})\rangle = \frac{E(k)}{8\pi\, k^2}\left\{\left[\delta_{ij} - \frac{k_i\, k_j}{k^2} + \frac{\mathrm{i}}{2k^2}\left(k_i \nabla_j - k_j \nabla_i\right)\right]\langle \boldsymbol{u}^2\rangle - \frac{\mathrm{i}}{k^2}\,\varepsilon_{ijp}\, k_p\, \langle \mathbf{u} \cdot (\nabla \times \mathbf{u})\rangle\right\},$$

where the energy spectrum function is $E(k) = \delta(k - k_0)$, the wave number $k_0 = 1/\ell_0$, the length ℓ_0 is the integral scale of random motions, u_0 is the characteristic turbulent velocity in the scale ℓ_0 and $\mathrm{Re} = u_0\, \ell_0/\nu \gg 1$. The function $\delta(y)$ is the Dirac's delta-function which has the following property: $\int_a^b f(y)\, \delta(y - c)\, dy = f(c)$, where $a < c < b$. Here ν is the kinematic viscosity of the fluid and η is the magnetic diffusion coefficient of the magnetic field caused by the electrical conductivity of the fluid.

6.1.4 Practice Problems for Chapter 4

52. To estimate the dissipation rate $\varepsilon_k = \nu\, \langle (\nabla_i u_j)^2\rangle$ of the second moment, $E_K = \frac{1}{2}\, \langle \boldsymbol{u}^2\rangle$. The spectrum function of velocity fluctuations $\boldsymbol{u}$ is a power-law function with the exponent $q < 3$, where ν is the kinematic viscosity.

53. To estimate the dissipation rate $\varepsilon_\theta = D^{(\theta)}\, \langle (\nabla \theta)^2\rangle$ of the second moment $E_\theta = \frac{1}{2}\, \langle \theta^2\rangle$. The spectrum function of temperature fluctuations θ is a power-law function with the exponent $q < 3$, where $D^{(\theta)}$ is the molecular diffusion coefficient of the temperature field.

54. To estimate the dissipation rate $\varepsilon_i^{(F)} = \left(\nu + D^{(\theta)}\right) \langle(\nabla_j \theta)(\nabla_j u_i)\rangle$ of the turbulent heat flux $F_i^{(\theta)} = \langle u_i \theta \rangle$. The spectrum function of the turbulent heat flux is a power-law function with the exponent $q < 3$, where ν is the kinematic viscosity and $D^{(\theta)}$ is the molecular diffusion coefficient of the temperature field.

55. To estimate the turbulent heat flux $\langle \boldsymbol{u}\theta \rangle$ and the turbulent diffusion tensor $D_{ij}^{(T)}$ of the mean temperature field using the steady-state solution of the budget equation for the turbulent heat flux for large Péclet numbers $\mathrm{Pe} \equiv u_0 \ell_0 / D^{(\theta)} \gg 1$. To estimate the turbulent kinetic energy $E_k = \langle \boldsymbol{u}^2 \rangle / 2$ for a forced turbulence, using the steady-state solution of the budget equation for the turbulent kinetic energy, where the length ℓ_0 is the integral scale of turbulence, u_0 is the characteristic turbulent velocity in the scale ℓ_0 and $D^{(\theta)}$ is the molecular diffusion coefficient of the temperature field.

56. To estimate the second moment $\langle \theta^2 \rangle$ of temperature fluctuations θ generated by tangling of the imposed mean temperature gradient $\boldsymbol{\nabla} \overline{T}$ by velocity fluctuations using the steady-state solution of the budget equation for the second moment $\langle \theta^2 \rangle$ of temperature fluctuations θ for large Péclet numbers $\mathrm{Pe} = u_0 \ell_0 / D^{(\theta)} \gg 1$. Here the length ℓ_0 is the integral scale of turbulence, u_0 is the characteristic turbulent velocity in the scale ℓ_0 and $D^{(\theta)}$ is the molecular diffusion coefficient of the temperature field.

57. To determine the dependence of potential temperature fluctuations $\langle \Theta'^2 \rangle$ on the gradient Richardson number Ri in a shear produced stably stratified homogeneous turbulence using the steady-state solution of the budget equation for the second moment $\langle \Theta'^2 \rangle$ of potential temperature fluctuations Θ' for large Péclet numbers $\mathrm{Pe} = u_0 \ell_0 / D^{(\theta)} \gg 1$. Here the length ℓ_0 is the integral scale of turbulence, u_0 is the characteristic turbulent velocity in the scale ℓ_0 and $D^{(\theta)}$ is the molecular diffusion coefficient of the temperature field.

6.1.5 Practice Problems for Chapter 5

58. To determine the tensor of turbulent diffusion $D_{mn}^{(T)}$ of temperature field for an anisotropic turbulence and the turbulent diffusion coefficient D_T for an isotropic turbulence applying the path-integral approach in a random delta-correlated in time incompressible fluid flow with a constant renovating time.

59. To determine tensor of turbulent diffusion $D_{mn}^{(T)}$ of particle number density and the effective particle velocity $\boldsymbol{V}^{\rm eff}$ applying the path-integral approach in a random delta-correlated in time density-stratified fluid flow with a constant renovating time.

60. To determine the effective diamagnetic velocity $V^{\rm eff}$ of the mean magnetic field, the turbulent magnetic diffusion coefficient $\eta_{_T}$, the α effect, the turbulent electromotive force (EMF) $\boldsymbol{\mathcal{E}}$, applying the path-integral approach in a random delta-correlated in time incompressible fluid flow with a constant renovating time.

6.2 Solutions to the Practice Problems on Turbulent Transport

6.2.1 Solutions of Practice Problems for Chapter 1

1. The spectrum for velocity fluctuations for fully developed homogeneous and isotropic hydrodynamic turbulence is given by $E_K(k) = \varepsilon^{2/3}\, k^{-5/3}$ in the inertial range of the turbulence, $k_0 \le k \le k_\nu$. The scale-dependence of the characteristic turbulent time $\tau(k)$ in the inertial range is determined by $\tau(k) = \varepsilon^{-1/3}\, k^{-2/3}$. The Kolmogorov viscous length-scale $\ell_\nu = k_\nu^{-1}$ is given by $\ell_\nu = \ell_0\, {\rm Re}^{-3/4}$ and the characteristic viscous time (the Kolmogorov time), τ_ν, is given by $\tau_\nu = \tau_0\, {\rm Re}^{-1/2}$, which are the functions of the Reynolds number, ${\rm Re} = u_0\, \ell_0/\nu$, where $\varepsilon = u_0^3/\ell_0$ is the dissipation rate of the turbulent kinetic energy density, the length $\ell_0 = k_0^{-1}$ is the integral scale of turbulence, u_0 is the characteristic turbulent velocity in the scale ℓ_0 and ν is the kinematic viscosity. The solution to this problem is similar to that discussed in Section 1.1.2 using Eqs. (1.9)–(1.21).

2. The molecular diffusion length-scale ℓ_D and the molecular diffusion time-scale τ_D of the temperature field versus Péclet number ${\rm Pe} = u_0\, \ell_0/D^{(\theta)}$, are given by $\ell_D = \ell_0\, {\rm Pe}^{-3/4}$ and $\tau_D = \tau_0\, {\rm Pe}^{-1/2}$, where u_0 is the characteristic turbulent velocity in the integral scale of turbulence ℓ_0, the characteristic turbulent time $\tau_0 = \ell_0/u_0$ and $D^{(\theta)}$ is the molecular diffusion coefficient of the temperature field. The solution to this problem is similar to that discussed in Section 1.2.1 using Eqs. (1.32)–(1.37).

3. The spectrum of isotropic temperature fluctuations produced in the inertial range $\ell_\nu \le \ell_D < \ell < \ell_0$ of the incompressible turbulent fluid flow is given by $\tilde{E}_\theta(k) \sim \varepsilon^{2/3}\, k^{-5/3}$. The spectrum function of temperature fluctuations is defined as $\left\langle \theta^2 \right\rangle_\ell = \int_{k_0}^{k} \tilde{E}_\theta(k')\, dk'$. The solution to this problem is similar to that discussed in Section 1.2.2 using Eqs. (1.40)–(1.45).

4. The spectrum of isotropic temperature fluctuations in the viscous range $\ell_D < \ell < \ell_\nu$ of the incompressible turbulent fluid flow is given by $\tilde{E}_\theta(k) \sim k^{-1}$. The spectrum function of temperature fluctuations is defined as $\left\langle \theta^2 \right\rangle_\ell = \int_{k_0}^{k} \tilde{E}_\theta(k')\, dk'$. The solution to this problem is similar to that discussed in Section 1.2.2 using Eqs. (1.43)–(1.44) and (1.46).

5. The spectrum of anisotropic temperature fluctuations produced by the tangling of the gradient of the mean temperature by velocity fluctuations in the inertial range

$\ell_\nu < \ell \ll \ell_D \ll \ell_0$ of the incompressible turbulent fluid flow for small Prandtl numbers is given by $\tilde{E}_\theta(k) \sim \varepsilon^{2/3} k^{-17/3} \left(\nabla \overline{T}/D^{(\theta)}\right)^2$. The spectrum function of temperature fluctuations is defined as $\left\langle\theta^2\right\rangle_\ell = \int_{k_0}^k \tilde{E}_\theta(k')\, dk'$. The solution to this problem is similar to that discussed in Section 1.2.3 using Eqs. (1.47)–(1.50).

6. The spectrum of anisotropic temperature fluctuations produced by the tangling of the gradient of the mean temperature by velocity fluctuations in the inertial range $\ell_\nu \ll \ell_D < \ell \ll \ell_0$ of the incompressible turbulent fluid flow is given by $\tilde{E}_\theta(k) \sim k^{-3} \left(\nabla \overline{T}\right)^2$. The solution to this problem is similar to that discussed in Section 1.2.4 using Eqs. (1.51)–(1.55).

7. The turbulent diffusion coefficient is $D_T = \frac{1}{3}\, \tau_0 \left\langle \boldsymbol{u}^2\right\rangle \mathrm{Pe}$, and the turbulent heat flux is $\langle u_i\, \theta\rangle = -D_T \nabla_i \overline{T}$. The solution to this problem is similar to that discussed in Sections 1.3.3–1.3.5 using Eqs. (1.57)–(1.58), (1.60) and (1.66)–(1.67).

8. The turbulent diffusion coefficient is $D_T = \frac{1}{3}\, \tau_0 \left\langle \boldsymbol{u}^2\right\rangle$, and the turbulent heat flux is $\langle u_i\, \theta\rangle = -D_T \nabla_i \overline{T}$. The solution to this problem is similar to that discussed in Sections 1.3.3–1.3.5 using Eqs. (1.57)–(1.58), (1.59) and (1.61)–(1.63).

9. The level, $\langle\theta^2\rangle$, of temperature fluctuations θ in a random incompressible fluid flow with an imposed mean temperature gradient, ∇T for small Péclet numbers is $\langle\theta^2\rangle = \frac{1}{3}\, \mathrm{Pe}^2\, \ell_0^2 \left(\nabla\overline{T}\right)^2$. The solution to this problem is similar to that discussed in Sections 1.3.3–1.3.5 using Eqs. (1.57)–(1.58), (1.60) and (1.68)–(1.69).

10. The level, $\langle\theta^2\rangle$, of temperature fluctuations θ in a turbulent incompressible flow with an imposed mean temperature gradient, ∇T, for large Péclet numbers is $\langle\theta^2\rangle = \frac{1}{3}\, \ell_0^2 \left(\nabla\overline{T}\right)^2$. The solution to this problem is similar to that discussed in Sections 1.3.3–1.3.5 using Eqs. (1.57)–(1.59) and (1.64)–(1.65).

11. The turbulent heat flux is $\langle u_i\, \theta\rangle = -D_T \nabla_i \overline{T}$, where the turbulent diffusion coefficient is $D_T = \frac{1}{12}\, u_0\, \ell_0\, \mathrm{Pe}$. The solution to this problem is similar to that discussed in Sections 1.5.1–1.5.4 using Eqs. (1.107)–(1.115), (1.117), (1.119)–(1.121), (1.131), (1.138), (1.133) and (1.135)–(1.136).

12. The dependence of the second moment, $\langle\theta^2\rangle$, of temperature fluctuations θ on the mean temperature gradient, ∇T, for small Péclet numbers is given by $\langle\theta^2\rangle = \frac{1}{21}\, \mathrm{Pe}^2\, \ell_0^2 \left(\nabla\overline{T}\right)^2$. The solution to this problem is similar to that discussed in Sections 1.5.1–1.5.4 using Eqs. (1.107)–(1.114), (1.116), (1.119)–(1.121), (1.132), (1.134) and (1.135), (1.137).

13. The turbulent heat flux is $\langle u_i\, \theta\rangle = -D_T \nabla_i \overline{T}$, where the turbulent diffusion coefficient is $D_T = \frac{1}{3}\, u_0\, \ell_0\, \mathrm{Pe}$. The solution to this problem is similar to that discussed in Sections 1.5.1–1.5.4 using Eqs. (1.107)–(1.115), (1.117), (1.119)–(1.121), (1.131) and (1.135).

14. The turbulent heat flux is $\langle u_i\,\theta\rangle = -D_T\nabla_i\overline{T}$, where the turbulent diffusion coefficient is $D_T = \frac{1}{3}\,u_0\,\ell_0$. The turbulent diffusion coefficient of the temperature field is independent of the Péclet number in this case. The solution to this problem is similar to that discussed in Sections 1.6.1–1.6.3 using Eqs. (1.141)–(1.142), (1.144), (1.146)–(1.147), (1.149), (1.151) (1.152), (1.156)–(1.157) and (1.154).

15. The dependence of the second moment, $\langle\theta^2\rangle$, of temperature fluctuations θ on the mean temperature gradient, $\boldsymbol{\nabla}T$, for large Péclet numbers is given by $\langle\theta^2\rangle = \frac{8}{9}\,\ell_0^2\,\left(\boldsymbol{\nabla}\overline{T}\right)^2$. The solution to this problem is similar to that discussed in Sections 1.6.1–1.6.3 using Eqs. (1.141)–(1.151), (1.153), (1.155) and (1.135).

16. The turbulent heat flux is $\langle u_i\,\theta\rangle = -D_T\nabla_i\overline{T}$, where the turbulent diffusion coefficient is $D_T = \frac{1}{3}\,\left\langle \boldsymbol{u}^2\right\rangle\,\tau(k_0)$, where $\tau(k_0) = \int_{k_0}^{k_\nu}\tau(k)\,\delta(k-k_0)\,dk$. The turbulent diffusion coefficient of the temperature field is independent of the Péclet number in this case. The solution to this problem is similar to that discussed in Sections 1.6.1–1.6.3 using Eqs. (1.141)–(1.142), (1.144), (1.146)–(1.147), (1.149), (1.151) and (1.152).

17. The dependence of the second moment, $\langle\theta^2\rangle$, of temperature fluctuations θ on the mean temperature gradient, $\boldsymbol{\nabla}T$, for large Péclet numbers is given by $\langle\theta^2\rangle = \frac{2}{3}\,\left\langle \boldsymbol{u}^2\right\rangle\,\tau^2(k_0)\,\left(\boldsymbol{\nabla}\overline{T}\right)^2$, where $\tau^2(k_0) = \int_{k_0}^{k_\nu}\tau^2(k)\,\delta(k-k_0)\,dk$. The solution to this problem is similar to that discussed in Sections 1.6.1–1.6.3 using Eqs. (1.141)–(1.151) and (1.153).

6.2.2 Solutions of Practice Problems for Chapter 2

18. The molecular diffusion length-scale $\ell_D^{(n)}$ and the molecular diffusion time-scale $\tau_D^{(n)}$ of the particle number density versus Péclet number, $\mathrm{Pe}^{(n)} = u_0\,\ell_0/D^{(n)}$, are given by $\ell_D^{(n)} = \ell_0\,\left(\mathrm{Pe}^{(n)}\right)^{-3/4}$ and $\tau_D^{(n)} = \tau_0\,\left(\mathrm{Pe}^{(n)}\right)^{-1/2}$, where u_0 is the characteristic turbulent velocity in the integral scale of turbulence ℓ_0, the characteristic turbulent time $\tau_0 = \ell_0/u_0$ and $D^{(n)}$ is the molecular (Brownian) diffusion coefficient of the particle number density. To obtain these results, we introduce the local Péclet number at the scale ℓ as $\mathrm{Pe}_\ell^{(n)} = \ell\,u_\ell/D^{(n)} = \varepsilon^{1/3}\ell^{4/3}/D^{(n)}$, where we take into account that in the inertial range of fully developed turbulence, $\ell_\nu \le \ell \le \ell_0$, the dissipation rate ε of the turbulent kinetic energy density is constant, so that $\varepsilon = u_0^3/\ell_0 = u_\ell^3/\ell = u_\nu^3/\ell_\nu = \mathrm{const}$, where u_ν is the turbulent velocity at the viscous Kolmogorov scale ℓ_ν; see Section 1.1.2. This implies that the ratio $\mathrm{Pe}_\ell^{(n)}/\mathrm{Pe}^{(n)} = (\ell/\ell_0)^{4/3}$. The diffusive scale, $\ell_D^{(n)}$, is defined as the scale in which the local Péclet number $\mathrm{Pe}_\ell^{(n)} = 1$. This implies that at the scale $\ell_D^{(n)}$, the nonlinear term in the equation for particle number density fluctuations equals the diffusion term. This yields $\ell_D^{(n)} = \ell_0 \left(\mathrm{Pe}^{(n)}\right)^{-3/4}$. We consider the case when the molecular diffusion length-scale $\ell_D^{(n)}$ is inside the inertial range of turbulence $\ell_\nu \le \ell_D^{(n)} \le \ell_0$. This implies that $u_D^{(n)} = \left(\varepsilon\, \ell_D^{(n)}\right)^{1/3}$. Substituting the expression for the diffusion scale $\ell_D^{(n)}$ into the formula for $u_D^{(n)}$, we obtain the characteristic velocity in the diffusion scale as $u_D^{(n)} = (\varepsilon\, \ell_0)^{1/3} \left(\mathrm{Pe}^{(n)}\right)^{-1/4} = u_0\, \mathrm{Pe}^{-1/4}$. Therefore, the characteristic diffusion time $\tau_D^{(n)} = \ell_D^{(n)}/u_D^{(n)}$ is given by $\tau_D^{(n)} = \tau_0 \left(\mathrm{Pe}^{(n)}\right)^{-1/2}$.

19. The turbulent diffusion coefficient of particles is $D_T = \frac{1}{3}\,\tau_0 \left\langle \boldsymbol{u}^2 \right\rangle \mathrm{Pe}^{(n)}$, the effective pumping velocity of non-inertial particles is $\boldsymbol{V}^{\rm eff} = D_T\, (\boldsymbol{\nabla}\overline{\rho})/\overline{\rho}$, and the turbulent particle flux is $\langle u_i\, n'\rangle = \boldsymbol{V}^{\rm eff}\, \overline{n} - D_T \nabla_i \overline{n}$. The solution to this problem is similar to that discussed in Sections 2.1.3–2.1.6 using Eqs. (2.12) and (2.25)–(2.28).

20. The turbulent diffusion coefficient of particles is $D_T = \frac{1}{3}\,\tau_0 \left\langle \boldsymbol{u}^2 \right\rangle$, the effective pumping velocity of non-inertial particles is $\boldsymbol{V}^{\rm eff} = D_T\, (\boldsymbol{\nabla}\overline{\rho})/\overline{\rho}$, and the turbulent particle flux is $\langle u_i\, n'\rangle = \boldsymbol{V}^{\rm eff}\, \overline{n} - D_T \nabla_i \overline{n}$. The solution to this problem is similar to that discussed in Sections 2.1.3–2.1.6 using Eqs. (2.8)–(2.11) and (2.13)–(2.19).

21. The level, $\langle n'^2\rangle$, of particle number density fluctuations n' in a random density-stratified fluid flow with an imposed mean particle number density gradient, $\boldsymbol{\nabla}\overline{n}$ for small Péclet numbers $\mathrm{Pe}^{(n)} = u_0\, \ell_0/D^{(n)} \ll 1$ is given by $\langle n'^2\rangle = \frac{1}{3} \left(\mathrm{Pe}^{(n)}\right)^2 \ell_0^2 \left[(\boldsymbol{\nabla}\overline{n})^2 + (\boldsymbol{\lambda}\cdot\boldsymbol{\nabla})\, \overline{n}^2 + \overline{n}^2\, \boldsymbol{\lambda}^2\right]$, where $\boldsymbol{\lambda} = -(\boldsymbol{\nabla}\overline{\rho})/\overline{\rho}$. The solution to this problem is similar to that discussed in Sections 2.1.4–2.1.6 using Eqs. (2.9)–(2.10), (2.12) and (2.29)–(2.32).

22. The level, $\langle n'^2\rangle$, of particle number density fluctuations n' in a turbulent density-stratified fluid flow with an imposed mean particle number density gradient, $\boldsymbol{\nabla}\overline{n}$ for large Péclet numbers $\mathrm{Pe}^{(n)} = u_0\, \ell_0/D^{(n)} \gg 1$ is given by $\left\langle n'^2\right\rangle = \frac{1}{3}\, \ell_0^2 \left[(\boldsymbol{\nabla}\overline{n})^2 + (\boldsymbol{\lambda}\cdot\boldsymbol{\nabla})\, \overline{n}^2 + \overline{n}^2\, \boldsymbol{\lambda}^2\right]$, where $\boldsymbol{\lambda} = -(\boldsymbol{\nabla}\overline{\rho})/\overline{\rho}$. The solution to this problem is similar to that discussed in Sections 2.1.4–2.1.6 using Eqs. (2.9)–(2.11), (2.21)–(2.23) and (2.24).

23. The turbulent diffusion coefficient of particles is $D_T = \frac{1}{12}\,\tau_0 \left\langle \boldsymbol{u}^2 \right\rangle \mathrm{Pe}^{(n)}$, the effective pumping velocity of non-inertial particles is $\boldsymbol{V}^{\rm eff} = D_T\, (\boldsymbol{\nabla}\overline{\rho})/\overline{\rho}$, and the turbulent particle flux is $\langle u_i\, n'\rangle = \boldsymbol{V}^{\rm eff}\, \overline{n} - D_T \nabla_i \overline{n}$. The spatial distribution of the mean number density of non-inertial particles in a steady state when the mean fluid density is the function of the vertical coordinate z is given by

$$\boxed{\frac{\overline{n}}{\overline{n}_*} = \left(\frac{\overline{\rho}}{\overline{\rho}_*}\right)^{\frac{D_T}{D^{(n)}+D_T}}},$$

where the subscripts $*$ represent the values of the mean particle number density $\overline{n}_*$ and the mean fluid density $\overline{\rho}_*$ at the boundary. The solution to this problem is similar to that discussed in Sections 2.2.1–2.2.4 using Eqs. (2.42)–(2.49), (2.51), (2.53), (2.54), (2.59), (2.61) and (2.63)–(2.64), (1.135); see also Eqs. (2.65)–(2.66) in Section 2.1.7.

24. The turbulent diffusion coefficient of particles is $D_T = \frac{1}{3}\,\tau(k_0)\,\langle \boldsymbol{u}^2\rangle\,\mathrm{Pe}^{(n)}$, the effective pumping velocity of non-inertial particles is $\boldsymbol{V}^{\rm eff} = D_T\,(\boldsymbol{\nabla}\overline{\rho})/\overline{\rho}$, and the turbulent particle flux is $\langle u_i\, n'\rangle = \boldsymbol{V}^{\rm eff}\,\overline{n} - D_T \nabla_i \overline{n}$, where $\tau(k_0) = \int_{k_0}^{k_\nu} \tau(k)\,\delta(k - k_0)\,dk$. The solution to this problem is similar to that discussed in Sections 2.2.1–2.2.4 using Eqs. (2.42)–(2.49), (2.51), (2.53), (2.54), (2.59) and (2.61).

25. The level of particle number density fluctuations, $\langle n'^2\rangle$, for small Péclet numbers is given by $\langle n'^2\rangle = \frac{1}{21}\,\left(\mathrm{Pe}^{(n)}\right)^2\,\ell_0^2\,\left[(\boldsymbol{\nabla}\overline{n})^2 + (\boldsymbol{\lambda}\cdot\boldsymbol{\nabla})\overline{n}^2\right]$, where $\boldsymbol{\lambda} = -(\boldsymbol{\nabla}\overline{\rho})/\overline{\rho}$. Solution of this problem is similar to that discussed in Sections 2.2.1–2.2.4 using Eqs. (2.42)–(2.48), (2.50), (2.53), (2.54), (2.60), (2.62) and (1.137).

26. The turbulent diffusion coefficient of particles is $D_T = \frac{1}{3}\,\tau_0\,\langle \boldsymbol{u}^2\rangle$, the effective pumping velocity of non-inertial particles is $\boldsymbol{V}^{\rm eff} = D_T\,(\boldsymbol{\nabla}\overline{\rho})/\overline{\rho}$, and the turbulent particle flux is $\langle u_i\, n'\rangle = \boldsymbol{V}^{\rm eff}\,\overline{n} - D_T \nabla_i \overline{n}$. The spatial distribution of the mean number density of non-inertial particles in a steady state when the mean fluid density is the function of the vertical coordinate z is given by

$$\boxed{\frac{\overline{n}}{\overline{n}_*} = \left(\frac{\overline{\rho}}{\overline{\rho}_*}\right)^{\frac{D_T}{D^{(n)}+D_T}},}$$

where the subscripts $*$ represent the values of the mean particle number density $\overline{n}_*$ and the mean fluid density $\overline{\rho}_*$ at the boundary. The solution to this problem is similar to that discussed in Sections 2.3.1–2.3.2 using Eqs. (2.82)–(2.87), (2.89), (2.90), (2.92), (2.94), (2.95), (2.97)–(2.101) and (2.103)–(2.105); see also Eqs. (2.65)–(2.66) in Section 2.1.7.

27. The turbulent diffusion coefficient of particles is $D_T = \frac{1}{3}\,\tau_0(k_0)\,\langle \boldsymbol{u}^2\rangle$, the effective pumping velocity of non-inertial particles is $\boldsymbol{V}^{\rm eff} = D_T\,(\boldsymbol{\nabla}\overline{\rho})/\overline{\rho}$, and the turbulent particle flux is $\langle u_i\, n'\rangle = \boldsymbol{V}^{\rm eff}\,\overline{n} - D_T \nabla_i \overline{n}$, where $\tau(k_0) = \int_{k_0}^{k_\nu} \tau(k)\,\delta(k - k_0)\,dk$. The solution to this problem is similar to that discussed in Sections 2.3.1–2.3.2 using Eqs. (2.82)–(2.87), (2.89), (2.90), (2.92), (2.94), (2.95) and (2.97)–(2.101).

28. The level of particle number density fluctuations, $\langle n'^2\rangle$, for large Péclet numbers is given by $\langle n'^2\rangle = \frac{8}{9}\,\ell_0^2\,\left[(\boldsymbol{\nabla}\overline{n})^2 + (\boldsymbol{\lambda}\cdot\boldsymbol{\nabla})\overline{n}^2\right]$, where the vector $\boldsymbol{\lambda} = -(\boldsymbol{\nabla}\overline{\rho})/\overline{\rho}$. Solution of this problem is similar to that discussed in Sections 2.3.1–2.3.2 using Eqs. (2.82)–(2.87), (2.89), (2.90), (2.92), (2.94), (2.95), (2.97)–(2.101) and (2.103)–(2.105); see also Eqs. (2.65)–(2.66) in Section 2.1.7.

29. The level of particle number density fluctuations, $\langle n'^2 \rangle$, for large Péclet numbers is given by $\langle n'^2 \rangle = \frac{2}{3} \langle \boldsymbol{u}^2 \rangle \, \tau_0^2(k_0) \, \left[(\nabla \overline{n})^2 + (\boldsymbol{\lambda} \cdot \nabla) \overline{n}^2 \right]$, where $\boldsymbol{\lambda} = -(\nabla \overline{\rho})/\overline{\rho}$ and $\tau^2(k_0) = \int_{k_0}^{k_\nu} \tau^2(k) \, \delta(k - k_0) \, dk$. The solution to this problem is similar to that discussed in Sections 2.3.1–2.3.2 using Eqs. (2.82)–(2.94), (2.96)–(2.100) and (2.102).

30. The effective pumping velocity of inertial particles is $\boldsymbol{V}^{\rm eff} = -\mu_p \, D_T \, \boldsymbol{\nabla} \ln \overline{T}$, where $\mu_p = 1 + 2 \, V_g \, L_P \, \ln {\rm Re}/(u_0 \, \ell_0)$, $\mathbf{V}_{\rm g}$ is the terminal fall velocity of particles, L_P is the pressure height scale, and Re is the Reynolds number. The spatial distribution of the mean number density of inertial particles in a steady state satisfying the boundary condition with a zero total particle flux at the boundary reads

$$\frac{\overline{n}(z)}{\overline{n}_*} = \left(\frac{\overline{T}(z)}{\overline{T}_*} \right)^{-\frac{\mu_p D_T}{D^{(n)} + D_T}} \exp \left[- \int_{z_0}^{z} \frac{V_{\rm g}}{D^{(n)} + D_T} \, dz' \right],$$

where the subscripts $*$ denote the values of the mean temperature and the mean particle number density at the boundaries. The solution to this problem is similar to that discussed in Section 2.4.3 using Eqs. (2.121)–(2.136).

31. The spectrum of isotropic fluctuations of particle number density produced in the inertial range $\ell_\nu \leq \ell_D^{(n)} < \ell < \ell_0$ of the incompressible turbulent fluid flow is given by $E_{\rm n}(k) \sim \varepsilon^{2/3} \, k^{-5/3}$. The spectrum function of particle number density fluctuations is defined as $\langle n'^2 \rangle_\ell = \int_{k_0}^{k} E_{\rm n}(k') \, dk'$. We assume that the rate of dissipation of isotropic fluctuations of particle number density $\langle n'^2 \rangle_\ell / \tau_\ell$ is constant inside the scale range $\ell_D^{(n)} < \ell < \ell_0$, i.e., $\langle n'^2 \rangle_\ell / \tau_\ell = E_{\rm n}(k) \, k / \tau(k) = {\rm const}$. We take into account that the rate of dissipation of the turbulent kinetic energy density is constant inside the inertial range of turbulence, $\varepsilon = u_\ell^2 / \tau_\ell = E_{\rm u}(k) \, k / \tau(k) =$ const. This yields the spectrum of isotropic fluctuations of particle number density $E_{\rm n}(k) \sim E_{\rm u}(k) \sim \varepsilon^{2/3} \, k^{-5/3}$ inside the scale subrange $\ell_D < \ell < \ell_0$.

32. The spectrum of isotropic fluctuations of particle number density produced in the viscous range $\ell_D^{(n)} < \ell < \ell_\nu$ of the incompressible turbulent fluid flow is given by $E_{\rm n}(k) \sim k^{-1}$. The spectrum function of particle number density fluctuations is defined as $\langle n'^2 \rangle_\ell = \int_{k_0}^{k} E_{\rm n}(k') \, dk'$. We assume that the rate of dissipation of isotropic fluctuations of particle number density $\langle n'^2 \rangle_\ell / \tau_\ell$ is constant inside the scale range $\ell_D^{(n)} < \ell < \ell_0$, i.e., $\langle n'^2 \rangle_\ell / \tau_\ell = E_{\rm n}(k) \, k / \tau(k) = {\rm const}$. The time $\tau(k)$ in the viscous sub-range of scales $\ell_D^{(n)} < \ell < \ell_\nu$ does not have a universal scaling. If we assume that $\tau(k) = {\rm const}$, the spectrum of isotropic fluctuations of particle number density produced in the viscous range $\ell_D^{(n)} < \ell < \ell_\nu$ is given by $E_{\rm n}(k) \sim k^{-1}$.

33. The spectrum of anisotropic fluctuations of particle number density produced by the tangling of the gradient of the mean particle number density by velocity fluctuations in the inertial range $\ell_\nu < \ell \ll \ell_D^{(n)} \ll \ell_0$ of the incompressible turbulent fluid flow for small Schmidt numbers $\mathrm{Sc} \equiv \nu/D^{(n)} \ll 1$ is given by $E_\mathrm{n}(k) \sim \varepsilon^{2/3}\, k^{-17/3}\, \left(\nabla\, \overline{n}/D^{(n)}\right)^2$. In the range of scales $\ell_\nu < \ell \ll \ell_D^{(n)}$, the molecular diffusion is a key effect. This implies that the molecular diffusion term, $D^{(n)}\Delta n'$, in the equation for particle number density fluctuations should be balanced by the source term, $(\boldsymbol{u} \cdot \nabla)\overline{n}$, i.e., $|D^{(n)}\Delta n'| \sim |(\boldsymbol{u} \cdot \nabla)\overline{n}|$. This yields $\langle n'^2 \rangle_\ell \sim u_\ell^2 \left(\ell^2 \nabla\, \overline{n}/D^{(n)}\right)^2$. In the $\boldsymbol{k}$ space, this equation can be rewritten as $E_\mathrm{n}(k) \sim E_\mathrm{u}(k)\, k^{-4}\, \left(\nabla\, \overline{n}/D^{(n)}\right)^2$, where the spectrum function of particle number density fluctuations is defined as $\langle n'^2 \rangle_\ell = \int_{k_0}^{k} E_\mathrm{n}(k')\, dk'$ and $E_\mathrm{u}(k)$ is the spectrum function of velocity fluctuations. Since the scale-subrange $\ell_\nu < \ell < \ell_D^{(n)}$ corresponds to the inertial range of turbulence, velocity fluctuations obey the Kolmogorov-Obukhov spectrum: $E_\mathrm{u}(k) \sim \varepsilon^{2/3}\, k^{-5/3}$. Substituting $E_\mathrm{u}(k)$ into the expression for $E_\mathrm{n}(k)$, we obtain the spectrum of anisotropic fluctuations of particle number in the subrange of scales $\ell_\nu < \ell \ll \ell_D^{(n)}$ as $E_\mathrm{n}(k) \sim \varepsilon^{2/3}\, k^{-17/3}\, \left(\nabla\, \overline{n}/D^{(n)}\right)^2$.

34. The spectrum of anisotropic fluctuations of particle number density produced by the tangling of the gradient of the mean particle number density by velocity fluctuations in the inertial range $\ell_\nu \ll \ell_D^{(n)} < \ell \ll \ell_0$ of the incompressible turbulent fluid flow is given by $E_\mathrm{n}(k) \sim k^{-3}\, (\nabla\, \overline{n})^2$. We average the equation for particle number density fluctuations over an ensemble up to the scale ℓ_* which is inside the subrange of scales: $\ell_D^{(n)} \ll \ell_* \ll \ell_0$. This yields the renormalized equation for particle number density fluctuations as $\partial n'/\partial t - D_{_T}(\ell)\, \Delta n' = -(\boldsymbol{u} \cdot \nabla)\overline{n}$, where $D_{_T}(\ell)$ is the scale-dependent turbulent diffusion coefficient which can be estimated as $D_{_T}(\ell) = \ell\, u_\ell$. Here we take into account that the main effect of turbulence on particle number density fluctuations in incompressible flow is the scale-dependent turbulent diffusion which is much larger than the molecular diffusion for large Péclet numbers. In this range of scales, the turbulent diffusion term, $D_{_T}(\ell)\Delta n'$, should be balanced by the source term, $(\boldsymbol{u} \cdot \nabla)\overline{n}$, i.e., $|D_{_T}(\ell)\, \Delta n'| \sim |(\boldsymbol{u} \cdot \nabla)\overline{n}|$. This implies that $\langle n'^2 \rangle_\ell \sim u_\ell^2 \left(\ell^2\, \nabla\, \overline{n}/D_{_T}(\ell)\right)^2 \sim u_\ell^2 \left(\ell^2\, \nabla\, \overline{n}/(\ell\, u_\ell)\right)^2 \sim (\ell\, \nabla\, \overline{n})^2$. This equation written in the $\boldsymbol{k}$ space yields the spectrum of anisotropic fluctuations of particle number density $E_\mathrm{n}(k) \sim k^{-3}\, (\nabla\, \overline{n})^2$ in the subrange of scales $\ell_D^{(n)} < \ell \ll \ell_0$.

6.2.3 *Solutions of Practice Problems for Chapter 3*

35. The molecular magnetic diffusion length-scale ℓ_η (the resistive scale) and time-scale τ_η of magnetic fluctuations versus magnetic Reynolds number, $\mathrm{Rm} = u_0\, \ell_0/\eta \gg 1$, are given by $\ell_\eta = \ell_0\, \mathrm{Rm}^{-3/4}$ and $\tau_\eta = \tau_0\, \mathrm{Rm}^{-1/2}$. The solution to this problem is similar to that discussed in Section 3.2.1 using Eqs. (3.23)–(3.37).

36. The spectrum of isotropic magnetic fluctuations produced in the inertial range $\ell_m \ll \ell_\nu < \ell < \ell_0$ of an incompressible turbulent fluid flow for magnetic Prandtl number $\mathrm{Pr}_m \equiv \nu/\eta > 1$ is given by $M_b(k) \sim \varepsilon^{2/3} k^{-5/3}$, where the spectrum function $M_b(k)$ for magnetic fluctuations is defined as $\langle \boldsymbol{b}^2 \rangle_\ell = \int_{k_0}^{k} M_b(k')\, dk'$. The solution to this problem is similar to that discussed in Section 3.2.2 using Eqs. (3.38)–(3.43).

37. The spectrum of isotropic magnetic fluctuations produced in the viscous range $\ell_m \ll \ell < \ell_\nu \ll \ell_0$ of an incompressible turbulent fluid flow for magnetic Prandtl number $\mathrm{Pr}_m \equiv \nu/\eta > 1$ is given by $M_b(k) \sim k^{-1}$, where the spectrum function $M_b(k)$ for magnetic fluctuations is defined as $\langle \boldsymbol{b}^2 \rangle_\ell = \int_{k_0}^{k} M_b(k')\, dk'$. The solution to this problem is similar to that discussed in Section 3.2.2 using Eqs. (3.42), (3.44).

38. The spectrum of anisotropic magnetic fluctuations produced by tangling of the mean magnetic field $\overline{\boldsymbol{B}}$ by velocity fluctuations in the inertial range of scales $\ell_\nu \ll \ell \le \ell_m \ll \ell_0$ of an incompressible turbulent fluid flow for small magnetic Prandtl numbers is given by $M_b(k) \sim \varepsilon^{2/3}\, k^{-11/3} \left(\overline{\boldsymbol{B}}/\eta\right)^2$, where the spectrum function $M_b(k)$ for magnetic fluctuations is defined as $\langle \boldsymbol{b}^2 \rangle_\ell = \int_{k_0}^{k} M_b(k')\, dk'$. The solution to this problem is similar to that discussed in Section 3.2.3 using Eqs. (3.45)–(3.48).

39. The spectrum of anisotropic magnetic fluctuations produced by tangling of the mean magnetic field by velocity fluctuations in the inertial range of scales $\ell_\nu \ll \ell_m \ll \ell \ll \ell_0$ of an incompressible turbulent fluid flow for small magnetic Prandtl numbers is given by $M_b(k) \sim k^{-1}\, \overline{\boldsymbol{B}}^2$, The solution to this problem is similar to that discussed in Section 3.2.4 using Eqs. (3.49)–(3.53).

40. The turbulent magnetic diffusion coefficient $\eta_{_T}$ for small magnetic Reynolds numbers $\mathrm{Rm} = u_0 \ell_0/\eta \ll 1$ is given by $\eta_{_T} = \frac{1}{3}\, \tau_0\, \langle \boldsymbol{u}^2 \rangle\, \mathrm{Rm}$; the α effect is given by $\alpha = -\frac{1}{3}\, \tau_0\, \langle \boldsymbol{u} \cdot (\boldsymbol{\nabla} \times \boldsymbol{u}) \rangle\, \mathrm{Rm}$; the effective diamagnetic velocity $\boldsymbol{V}^{\mathrm{eff}}$ of the mean magnetic field is given by $\boldsymbol{V}^{\mathrm{eff}} = -\frac{1}{2}\, \boldsymbol{\nabla} \eta_{_T}$; and the turbulent electromotive force (EMF) $\boldsymbol{\mathcal{E}}$ is given by $\boldsymbol{\mathcal{E}} = \boldsymbol{\alpha}\, \overline{\boldsymbol{B}} + \boldsymbol{V}^{\mathrm{eff}} \times \overline{\boldsymbol{B}} - \eta_{_T} (\boldsymbol{\nabla} \times \overline{\boldsymbol{B}})$. The solution to this problem is similar to that discussed in Sections 3.3.2–3.3.4 using Eqs. (3.56), (3.58), (3.59)–(3.62) and (3.78)–(3.80).

41. The turbulent magnetic diffusion coefficient $\eta_{_T}$ for large magnetic Reynolds numbers $\mathrm{Rm} = u_0 \ell_0/\eta \gg 1$ is given by $\eta_{_T} = \frac{1}{3}\, \tau_0\, \langle \boldsymbol{u}^2 \rangle$; the α effect is given by $\alpha = -\frac{1}{3}\, \tau_0\, \langle \boldsymbol{u} \cdot (\boldsymbol{\nabla} \times \boldsymbol{u}) \rangle$; the effective diamagnetic velocity $\boldsymbol{V}^{\mathrm{eff}}$ of the mean magnetic field is given by $\boldsymbol{V}^{\mathrm{eff}} = -\frac{1}{2}\, \boldsymbol{\nabla} \eta_{_T}$; and the turbulent electromotive force (EMF) $\boldsymbol{\mathcal{E}}$ is given by $\boldsymbol{\mathcal{E}} = \boldsymbol{\alpha}\, \overline{\boldsymbol{B}} + \boldsymbol{V}^{\mathrm{eff}} \times \overline{\boldsymbol{B}} - \eta_{_T} (\boldsymbol{\nabla} \times \overline{\boldsymbol{B}})$. The solution to this problem is similar to that discussed in Sections 3.3.2–3.3.4 using Eqs. (3.56)–(3.57) and (3.59)–(3.74).

42. The level, $\langle \boldsymbol{b}^2 \rangle$, of magnetic fluctuations $\boldsymbol{b}$ for small magnetic Reynolds numbers $\mathrm{Rm} = u_0 \ell_0/\eta \ll 1$ is given by $\langle \boldsymbol{b}^2 \rangle = \mathrm{Rm}^2\, \overline{\boldsymbol{B}}^2$. The solution to this problem is similar to that discussed in Sections 3.3.2–3.3.4 using Eqs. (3.56), (3.58) and (3.81)–(3.82).

43. The level, $\langle \boldsymbol{b}^2 \rangle$, of magnetic fluctuations $\boldsymbol{b}$ for large magnetic Reynolds numbers $\mathrm{Rm} = u_0 \ell_0 / \eta \gg 1$ is given by $\left\langle \boldsymbol{b}^2 \right\rangle = (\ln \mathrm{Rm})\, \overline{\boldsymbol{B}}^2$. The solution to this problem is similar to that discussed in Sections 3.3.2–3.3.4 using Eqs. (3.56)–(3.57) and (3.75)–(3.77).

44. The turbulent magnetic diffusion coefficient η_T for small magnetic Reynolds numbers $\mathrm{Rm} = u_0 \ell_0 / \eta \ll 1$ is given by $\eta_T = \frac{1}{12}\, \tau_0 \left\langle \boldsymbol{u}^2 \right\rangle \mathrm{Rm}$; the α effect is given by $\alpha = -\frac{1}{12}\, \tau_0 \, \langle \boldsymbol{u} \cdot (\nabla \times \boldsymbol{u}) \rangle \, \mathrm{Rm}$; the effective diamagnetic velocity $\boldsymbol{V}^{\mathrm{eff}}$ of the mean magnetic field is given by $\boldsymbol{V}^{\mathrm{eff}} = -\frac{1}{2}\, \nabla \eta_T$; and the turbulent electromotive force (EMF) $\boldsymbol{\mathcal{E}}$ is given by $\boldsymbol{\mathcal{E}} = \boldsymbol{\alpha} \overline{\boldsymbol{B}} + \boldsymbol{V}^{\mathrm{eff}} \times \overline{\boldsymbol{B}} - \eta_T (\nabla \times \overline{\boldsymbol{B}})$. The solution to this problem is similar to that discussed in Sections 3.4.1–3.4.4 using Eqs. (3.84)–(3.97), (3.99), (3.101), (3.105) and (3.107)–(3.110).

45. The turbulent magnetic diffusion coefficient η_T for small magnetic Reynolds numbers $\mathrm{Rm} = u_0 \ell_0 / \eta \ll 1$ is given by $\eta_T = \frac{1}{3}\, \tau_0 \left\langle \boldsymbol{u}^2 \right\rangle \mathrm{Rm}$; the α effect is given by $\alpha = -\frac{1}{3}\, \tau_0 \, \langle \boldsymbol{u} \cdot (\nabla \times \boldsymbol{u}) \rangle \, \mathrm{Rm}$; the effective diamagnetic velocity $\boldsymbol{V}^{\mathrm{eff}}$ of the mean magnetic field is given by $\boldsymbol{V}^{\mathrm{eff}} = -\frac{1}{2}\, \nabla \eta_T$; and the turbulent electromotive force (EMF) $\boldsymbol{\mathcal{E}}$ is given by $\boldsymbol{\mathcal{E}} = \boldsymbol{\alpha} \overline{\boldsymbol{B}} + \boldsymbol{V}^{\mathrm{eff}} \times \overline{\boldsymbol{B}} - \eta_T (\nabla \times \overline{\boldsymbol{B}})$. The solution to this problem is similar to that discussed in Sections 3.4.1–3.4.4 using Eqs. (3.84)–(3.97), (3.99), (3.101) and (3.105).

46. The level, $\langle \boldsymbol{b}^2 \rangle$, of magnetic fluctuations $\boldsymbol{b}$ for small magnetic Reynolds numbers $\mathrm{Rm} = u_0 \ell_0 / \eta \ll 1$ is given by $\left\langle \boldsymbol{b}^2 \right\rangle = \frac{1}{12}\, \mathrm{Rm}^2\, \overline{\boldsymbol{B}}^2$. Solution of this problem is similar to that discussed in Sections 3.4.1–3.4.4 using Eqs. (3.84), (3.89)–(3.96), (3.98), (3.100), (3.101), (3.106) and (3.111).

47. The level, $\langle \boldsymbol{b}^2 \rangle$, of magnetic fluctuations $\boldsymbol{b}$ for small magnetic Reynolds numbers $\mathrm{Rm} = u_0 \ell_0 / \eta \ll 1$ is given by $\left\langle \boldsymbol{b}^2 \right\rangle = \frac{1}{3}\, \mathrm{Rm}^2\, \overline{\boldsymbol{B}}^2$. Solution of this problem is similar to that discussed in Sections 3.4.1–3.4.4 using Eqs. (3.84), (3.89)–(3.96), (3.98), (3.100), (3.101) and (3.106).

48. The turbulent magnetic diffusion coefficient η_T for large magnetic Reynolds numbers $\mathrm{Rm} = u_0 \ell_0 / \eta \gg 1$ is given by $\eta_T = \frac{1}{3}\, \tau_0 \left\langle \boldsymbol{u}^2 \right\rangle$; the α effect is given by $\alpha = -\frac{1}{3}\, \tau_0 \, \langle \boldsymbol{u} \cdot (\nabla \times \boldsymbol{u}) \rangle$; the effective diamagnetic velocity $\boldsymbol{V}^{\mathrm{eff}}$ of the mean magnetic field is given by $\boldsymbol{V}^{\mathrm{eff}} = -\frac{1}{6}\, \tau_0 \nabla \left\langle \boldsymbol{u}^2 \right\rangle$; and the turbulent electromotive force (EMF) $\boldsymbol{\mathcal{E}}$ is given by $\boldsymbol{\mathcal{E}} = \boldsymbol{\alpha} \overline{\boldsymbol{B}} + \boldsymbol{V}^{\mathrm{eff}} \times \overline{\boldsymbol{B}} - \eta_T (\nabla \times \overline{\boldsymbol{B}})$. Solution of this problem is similar to that discussed in Sections 3.5.1–3.4.4 using Eqs. (3.133)–(3.135), (3.137)–(3.153), (3.155), (3.156) and (3.158)–(3.161).

49. The turbulent magnetic diffusion coefficient η_T for large magnetic Reynolds numbers $\mathrm{Rm} = u_0 \ell_0 / \eta \gg 1$ is given by $\eta_T = \frac{1}{3}\, \tau(k_0) \left\langle \boldsymbol{u}^2 \right\rangle$; the α effect is given by

$\alpha = -\frac{1}{3}\,\tau(k_0)\,\langle \boldsymbol{u} \cdot (\nabla \times \boldsymbol{u})\rangle$; the effective diamagnetic velocity $\boldsymbol{V}^{\rm eff}$ of the mean magnetic field is given by $\boldsymbol{V}^{\rm eff} = -\frac{1}{6}\,\tau(k_0)\nabla\left\langle \boldsymbol{u}^2\right\rangle$; and the turbulent electromotive force (EMF) $\boldsymbol{\mathcal{E}}$ is given by $\boldsymbol{\mathcal{E}} = \boldsymbol{\alpha}\,\overline{\boldsymbol{B}} + \boldsymbol{V}^{\rm eff} \times \overline{\boldsymbol{B}} - \eta_{_T}(\nabla \times \overline{\boldsymbol{B}})$. The solution to this problem is similar to that discussed in Sections 3.5.1–3.5.6 using Eqs. (3.133)–(3.135), (3.137)–(3.153), (3.155), (3.156) and (3.158)–(3.161).

50. The level, $\langle \boldsymbol{b}^2\rangle$, of magnetic fluctuations $\boldsymbol{b}$ for large magnetic Reynolds numbers ${\rm Rm} = u_0\ell_0/\eta \gg 1$ is given by $\left\langle \boldsymbol{b}^2\right\rangle = \frac{4}{9}\,(\ln {\rm Re})\;\overline{\boldsymbol{B}}^2$. Solution of this problem is similar to that discussed in Sections 3.5.1–3.5.6 using Eqs. (3.134)–(3.139), (3.152), (3.154), (3.155) and (3.157).

51. The level, $\langle \boldsymbol{b}^2\rangle$, of magnetic fluctuations $\boldsymbol{b}$ for large magnetic Reynolds numbers ${\rm Rm} = u_0\ell_0/\eta \gg 1$ is given by $\left\langle \boldsymbol{b}^2\right\rangle = \frac{2}{3}\left(\tau^2(k_0)/\tau_0^2\right)\overline{\boldsymbol{B}}^2$. The solution to this problem is similar to that discussed in Sections 3.5.1–3.5.6 using Eqs. (3.134)–(3.139), (3.152), (3.154), (3.155) and (3.157).

6.2.4 Solutions of Practice Problems for Chapter 4

52. The dissipation rate $\varepsilon_k = \nu\,\langle(\nabla_i u_j)^2\rangle$ of the second moment, $E_K = \frac{1}{2}\,\langle \boldsymbol{u}^2\rangle$ is given by $\varepsilon_k = E_K/\tau_0$. The solution to this problem is similar to that discussed in Section 4.1.1 using Eqs. (4.10)–(4.17).

53. The dissipation rate $\varepsilon_\theta = D^{(\theta)}\,\langle(\nabla\theta)^2\rangle$ of the second moment, $E_\theta = \frac{1}{2}\,\langle\theta^2\rangle$ is given by $\varepsilon_\theta = E_\theta/\tau_0$. The solution to this problem is similar to that discussed in Section 4.1.2 using Eqs. (4.28)–(4.32).

54. The dissipation rate $\varepsilon_i^{(F)} = \left(\nu + D^{(\theta)}\right)\langle(\nabla_j\theta)(\nabla_j u_i)\rangle$ of the turbulent heat flux, $F_i^{(\theta)} = \langle u_i\theta\rangle$ is given by $\varepsilon_i^{(F)} = \left(1 + {\rm Pr}^{-1}\right)F_i^{(\theta)}/2\tau_0$. The solution to this problem is similar to that discussed in Section 4.1.3 using Eqs. (4.41)–(4.45).

55. The turbulent heat flux, $F_i^{(\theta)} = \langle u_i\theta\rangle$ is given by $F_i^{(\theta)} = -D_{ij}^{(T)}\,\nabla_j\overline{T}$, where the turbulent diffusion tensor $D_{ij}^{(T)}$ of the mean temperature field for large Péclet numbers is given by $D_{ij}^{(T)} = 2\tau_0\langle u_i u_j\rangle\left(1 + {\rm Pr}^{-1}\right)^{-1}$. The turbulent kinetic energy, $E_k = \frac{1}{2}\langle \boldsymbol{u}^2\rangle$ for a forced turbulence is given by $E_K = \tau_0\,\rho\,\langle \boldsymbol{u} \cdot \boldsymbol{f}\rangle$. The solution to this problem is similar to that discussed in Sections 4.1.3–4.1.4 using Eqs. (4.37), (4.46) and (4.48)–(4.49).

56. The second moment, $\langle\theta^2\rangle$, of temperature fluctuations θ generated by tangling of the imposed mean temperature gradient $\nabla\overline{T}$ by velocity fluctuations in a steady state is given by $\langle\theta^2\rangle \approx \ell_0^2(\nabla\overline{T})^2\left(1 + {\rm Pr}^{-1}\right)^{-1}$. The solution to this problem is similar to that discussed in Sections 4.1.2 and 4.1.4 using Eqs. (4.24), (4.46) and (4.47)–(4.53).

57. The dependence of potential temperature fluctuations $\langle \Theta'^2 \rangle$ on the gradient Richardson number Ri in a shear produced stably stratified homogeneous turbulence is given by $\langle \Theta'^2 \rangle = 2\ell_0^2 \left(\nabla \overline{\Theta}\right)^2 \left(3 + 4\tau_0^2\, S^2\, \mathrm{Ri}^2\right)^{-1}$, where Θ' are potential temperature fluctuations, the length ℓ_0 is the integral scale of turbulence, S is the mean shear and τ_0 is the characteristic turbulent time. The solution to this problem is similar to that discussed in Sections 4.2.3–4.2.4 using Eqs. (4.83)– (4.85), (4.88) and (4.91)–(4.92).

6.2.5 *Solutions of Practice Problems for Chapter 5*

58. The tensor of turbulent diffusion $D_{mn}^{(T)}$ of the mean temperature field for a developed random flow is given by $D_{mn}^{(T)} = \langle \tau u_m\, u_n \rangle$, and the turbulent diffusion coefficient, D_T, for an isotropic developed turbulence is $D_T = \frac{1}{3}\,\tau\,\langle \boldsymbol{u}^2 \rangle$. The solution to this problem is similar to that discussed in Sections 5.1.1–5.1.2 using Eqs. (5.1)–(5.4) and (5.13)–(5.23).

59. The tensor of turbulent diffusion $D_{mn}^{(T)}$ of the mean particle number density is given by $D_{mn}^{(T)} = \langle \tau u_m\, u_n \rangle$, and the effective particle velocity, $\boldsymbol{V}^{\mathrm{eff}}$, is given by $\boldsymbol{V}^{\mathrm{eff}} = -\langle \tau \boldsymbol{u}\, (\boldsymbol{\nabla} \cdot \boldsymbol{u}) \rangle$. The solution to this problem is similar to that discussed in Sections 5.2.2–5.2.3 using Eqs. (5.25)–(5.29) and (5.41)–(5.54).

60. The effective diamagnetic velocity $\boldsymbol{V}^{\mathrm{eff}}$ of the mean magnetic field is given $V_m^{\mathrm{eff}} = -\frac{1}{2}\,\nabla_n \eta_{nm}^{(T)}$, the turbulent magnetic diffusion is given by $\eta_{_T} = \ell_0\, u_0/3$, the α effect $\alpha = -\frac{1}{3}\,\langle \tau \boldsymbol{u} \cdot (\boldsymbol{\nabla} \times \boldsymbol{u}) \rangle$. The solution to this problem is similar to that discussed in Sections 5.3.1–5.3.3 using Eqs. (5.71)–(5.99).

Notations and Definitions

$\boldsymbol{A}$ is the magnetic vector potential;
$\overline{\boldsymbol{A}}$ is the mean magnetic vector potential;
$\boldsymbol{B}$ is the magnetic field;
$\overline{\boldsymbol{B}}$ is the mean magnetic field;
C is the electric capacitance;
$C_{\rm p} = n\, m_{\rm p}/\rho$ is the mass concentration of particles or gaseous admixtures;
$\overline{C}_{\rm p}$ is the mean mass concentration of particles or gaseous admixtures;
$C'_{\rm p}$ are the mass concentration fluctuations of particles or gaseous admixtures;
$D^{(\theta)}$ is the coefficient of the molecular diffusion of the temperature field;
$D^{(n)} = k_B\, T/(6\pi\rho\, \nu\, a_p)$ is the coefficient of the molecular (Brownian) diffusion of particles having the radius a_p;
$D_{ij}^{(T)} = \tau_0\, \langle u_i u_j \rangle$ is the turbulent diffusion tensor;
$D_{_T}$ is the coefficient of turbulent diffusion of the temperature field;
$D_{_T}(\ell)$ is the scale-dependent turbulent diffusion coefficient;
$D_* = D^{(\theta)} + D_{_T}$ is the total diffusion coefficient of the temperature field;
$E_K = \rho\, \langle \boldsymbol{u}^2 \rangle/2$ is the density of turbulent kinetic energy;
$E_\theta = \langle \theta^2 \rangle/2$ is the second moment of temperature fluctuations;
$E_\Theta = \langle \Theta'^2 \rangle/2$ is the second moment of the potential temperature fluctuations;
$E(k) = k_0^{-1}\, (2/3)\, (k/k_0)^{-5/3}$ is the Kolmogorov kinetic energy spectrum function;
$\tilde{E}_\theta(k)$ is the spectrum function of temperature fluctuations;
$E_F(k)$ is the turbulent heat flux spectrum function;
$E_P = (\beta^2/N^2) E_\Theta$ is the density of turbulent potential energy;
$E_T = E_K + E_P$ is the density of the total turbulent energy;
$E_{\rm u}(k)$ is the spectrum function of velocity fluctuations;
$\boldsymbol{E}$ is the electric field;
$G_D(\omega, \boldsymbol{k}) = \left(D^{(n)} k^2 + i\omega\right)^{-1}$ is the Green function;
$\boldsymbol{F}^{(\theta)} \equiv \langle \boldsymbol{u}\theta \rangle$ is the turbulent heat flux or turbulent flux of temperature field;

$\boldsymbol{F}^{(\Theta)} = \langle \boldsymbol{u}\Theta' \rangle$ is the turbulent flux of the potential temperature;
$\boldsymbol{F}^{(n)} \equiv \langle \boldsymbol{u}n' \rangle$ is the turbulent flux of particles;
I is the electric current;
$H_\rho = 1/|\lambda|$ is the density stratification length;
$N = \left|\beta \nabla_z \overline{\Theta}\right|^{1/2}$ is the Brunt Väisälä frequency or the frequency of the internal gravity waves;
N_A is the Avogadro number;
L is inductance;
L_x, L_y, L_z are the sizes of the box along x, y, z directions;
L_n is the characteristic spatial scale of variations of the mean particle number density;
$L_P = |\nabla_z \overline{P}/\overline{P}|^{-1}$ is the pressure height scale;
L_T is the characteristic spatial scale of variations of the mean temperature field;
M_b is the spectrum function for magnetic fluctuations;
$\mathrm{Ma} = u/c_\mathrm{s}$ is the Mach number;
P is the fluid pressure;
$\overline{P} = \langle P \rangle$ is the mean fluid pressure;
P_0 is the fluid pressure in the basic reference state;
$\tilde{P}$ are the deviations of the fluid pressure from the basic reference state;
$\mathrm{Pr} = \nu/D^{(\theta)}$ is the Prandtl number;
$\mathrm{Pr}_T = \nu_T/D_T$ is the turbulent Prandtl number;
$\mathrm{Pe} = \ell_0 u_0/D^{(\theta)}$ is the Péclet number for temperature fluctuations;
$\mathrm{Pe}^{(n)} = \ell_0 u_0/D^{(n)}$ is the Péclet number for particle number density fluctuations;
Q are the third-order moment terms appearing due to the nonlinear terms in an equation for temperature fluctuations;
R is the gas constant;
$\mathrm{Re} = \ell_0 u_0/\nu$ is the Reynolds number;
$\mathrm{Re}_\ell = \ell u_\ell/\nu$ is the local Reynolds number in the scale ℓ;
$\mathrm{Ri} = N^2/S^2$ is the gradient Richardson number;
$\mathrm{St} = \tau_p/\tau_\eta$ is the Stokes number;
$S = \partial \overline{U}_x/\partial z$ is the large-scale shear;
$T(t, \boldsymbol{x})$ is the instantaneous fluid temperature field;
$\overline{T} = \langle T \rangle$ is the mean fluid temperature;
$\tilde{T}$ are the deviations of the fluid temperature from the basic reference state;
$\boldsymbol{\nabla}\overline{T}$ is the gradient of the mean temperature;
$\boldsymbol{U}$ is the fluid velocity;
$\overline{\boldsymbol{U}} = \langle \boldsymbol{U} \rangle$ is the mean fluid velocity;
$\boldsymbol{V}^{\mathrm{eff}} = -D_T \boldsymbol{\lambda} = D_T \frac{\nabla \overline{\rho}}{\overline{\rho}}$ is the effective pumping velocity of particles;
$V_\mathrm{g} = \tau_p \boldsymbol{g}$ is the terminal fall velocity of particles due to the gravity field;
U is the electric potential;

W is the Wolf number;
a_p is the particle radius;
$\boldsymbol{b}$ are the fluctuations of the magnetic field;
c is the speed of light;
c_s is the sound speed;
$\boldsymbol{e}_x, \boldsymbol{e}_y, \boldsymbol{e}_z$ are unit vectors along the x-, y-, z-axis of the Cartesian coordinates;
$\boldsymbol{j}$ is the density of electric current;
$f_{ij} = \langle u_i \, u_j \rangle$ is the Reynolds stress;
$\boldsymbol{f} \rho$ is the external force, that creates turbulent random velocity field;
f_w is the total number of sunspots in the visible part of the sun;
$\boldsymbol{g}$ is the acceleration caused by the gravity field;
g_w is the number of sunspot groups;
k_B is the Boltzmann constant;
$\boldsymbol{k}$ is the wave vector;
$k_0 = \ell_0^{-1}$ is the wave number in the integral scale ℓ_0;
$k = \ell^{-1}$ is the wave number in the scale ℓ;
$\ell_D = \ell_0 \mathrm{Pe}^{-3/4}$ is the molecular diffusion scale of temperature field;
$\ell_D^{(n)} = \ell_0 \left(\mathrm{Pe}^{(n)}\right)^{-3/4}$ is the molecular diffusion scale of particle number density;
$\ell_\nu = \ell_0 \mathrm{Re}^{-3/4}$ is the Kolmogorov (viscous) scale;
ℓ_0 is the integral scale of turbulence;
$m_p = (4\pi/3) \; \rho_p \, a_p^3$ is the mass of a spherical particle;
m_μ is the mass of the molecules of the surrounding;
$n(t, \boldsymbol{x})$ is the instantaneous particle number density;
$\overline{n} = \langle n \rangle$ is the mean particle number density;
n' are the particle number density fluctuations;
$q = \int \rho_q \, d\boldsymbol{r}$ is the electric charge;
$\boldsymbol{u}$ are the velocity fluctuations;
$u_D = u_0 \mathrm{Pe}^{-1/4}$ is the characteristic velocity in the diffusion scale;
u_ℓ is the characteristic turbulent velocity in the scale ℓ inside the inertial range of turbulence;
u_ν is the characteristic velocity in the viscous scale ℓ_ν;
$u_{\mathrm{rms}} = \sqrt{\langle \boldsymbol{u}^2 \rangle}$ is the r.m.s. velocity (the characteristic turbulent velocity);
$u_T = (k_B T_* / m_\mu)^{1/2}$ is the thermal velocity,
s is the entropy;
$\tilde{s}$ are the deviations of the non-dimensional entropy from the basic reference state;
s' are fluctuations of entropy;
$\overline{s}$ is the mean entropy;
t_n is the characteristic time-scale of variations of the mean particle number density;
t_M is the total time of measurements;

t_T is the characteristic time-scale of variations of the mean temperature field;
$\mathrm{w}_j(t)$ is the Wiener random process;
$\boldsymbol{\mathcal{E}} \equiv \langle \boldsymbol{u} \times \boldsymbol{b} \rangle$ is the turbulent electromotive force or turbulent EMF;
Θ is the potential temperature;
Θ_0 is the potential temperature in the basic reference state;
$\tilde{\Theta}$ are deviations of potential temperature from the basic reference state Θ_0;
$\overline{\Theta} = \langle \Theta \rangle$ is the mean potential temperature;
$\Pi_K = \nu_T S^2$ is the production rate of the density of turbulent kinetic energy caused by the large-scale shear
Π_θ is the production rate of the second moment, E_θ, of the temperature fluctuations;
$\Pi_i^{(F)}$ is the production rate of the turbulent heat flux;
$\Phi(\omega)$ is the frequency function;
$\boldsymbol{\Phi}_K$ is the flux of the density of turbulent kinetic energy;
$\boldsymbol{\Phi}_\theta$ is the flux of E_θ;
$\boldsymbol{\Phi}_P$ is the flux of E_P;
$\boldsymbol{\Phi}_{ij}^{(F)}$ is the flux of $F_i^{(\theta)}$;
Ω is the frequency;
α is the α-effect;
α_{ij} is the tensor that determines α-effect;
α_{TE} is the temperature expansion coefficient;
$\beta = |\boldsymbol{g}|/T_0$ is the buoyancy parameter;
δ_{ij} is the Kronecker tensor (or the unit matrix), that is defined as follows: $\delta_{ij} = 1$ for $i = j$, and $\delta_{ij} = 0$ for $i \neq j$.
$\delta(\boldsymbol{r})$ is the Dirac delta function defined by the following properties: $\delta(\boldsymbol{r}) = 0$ when $\boldsymbol{r} \neq 0$; $\delta(\boldsymbol{r}) = \infty$ when $\boldsymbol{r} = 0$; $\int_{-\infty}^{\infty} \delta(x)\, dx = 1$ and $\int f(\boldsymbol{r})\delta(\boldsymbol{r} - \boldsymbol{r}_0)\, d\boldsymbol{r} = f(\boldsymbol{r}_0)$;
$\gamma = c_\mathrm{p}/c_\mathrm{v}$ is the ratio of specific heats;
ε_{ijm} is the fully antisymmetric Levi-Civita tensor;
ε, ε_K is the rate of the dissipation of the turbulent kinetic energy;
ε_θ is the dissipation rate of the second moment E_θ of the temperature fluctuations;
$\varepsilon_i^{(F)}$ is the dissipation rate of the turbulent heat flux;
$\eta = c^2/4\pi\sigma$ is the coefficient of magnetic diffusion due to the electrical conductivity of the fluid;
$\eta_{ij}^{(T)}$ is the turbulent magnetic diffusion tensor;
$\boldsymbol{\lambda} = -\boldsymbol{\nabla}\overline{\rho}/\overline{\rho}$;
ϑ, φ are the angles in spherical coordinates;
$\mu = m_\mu N_A$ is the molar mass;
μ_p is the turbulent thermal diffusion coefficient;
ν is the kinematic viscosity;
ν_T is the turbulent viscosity;

$\boldsymbol{\xi}(t, \boldsymbol{x}|s)$ is the Wiener trajectory;
ρ is the instantaneous fluid density;
$\overline{\rho}$ is the mean fluid density;
ρ_p is the material density of inertial particles;
ρ_0 is the fluid density in the basic reference state;
ρ_q is the density of the electric charge;
$\tilde{\rho}$ are the deviations of the fluid density from the basic reference state;
σ is the electric conductivity;
$\tau_{\rm diffusion} = L_T^2/D_*$ is the characteristic diffusion time;
τ_c is the correlation time of a random velocity field;
$\tau_D = \ell_D/u_D$ is the characteristic molecular diffusion time;
$\tau_0 = \ell_0/u_0$ is the characteristic turbulent time in the integral turbulent scale ℓ_0;
$\tau_\ell = \ell/u_\ell$ is the characteristic time in the inertial range of turbulence;
$\tau(k) = 2\,\tau_0\,(k/k_0)^{-2/3}$ is the scale-dependent turbulent correlation time (Kolmogorov scaling);
$\tau_{n'}$ is the characteristic time of the particle number density fluctuations that can be identified with the correlation time τ_0 of the turbulent velocity field for large Reynolds numbers;
$\tau_p = m_p/6\pi\rho\,\nu\,a_p$ is the Stokes time of particles;
$\tau_r(k)$ is the scale-dependent relaxation time, which can be identified with the correlation time $\tau(k)$ of the turbulent velocity field for large Reynolds and Péclet numbers;
τ_θ is the characteristic time of the temperature fluctuations;
θ are the temperature fluctuations;
ω is the frequency.

References

Acheson, D. J. 1978. On the instability of toroidal magnetic fields and differential rotation in stars. *Philos. Trans. R. Soc. London A*, **289**, 459–500.

Afonso, M. M., Mitra, Dh., and Vincenzi, D. 2019. Kazantsev dynamo in turbulent compressible flows. *Proc. R. Soc. A*, **475**, 20180591.

Ahlers, G., Grossmann, S., and Lohse, D. 2009. Heat transfer and large scale dynamics in turbulent Rayleigh-Bénard convection. *Rev. Mod. Phys.*, **81**(2), 503.

Akasofu, S.-I., and Chapman, S. 1972. *Solar-Terrestrial Physics*. Oxford: Clarendon.

Akhiezer, A. I., and Peletminsky, S. V. 1981. *Methods of Statistical Physics*. Oxford: Pergamon.

Aliseda, A., Cartellier, A., Hainaux, F., and Lasheras, J. C. 2002. Effect of preferential concentration on the settling velocity of heavy particles in homogeneous isotropic turbulence. *J. Fluid Mech.*, **468**, 77–105.

Amir, G., Bar, N., Eidelman, A., Elperin, T., Kleeorin, N., and Rogachevskii, I. 2017. Turbulent thermal diffusion in strongly stratified turbulence: theory and experiments. *Phys. Rev. Fluids*, **2**(6), 064605.

Armitage, P. J. 2010. *Astrophysics of Planet Formation*. Cambridge: Cambridge University Press.

Arnol'd, V. I. 1965. Sur la topologie des écoulements stationnaires des fluides parfaits. *C. R. Acad. Sci. Paris*, **261**, 17–20.

Arponen, H., and Horvai, P. 2007. Dynamo effect in the Kraichnan magnetohydrodynamic turbulence. *J. Statist. Phys.*, **129**(2), 205–239.

Atkinson, B. W., and Wu Zhang, J. 1996. Mesoscale shallow convection in the atmosphere. *Rev. Geophys.*, **34**(4), 403–431.

Avellaneda, M., and Majda, A. J. 1990. Mathematical models with exact renormalization for turbulent transport. *Commun. Math. Phys.*, **131**(2), 381–429.

Baines, P. G. 1997. *Topographic Effects in Stratified Flows*. Cambridge: Cambridge University Press.

Balachandar, S., and Eaton, J. K. 2010. Turbulent dispersed multiphase flow. *Annu. Rev. Fluid Mech.*, **42**, 111–133.

Bałdyga, J., and Bourne, J. R. 1999. *Turbulent Mixing and Chemical Reactions*. Chichester: John Wiley.

Balkovsky, E., Falkovich, G., and Fouxon, A. 2001. Intermittent distribution of inertial particles in turbulent flows. *Phys. Rev. Lett.*, **86**(13), 2790.

Barel, L., Eidelman, A., Elperin, T., Fleurov, G., Kleeorin, N., Levy, A., Rogachevskii, I., and Shildkrot, O. 2020. Detection of standing internal gravity waves in experiments with convection over a wavy heated wall. *Phys. Fluids*, **32**(9), 095105.

Batchelor, G. K. 1950. On the spontaneous magnetic field in a conducting liquid in turbulent motion. *Proc. R. Soc. Lond.*, **201**, 405–416.

Batchelor, G. K. 1953. *The Theory of Homogeneous Turbulence*. Cambridge: Cambridge University Press.

Batchelor, G. K. 1959. Small scale variation of convected quantities like temperature in turbulent fluid. Part 1. General discussion and the case of small conductivity. *J. Fluid Mech.*, **5**, 113–134.

Batchelor, G. K. 1967. *An Introduction to Fluid Dynamics*. Cambridge: Cambridge University Press.

Batchelor, G. K., Howells, I. D., and Townsend, A. 1959. Small scale variation of convected quantities like temperature in turbulent fluid. Part 2. The case of large conductivity. *J. Fluid Mech.*, **5**, 135–139.

Bec, J., Biferale, L., Cencini, M., Lanotte, A., Musacchio, S., and Toschi, F. 2007. Heavy particle concentration in turbulence at dissipative and inertial scales. *Phys. Rev. Lett.*, **98**(8), 084502.

Beer, T. 1974. *Atmospheric Waves*. New York: Wiley.

Bell, A. R. 2004. Turbulent amplification of magnetic field and diffusive shock acceleration of cosmic rays. *Mon. Not. R. Astron. Soc.*, **353**(2), 550–558.

Bell, A. R. 2005. The interaction of cosmic rays and magnetized plasma. *Mon. Not. R. Astron. Soc.*, **358**(1), 181–187.

Benzi, R., and Nelson, D. R. 2009. Fisher equation with turbulence in one dimension. *Physica. D.*, **238**(19), 2003–2015.

Beresnyak, A., and Lazarian, A. 2019. *Turbulence in Magnetohydrodynamics*. Berlin: Walter de Gruyter.

Beresnyak, A., and Li, H. 2014. Basic Bell-MHD turbulence. *Astrophys. J.*, **788**(2), 107.

Bernard, P. S., and Wallace, J. M. 2002. *Turbulent Flow: Analysis, Measurement, and Prediction*. Hoboken: John Wiley & Sons.

Bertin, F., Barat, J., and Wilson, R. 1997. Energy dissipation rates, eddy diffusivity, and the Prandtl number: an in situ experimental approach and its consequences on radar estimate of turbulent parameters. *Radio Science*, **32**(2), 791–804.

Bhat, P., and Subramanian, K. 2014. Fluctuation dynamo at finite correlation times and the Kazantsev spectrum. *Astrophys. J. Lett.*, **791**(2), L34.

Biskamp, D. 2003. *Magnetohydrodynamic Turbulence*. Cambridge: Cambridge University Press.

Blackadar, A. K. 2012. *Turbulence and Diffusion in the Atmosphere*. Berlin: Springer.

Blackman, E. G., and Brandenburg, A. 2002. Dynamic nonlinearity in large-scale dynamos with shear. *Astrophys. J.*, **579**(1), 359.

Blackman, E. G., and Field, G. B. 2000. Constraints on the magnitude of α in dynamo theory. *Astrophys. J.*, **534**(2), 984.

Blackman, E. G., and Field, G. B. 2003. A new approach to turbulent transport of a mean scalar. *Phys. Fluids*, **15**(11), L73–L76.

Boffetta, G., De Lillo, F., and Gamba, A. 2004. Large scale inhomogeneity of inertial particles in turbulent flows. *Phys. Fluids*, **16**(4), L20–L23.

Boldyrev, S., and Cattaneo, F. 2004. Magnetic-field generation in Kolmogorov turbulence. *Phys. Rev. Lett.*, **92**(14), 144501.

Bradshaw, P. 1971. *Introduction to Turbulence and its Measurement*. Oxford: Pergamon Press.

Braginsky, S. I., and Roberts, P. H. 1995. Equations governing convection in Earth's core and the geodynamo. *Geophys. Astrophys. Fluid Dyn.*, **79**, 1–97.

Brandenburg, A. 2011. Nonlinear small-scale dynamos at low magnetic Prandtl numbers. *Astrophys. J.*, **741**(2), 92.

Brandenburg, A., Jennings, R. L., Nordlund, Å, Rieutord, M., Stein, R. F., and Tuominen, I. 1996. Magnetic structures in a dynamo simulation. *J. Fluid Mech.*, **306**, 325–352.

Brandenburg, A., and Dobler, W. 2002. Hydromagnetic turbulence in computer simulations. *Comput. Phys. Commun.*, **147**, 471–475.

Brandenburg, A., Käpylä, P. J., and Mohammed, A. 2004. Non-Fickian diffusion and tau approximation from numerical turbulence. *Phys. Fluids*, **16**(4), 1020–1027.

Brandenburg, A., and Subramanian, K. 2005a. Astrophysical magnetic fields and nonlinear dynamo theory. *Physics Reports*, **417**, 1–209.

Brandenburg, A., and Subramanian, K. 2005b. Minimal tau approximation and simulations of the alpha effect. *Astron. Astrophys.*, **439**(3), 835–843.

Brandenburg, A., Rädler, K.-H., Rheinhardt, M., and Käpylä, P. J. 2008. Magnetic diffusivity tensor and dynamo effects in rotating and shearing turbulence. *Astrophys. J.*, **676**(1), 740.

Brandenburg, A., Kleeorin, N., and Rogachevskii, I. 2010. Large-scale magnetic flux concentrations from turbulent stresses. *Astron. Nachr.*, **331**(1), 5–13.

Brandenburg, A., Kemel, K., Kleeorin, N., Mitra, Dh., and Rogachevskii, I. 2011a. Detection of negative effective magnetic pressure instability in turbulence simulations. *Astrophys. J. Lett.*, **740**(2), L50.

Brandenburg, A., Haugen, N. E. L., and Babkovskaia, N. 2011b. Turbulent front speed in the Fisher equation: dependence on Damköhler number. *Phys. Rev. E*, **83**(1), 016304.

Brandenburg, A., Rädler, K.-H., and Kemel, K. 2012a. Mean-field transport in stratified and/or rotating turbulence. *Astron. Astroph.*, **539**, A35.

Brandenburg, A., Kemel, K., Kleeorin, N., and Rogachevskii, I. 2012b. The negative effective magnetic pressure in stratified forced turbulence. *Astrophys. J.*, **749**(2), 179.

Brandenburg, A., Gressel, O., Käpylä, P. J., Kleeorin, N., Mantere, M. J., and Rogachevskii, I. 2013a. New scaling for the alpha effect in slowly rotating turbulence. *Astrophys. J.*, **762**(2), 127.

Brandenburg, A., Kleeorin, N., and Rogachevskii, I. 2013b. Self-assembly of shallow magnetic spots through strongly stratified turbulence. *Astrophys. J. Lett.*, **776**(2), L23.

Brandenburg, A., Gressel, O., Jabbari, S., Kleeorin, N., and Rogachevskii, I. 2014. Mean-field and direct numerical simulations of magnetic flux concentrations from vertical field. *Astron. Astrophys.*, **562**, A53.

Brandenburg, A., Rogachevskii, I., and Kleeorin, N. 2016. Magnetic concentrations in stratified turbulence: the negative effective magnetic pressure instability. *New Journal of Physics*, **18**(12), 125011.

Brandenburg, A., Schober, J., and Rogachevskii, I. 2017a. The contribution of kinetic helicity to turbulent magnetic diffusivity. *Astron. Nachr.*, **338**(7), 790–793.

Brandenburg, A., Schober, J., Rogachevskii, I., Kahniashvili, T., Boyarsky, A., Fröhlich, J., Ruchayskiy, O., and Kleeorin, N. 2017b. The turbulent chiral magnetic cascade in the early universe. *Astrophys. J. Lett.*, **845**(2), L21.

Bray, K. N. C. 1995. Turbulent transport in flames. *Proc. R. Soc. London A*, **451**, 231–256.

Britter, R. E., and Hanna, S. R. 2003. Flow and dispersion in urban areas. *Annu. Rev. Fluid Mech.*, **35**(1), 469–496.

Buchholz, J., Eidelman, A., Elperin, T., Grünefeld, G., Kleeorin, N., Krein, A., and Rogachevskii, I. 2004. Experimental study of turbulent thermal diffusion in oscillating grids turbulence. *Experim. Fluids*, **36**(6), 879–887.

Bukai, M., Eidelman, A., Elperin, T., Kleeorin, N., Rogachevskii, I., and Sapir-Katiraie, I. 2009. Effect of large-scale coherent structures on turbulent convection. *Phys. Rev. E*, **79**(6), 066302.

Bukai, M., Eidelman, A., Elperin, T., Kleeorin, N., Rogachevskii, I., and Sapir-Katiraie, I. 2011. Transition phenomena in unstably stratified turbulent flows. *Phys. Rev. E*, **83**(3), 036302.

Busse, F. H. 1983. Generation of mean flows by thermal convection. *Physica. D.*, **9**(3), 287–299.

Busse, F. H., and Whitehead, J. A. 1971. Instabilities of convection rolls in a high Prandtl number fluid. *J. Fluid Mech.*, **47**(2), 305–320.

Busse, F. H., and Whitehead, J. A. 1974. Oscillatory and collective instabilities in large Prandtl number convection. *J. Fluid Mech.*, **66**(1), 67–79.

Bykov, A. M., Osipov, S. M., and Ellison, D. C. 2011. Cosmic ray current driven turbulence in shocks with efficient particle acceleration: the oblique, long-wavelength mode instability. *Monthly Notices of the Royal Astronomical Society*, **410**(1), 39–52.

Canuto, V. M., and Minotti, F. 1993. Stratified turbulence in the atmosphere and oceans: a new subgrid model. *J. Atmos. Sci.*, **50**(13), 1925–1935.

Canuto, V. M., Cheng, Y., Howard, A. M., and Esau, I. N. 2008. Stably stratified flows: a model with no Ri (cr). *J. Atmos. Sci.*, **65**(7), 2437–2447.

Caporaloni, M., Tampieri, F., Trombetti, F., and Vittori, O. 1975. Transfer of particles in nonisotropic air turbulence. *J. Atmosph. Sci.*, **32**(3), 565–568.

Cardy, J., Falkovich, G., and Gawedzki, K. 2008. *Non-Equilibrium Statistical Mechanics and Turbulence*. Cambridge: Cambridge University Press.

Cattaneo, F. 1999. On the origin of magnetic fields in the quiet photosphere. *Astrophys. J. Lett.*, **515**(1), L39.

Chandrasekhar, S. 1943. Stochastic problems in physics and astronomy. *Rev. Mod. Phys.*, **15**(1), 1.

Chandrasekhar, S. 2013. *Hydrodynamic and Hydromagnetic Stability*. New York: Dover.

Chapman, S. 1912. The kinetic theory of a gas constituted of spherically symmetrical molecules. *Phil. Trans. Roy. Soc. London A*, **211**(471-483), 433–483.

Chapman, S. 1916. On the law of distribution of molecular velocities, and on the theory of viscosity and thermal conduction, in a non-uniform simple monatomic gas. *Phil. Trans. Roy. Soc. London A*, **216**, 279–348.

Chapman, S., and Cowling, T. G. 1970. *The Mathematical Theory of Non-Uniform Gases*. Cambridge: Cambridge University Press.

Chapman, S., and Dootson, F. W. 1917. A note on thermal diffusion. *Phil. Mag. and J. Sci.*, **33**, 248–253.

Chassaing, P., Antonia, R. A., Anselmet, F., Joly, L., and Sarkar, S. 2002. *Variable Density Fluid Turbulence*. Dordrecht: Kluwer Academic.

Chen, F. F. 1984. *Introduction to Plasma Physics and Controlled Fusion*. New York: Plenum Press.

Chertkov, M., and Falkovich, G. 1996. Anomalous scaling exponents of a white-advected passive scalar. *Phys. Rev. Lett.*, **76**(15), 2706.

Childress, S. 1970. New solutions of the kinematic dynamo problem. *J. Mathemat. Phys.*, **11**(10), 3063–3076.

Childress, S., and Gilbert, A. D. 1995. *Stretch, Twist, Fold: The Fast Dynamo*. Vol. 37. Berlin: Springer.

Chillà, F., and Schumacher, J. 2012. New perspectives in turbulent Rayleigh-Bénard convection. *Eur. Phys. J. E*, **35**(7), 1–25.

Ching, E. S. C. 2014. *Statistics and Scaling in Turbulent Rayleigh-Bénard Convection*. Berlin: Springer.

Chorin, A. J. 1994. *Vorticity and Turbulence*. Berlin: Springer.

Chung, D., and Matheou, G. 2012. Direct numerical simulation of stationary homogeneous stratified sheared turbulence. *J. Fluid Mech.*, **696**, 434.

Clark di Leoni, P., and Mininni, P. D. 2015. Absorption of waves by large-scale winds in stratified turbulence. *Phys. Rev. E*, **91**(3), 033015.

Cohen, N., Eidelman, A., Elperin, T., Kleeorin, N., and Rogachevskii, I. 2014. Sheared stably stratified turbulence and large-scale waves in a lid driven cavity. *Phys. Fluids*, **26**(10), 105106.

Collins, L. R., and Keswani, A. 2004. Reynolds number scaling of particle clustering in turbulent aerosols. *New Journal of Physics*, **6**(1), 119.

Connaughton, C., Nazarenko, S., and Quinn, B. 2015. Rossby and drift wave turbulence and zonal flows: the Charney–Hasegawa–Mima model and its extensions. *Phys. Reports*, **604**, 1–71.

Corrsin, S. 1951. On the spectrum of isotropic temperature fluctuations in an isotropic turbulence. *J. Appl. Phys.*, **22**(4), 469–473.

Courvoisier, A., Hughes, D. W., and Tobias, S. M. 2006. α effect in a family of chaotic flows. *Phys. Rev. Lett.*, **96**(3), 034503.

Covas, E., Tworkowski, A., Tavakol, R., and Brandenburg, A. 1997. Robustness of truncated $\alpha\ \Omega$ dynamos with a dynamic α. *Solar Phys.*, **172**(1–2), 3–9.

Covas, E., Tavakol, R., Tworkowski, A., and Brandenburg, A. 1998. Axisymmetric mean field dynamos with dynamic and algebraic $alpha$–quenchings. *Astron. Astrophys.*, **329**, 350–360.

Crowe, C. T., Schwarzkopf, J. D., Sommerfeld, M., and Tsuji, Y. 2011. *Multiphase Flows with Droplets and Particles*. New York: CRC Press.

Csanady, G. T. 2012. *Turbulent Diffusion in the Environment*. Vol. 3. Berlin: Springer.

Dalaudier, F., and Sidi, C. 1987. Evidence and interpretation of a spectral gap in the turbulent atmospheric temperature spectra. *J. Atmos. Sci.*, **44**(20), 3121–3126.

Davidson, P. A. 2013. *Turbulence in Rotating, Stratified and Electrically Conducting Fluids*. Cambridge: Cambridge University Press.

Davidson, P. A. 2015. *Turbulence. An Introduction for Scientists and Engineers*. New York: Oxford University Press.

De Groot, S. R., and Mazur, P. 1984. *Non-Equilibrium Thermodynamics*. New York: Dover.

De Lillo, F., Cencini, M., Musacchio, S., and Boffetta, G. 2016. Clustering and turbophoresis in a shear flow without walls. *Phys. Fluids*, **28**(3), 035104.

Del Sordo, F., Guerrero, G., and Brandenburg, A. 2013. Turbulent dynamos with advective magnetic helicity flux. *Monthly Notices of the Royal Astronomical Society*, **429**(2), 1686–1694.

Devlen, E., Brandenburg, A., and Mitra, Dh. 2013. A mean field dynamo from negative eddy diffusivity. *Monthly Notices of the Royal Astronomical Society*, **432**(2), 1651–1657.

Diamond, P. H., Itoh, S.-I., and Itoh, K. 2010. *Modern Plasma Physics: Physical Kinetics of Turbulent Plasmas*. Cambridge: Cambridge University Press.

Dikpati, M., and Gilman, P. A. 2006. Simulating and predicting solar cycles using a flux-transport dynamo. *Astrophys. J.*, **649**(1), 498.

Dimotakis, P. E. 2005. Turbulent mixing. *Annu. Rev. Fluid Mech.*, **37**, 329–356.

Dombre, T., Frisch, U., Greene, J. M., Hénon, M., Mehr, A., and Soward, A. M. 1986. Chaotic streamlines in the ABC flows. *J. Fluid Mech.*, **167**, 353–391.

Drobyshevski, E. M., and Yuferev, V. S. 1974. Topological pumping of magnetic flux by three-dimensional convection. *J. Fluid Mech.*, **65**(1), 33–44.

Druzhinin, O. A., and Ostrovsky, L. A. 2015. Dynamics of turbulence under the effect of stratification and internal waves. *Nonlinear Processes in Geophysics*, **22**, 337.

Druzhinin, O. A., Ostrovsky, L. A., and Zilitinkevich, S. S. 2013. The study of the effect of small-scale turbulence on internal gravity waves propagation in a pycnocline. *Nonlinear Processes in Geophysics*, **20**, 977.

Duncan, K., Mehlig, B., Östlund, S., and Wilkinson, M. 2005. Clustering by mixing flows. *Phys. Rev. Lett.*, **95**(24), 240602.

Eaton, J. K., and Fessler, J. R. 1994. Preferential concentration of particles by turbulence. *Int. J. Multiphase Flow*, **20**, 169–209.

Echekki, T., and Mastorakos, E. 2010. *Turbulent Combustion Modeling: Advances, New Trends and Perspectives*. Berlin: Springer.

Eidelman, A., Elperin, T., Kleeorin, N., Krein, A., Rogachevskii, I., Buchholz, J., and Grünefeld, G. 2004. Turbulent thermal diffusion of aerosols in geophysics and laboratory experiments. *Nonlinear Processes in Geophysics*, **11**, 343–350.

Eidelman, A., Elperin, T., Kleeorin, N., Markovich, A., and Rogachevskii, I. 2006a. Experimental detection of turbulent thermal diffusion of aerosols in non-isothermal flows. *Nonlinear Processes in Geophysics*, **13**, 109–117.

Eidelman, A., Elperin, T., Kleeorin, N., Markovich, A., and Rogachevskii, I. 2006b. Hysteresis phenomenon in turbulent convection. *Experiments in Fluids*, **40**(5), 723–732.

Eidelman, A., Elperin, T., Kleeorin, N., Rogachevskii, I., and Sapir-Katiraie, I. 2006c. Turbulent thermal diffusion in a multi-fan turbulence generator with imposed mean temperature gradient. *Experiments in Fluids*, **40**(5), 744–752.

Eidelman, A., Elperin, T., Kleeorin, N., Hazak, G., Rogachevskii, I., Sadot, O., and Sapir-Katiraie, I. 2009. Mixing at the external boundary of a submerged turbulent jet. *Phys. Rev. E*, **79**(2), 026311.

Eidelman, A., Elperin, T., Kleeorin, N., Melnik, B., and Rogachevskii, I. 2010. Tangling clustering of inertial particles in stably stratified turbulence. *Phys. Rev. E*, **81**(5), 056313.

Eidelman, A., Elperin, T., Gluzman, I., Kleeorin, N., and Rogachevskii, I. 2013. Experimental study of temperature fluctuations in forced stably stratified turbulent flows. *Phys. Fluids*, **25**(1), 015111.

Elperin, T., Kleeorin, N., and Rogachevskii, I. 1995. Dynamics of the passive scalar in compressible turbulent flow: large-scale patterns and small-scale fluctuations. *Phys. Rev. E*, **52**(3), 2617.

Elperin, T., Kleeorin, N., and Rogachevskii, I. 1996a. Isotropic and anisotropic spectra of passive scalar fluctuations in turbulent fluid flow. *Phys. Rev. E*, **53**(4), 3431.

Elperin, T., Kleeorin, N., and Rogachevskii, I. 1996b. Self-excitation of fluctuations of inertial particle concentration in turbulent fluid flow. *Phys. Rev. Lett.*, **77**(27), 5373.

Elperin, T., Kleeorin, N., and Rogachevskii, I. 1996c. Turbulent thermal diffusion of small inertial particles. *Phys. Rev. Lett.*, **76**(2), 224.

Elperin, T., Kleeorin, N., Podolak, M., and Rogachevskii, I. 1997a. A mechanism for the formation of aerosol concentrations in the atmosphere of Titan. *Planetary and Space Science*, **45**(8), 923–929.

Elperin, T., Kleeorin, N., and Rogachevskii, I. 1997b. Temperature fluctuations and anomalous scaling in low-Mach-number compressible turbulent flow. *Phys. Rev. E*, **55**(6), 7043.

Elperin, T., Kleeorin, N., and Rogachevskii, I. 1997c. Turbulent barodiffusion, turbulent thermal diffusion, and large-scale instability in gases. *Phys. Rev. E*, **55**(3), 2713.

Elperin, T., Kleeorin, N., and Rogachevskii, I. 1998a. Effect of chemical reactions and phase transitions on turbulent transport of particles and gases. *Phys. Rev. Lett.*, **80**(1), 69.

Elperin, T., Kleeorin, N., and Rogachevskii, I. 1998b. Formation of inhomogeneities in two-phase low-Mach-number compressible turbulent fluid flows. *Int. J. Multiphase Flow*, **24**(7), 1163–1182.

Elperin, T., Kleeorin, N., and Rogachevskii, I. 2000a. Mechanisms of formation of aerosol and gaseous inhomogeneities in the turbulent atmosphere. *Atmosph. Res.*, **53**, 117–129.

Elperin, T., Kleeorin, N., Rogachevskii, I., and Sokoloff, D. 2000b. Passive scalar transport in a random flow with a finite renewal time: mean-field equations. *Phys. Rev. E*, **61**(3), 2617.

Elperin, T., Kleeorin, N., Rogachevskii, I., and Sokoloff, D. 2000c. Turbulent transport of atmospheric aerosols and formation of large-scale structures. *Phys. Chem. Earth*, **25**(12), 797–803.

Elperin, T., Kleeorin, N., Rogachevskii, I., and Sokoloff, D. 2001a. Mean-field theory for a passive scalar advected by a turbulent velocity field with a random renewal time. *Phys. Rev. E*, **64**(2), 026304.

Elperin, T., Kleeorin, N., Rogachevskii, I., and Sokoloff, D. 2001b. Strange behavior of a passive scalar in a linear velocity field. *Phys. Rev. E*, **63**(4), 046305.

Elperin, T., Kleeorin, N., L'vov, V. S., Rogachevskii, I., and Sokoloff, D. 2002a. Clustering instability of the spatial distribution of inertial particles in turbulent flows. *Phys. Rev. E*, **66**(3), 036302.

Elperin, T., Kleeorin, N., Rogachevskii, I., and Zilitinkevich, S. 2002b. Formation of large-scale semi-organized structures in turbulent convection. *Phys. Rev. E*, **66**, 066305.

Elperin, T., Kleeorin, N., and Rogachevskii, I. 2003. Generation of large-scale vorticity in a homogeneous turbulence with a mean velocity shear. *Phys. Rev. E*, **68**(1), 016311.

Elperin, T., Golubev, I., Kleeorin, N., and Rogachevskii, I. 2005. Excitation of large-scale inertial waves in a rotating inhomogeneous turbulence. *Phys. Rev. E*, **71**(3), 036302.

Elperin, T., Golubev, I., Kleeorin, N., and Rogachevskii, I. 2006a. Large-scale instabilities in a nonrotating turbulent convection. *Phys. Fluids*, **18**(12), 126601.

Elperin, T., Kleeorin, N., Rogachevskii, I., and Zilitinkevich, S. 2006b. Tangling turbulence and semi-organized structures in convective boundary layers. *Boundary-Layer Meteorol.*, **119**(3), 449–472.

Elperin, T., Kleeorin, N., Liberman, M., L'vov, V. S., and Rogachevskii, I. 2007a. Clustering of aerosols in atmospheric turbulent flow. *Environ. Fluid Mech.*, **7**(2), 173–193.

Elperin, T., Golubev, I., Kleeorin, N., and Rogachevskii, I. 2007b. Large-scale instability in a sheared nonhelical turbulence: formation of vortical structures. *Phys. Rev. E*, **76**(6), 066310.

Elperin, T., Kleeorin, N., Liberman, M., and Rogachevskii, I. 2013. Tangling clustering instability for small particles in temperature stratified turbulence. *Phys. Fluids*, **25**(8), 085104.

Elperin, T., Kleeorin, N., Liberman, M., and Rogachevskii, I. 2014. Turbulent diffusion of chemically reacting gaseous admixtures. *Phys. Rev. E*, **90**(5), 053001.

Elperin, T., Kleeorin, N., Krasovitov, B., Kulmala, M., Liberman, M., Rogachevskii, I., and Zilitinkevich, S. 2015. Acceleration of raindrop formation due to the tangling-clustering instability in a turbulent stratified atmosphere. *Phys. Rev. E*, **92**(1), 013012.

Elperin, T., Kleeorin, N., and Rogachevskii, I. 2017a. Three-dimensional slow Rossby waves in rotating spherical density-stratified convection. *Phys. Rev. E*, **96**(3), 033106.

Elperin, T., Kleeorin, N., Liberman, M., Lipatnikov, A. N., Rogachevskii, I., and Yu, R. 2017b. Turbulent diffusion of chemically reacting flows: theory and numerical simulations. *Phys. Rev. E*, **96**(5), 053111.

Enskog, D. 1911. Bemerkungen zu einer Fundamentalgleichung in der kinetischen Gastheorie. *Physik. Zs. Leipzig*, **12**, 533–539.

Enskog, D. 1912. Zur Elektronentheorie der Dispersion und Absorption der Metalle. *Annalen der Physik*, **343**(9), 731–763.

Etling, D., and Brown, R. A. 1993. Roll vortices in the planetary boundary layer: a review. *Boundary-Layer Meteorol.*, **65**(3), 215–248.

Faber, T. E. 1995. *Fluid Dynamics for Physicists*. Cambridge: Cambridge University Press.

Falkovich, G. 2011. *Fluid Mechanics: A Short Course for Physicists*. Cambridge: Cambridge University Press.

Falkovich, G., Gawedzki, K., and Vergassola, M. 2001. Particles and fields in fluid turbulence. *Rev. Mod. Phys.*, **73**(4), 913.

Falkovich, G., Fouxon, A., and Stepanov, M. G. 2002. Acceleration of rain initiation by cloud turbulence. *Nature*, **419**(6903), 151–154.

Falkovich, G., and Pumir, A. 2007. Sling effect in collisions of water droplets in turbulent clouds. *J. Atmosph. Sci.*, **64**(12), 4497–4505.

Federrath, C. 2013. On the universality of supersonic turbulence. *Mon. Not. R. Astron. Soc.*, **436**(2), 1245–1257.

Federrath, C. 2016. Magnetic field amplification in turbulent astrophysical plasmas. *J. Plasma Phys.*, **82**(6), 535820601.

Federrath, C., Schober, J., Bovino, S., and Schleicher, D. R. G. 2014. The turbulent dynamo in highly compressible supersonic plasmas. *Astrophys. J. Lett.*, **797**(2), L19.

Feynman, R. P., and Hibbs, A. R. 1965. *Quantum Mechanics and Path Integrals*. New York: McGraw-Hill.

Field, G. B., Blackman, E. G., and Chou, H. 1999. Nonlinear α-effect in dynamo theory. *Astrophys. J.*, **513**(2), 638.

Finn, J. M., and Ott, E. 1988. Chaotic flows and fast magnetic dynamos. *Phys. Fluids*, **31**(10), 2992–3011.

Finn, J. M., and Ott, E. 1990. The fast kinematic magnetic dynamo and the dissipationless limit. *Phys. Fluids B: Plasma Physics*, **2**(5), 916–926.

Fisher, R. A. 1937. The wave of advance of advantageous genes. *Annals of Eugenics*, **7**(4), 355–369.

Fitzjarrald, D. E. 1976. An experimental study of turbulent convection in air. *J. Fluid Mech.*, **73**(4), 693–719.

Fox, R. O. 2003. *Computational Models for Turbulent Reacting Flows*. Cambridge: Cambridge University Press.

Frisch, U. 1995. *Turbulence: the Legacy of A. N. Kolmogorov*. Cambridge: Cambridge University Press.

Galanti, B., Sulem, P.-L., and Pouquet, A. 1992. Linear and non-linear dynamos associated with ABC flows. *Geophys. Astrophys. Fluid Dynamics*, **66**(1–4), 183–208.

Galanti, B., Kleeorin, N., and Rogachevskii, I. 1994. Nonlinear dynamo in ABC flow: the Hall effect. *Phys. Plasmas*, **1**(12), 3843–3849.

Galanti, B., Kleeorin, N., and Rogachevskii, I. 1995. Ampère and Hall nonlinearities in the magnetic dynamo in the Arnol'd–Beltrami–Childress (ABC) flow. *Phys. Plasmas*, **2**(11), 4161–4168.

Galeev, A. A., Rosenbluth, M. N., Sagdeev, R. Z., and Sudan, R. N. 1983. *Handbook of Plasma Physics*. Amsterdam: North-Holland.

Galloway, D., and Frisch, U. 1986. Dynamo action in a family of flows with chaotic streamlines. *Geophys. Astrophys. Fluid Dynamics*, **36**(1), 53–83.

Galloway, D. J., and Proctor, M. R. E. 1983. The kinematics of hexagonal magnetoconvection. *Geophys. Astrophys. Fluid Dyn.*, **24**(2), 109–136.

Galloway, D. J., and Proctor, M. R. E. 1992. Numerical calculations of fast dynamos in smooth velocity fields with realistic diffusion. *Nature*, **356**(6371), 691–693.

Galperin, B., Sukoriansky, S., and Anderson, P. S. 2007. On the critical Richardson number in stably stratified turbulence. *Atmosph. Sci. Lett.*, **8**(3), 65–69.

Gawedzki, K., and Kupiainen, A. 1995. Anomalous scaling of the passive scalar. *Phys. Rev. Lett.*, **75**(21), 3834.

Gibson, E. G. 1973. *The Quiet Sun*. Vol. 303. Washington: NASA.

Gilbert, A. D., and Childress, S. 1990. Evidence for fast dynamo action in a chaotic web. *Phys. Rev. Lett.*, **65**(17), 2133.

Gill, A. E. 2016. *Atmosphere–Ocean Dynamics*. New York: Academic Press.

Gilman, P. A. 1970. Instability of magnetohydrostatic stellar interiors from magnetic buoyancy. *Astrophys. J.*, **162**, 1019.

Golitsyn, G. S. 1960. Fluctuations of the magnetic field and current density in a turbulent flow of a weakly conducting fluid. *Sov. Phys. Doklady*, **5**, 536.

Grabowski, W. W., and Wang, L.-P. 2013. Growth of cloud droplets in a turbulent environment. *Annu. Rev. Fluid Mech.*, **45**, 293–324.

Grachev, A. A., Andreas, E. L., Fairall, C. W., Guest, P. S., and Persson, P. O. G. 2007. On the turbulent Prandtl number in the stable atmospheric boundary layer. *Boundary-Layer Meteorol.*, **125**(2), 329–341.

Greenspan, HP. 1990. *The Theory of Rotating Fluids*. Brookline: Breukelen Press.

Grossmann, S., and Lohse, D. 2011. Multiple scaling in the ultimate regime of thermal convection. *Phys. Fluids*, **23**(4), 045108.

Gruzinov, A. V., and Diamond, P. H. 1994. Self-consistent theory of mean-field electrodynamics. *Phys. Rev. Lett.*, **72**(11), 1651.

Guervilly, C., Hughes, D. W., and Jones, C. A. 2014. Large-scale vortices in rapidly rotating Rayleigh-Bénard convection. *J. Fluid Mech.*, **758**, 407–435.

Guha, A. 1997. A unified Eulerian theory of turbulent deposition to smooth and rough surfaces. *J. Aerosol Sci.*, **28**(8), 1517–1537.

Guha, A. 2008. Transport and deposition of particles in turbulent and laminar flow. *Annu. Rev. Fluid Mech.*, **40**, 311–341.

Gustavsson, K., and Mehlig, B. 2014. Relative velocities of inertial particles in turbulent aerosols. *Journal of Turbulence*, **15**(1), 34–69.

Gustavsson, K., and Mehlig, B. 2016. Statistical models for spatial patterns of heavy particles in turbulence. *Advances in Physics*, **65**(1), 1–57.

Gustavsson, K., Mehlig, B., and Wilkinson, M. 2008. Collisions of particles advected in random flows. *New Journal of Physics*, **10**(7), 075014.

Gvaramadze, V. V., Lominadze, J. G., Ruzmaikin, A. A., Sokoloff, D., and Shukurov, A. M. 1988. Turbulent generation of magnetic fields in astrophysical jets. *Astrophys. Space Science*, **140**(1), 165–174.

Hanazaki, H., and Hunt, J. C. R. 1996. Linear processes in unsteady stably stratified turbulence. *J. Fluid Mech.*, **318**, 303–337.

Hartlep, T., Tilgner, A., and Busse, F. H. 2003. Large scale structures in Rayleigh-Bénard convection at high Rayleigh numbers. *Phys. Rev. Lett.*, **91**(6), 064501.

Haugen, N. E. L., Brandenburg, A., and Dobler, W. 2004. Simulations of nonhelical hydromagnetic turbulence. *Phys. Rev. E*, **70**(1), 016308.

Haugen, N. E. L., Kleeorin, N., Rogachevskii, I., and Brandenburg, A. 2012. Detection of turbulent thermal diffusion of particles in numerical simulations. *Phys. Fluids*, **24**(7), 075106.

Hinze, J. O. 1959. *Turbulence. An Introduction to Its Mechanism and Theory*. New York: McGraw-Hill.

Hubbard, A. 2016. Turbulent thermal diffusion: a way to concentrate dust in protoplanetary discs. *Monthly Notices of the Royal Astronomical Society*, **456**(3), 3079–3089.

Hubbard, A., and Brandenburg, A. 2010a. Magnetic helicity flux in the presence of shear. *Astrophys. J.*, **727**(1), 11.

Hubbard, A., and Brandenburg, A. 2010b. Magnetic helicity fluxes in an α^2 dynamo embedded in a halo. *Geophys. Astrophys. Fluid Dyn.*, **104**(5-6), 577–590.

Hubbard, A., and Brandenburg, A. 2012. Catastrophic quenching in $\alpha\Omega$ dynamos revisited. *Astrophys. J.*, **748**(1), 51.

Iskakov, A. B., Schekochihin, A. A., Cowley, S. C., McWilliams, J. C., and Proctor, M. R. E. 2007. Numerical demonstration of fluctuation dynamo at low magnetic Prandtl numbers. *Phys. Rev. Lett.*, **98**(20), 208501.

Jabbari, S., Brandenburg, A., Kleeorin, N., Mitra, Dh., and Rogachevskii, I. 2013. Surface flux concentrations in a spherical α^2 dynamo. *Astron. Astrophys.*, **556**, A106.

Jabbari, S., Brandenburg, A., Losada, I. R., Kleeorin, N., and Rogachevskii, I. 2014. Magnetic flux concentrations from dynamo-generated fields. *Astron. Astrophys.*, **568**, A112.

Jabbari, S., Brandenburg, A., Kleeorin, N., Mitra, Dh., and Rogachevskii, I. 2015. Bipolar magnetic spots from dynamos in stratified spherical shell turbulence. *Astrophys. J.*, **805**(2), 166.

Jabbari, S., Brandenburg, A., Mitra, Dh., Kleeorin, N., and Rogachevskii, I. 2016. Turbulent reconnection of magnetic bipoles in stratified turbulence. *Monthly Notices of the Royal Astronomical Society*, **459**(4), 4046–4056.

Jabbari, S., Brandenburg, A., Kleeorin, N., and Rogachevskii, I. 2017. Sharp magnetic structures from dynamos with density stratification. *Monthly Notices of the Royal Astronomical Society*, **467**(3), 2753–2765.

Jones, C. A., Thompson, M. J., and Tobias, S. M. 2010. The solar dynamo. *Space Sci. Rev.*, **152**, 591–616.

Jovanovic, J. 2004. *The Statistical Dynamics of Turbulence*. Berlin: Springer.

Kac, M. 1959. *Probability and Related Topics in Physical Sciences*. Vol. 1. New York: Interscience.

Kaftory, D., Hetsroni, G., and Banerjee, S. 1995a. Particle behaviour in the turbulent boundary layer. I. Motion, deposition, and entrainment. *Phys. Fluids*, **7**, 1095–1106.

Kaftory, D., Hetsroni, G., and Banerjee, S. 1995b. Particle behaviour in the turbulent boundary layer. II. Velocity and distribution profiles. *Phys. Fluids*, **7**, 1107–1121.

Kaimal, J. C., and Finnigan, J. J. 1994. *Atmospheric Boundary Layer Flows: Their Structure and Measurement*. New York: Oxford University Press.

Käpylä, P. J., Mitra, Dh., and Brandenburg, A. 2009. Numerical study of large-scale vorticity generation in shear-flow turbulence. *Phys. Rev. E*, **79**(1), 016302.

Käpylä, P. J., Mantere, M. J., and Hackman, T. 2011. Starspots due to large-scale vortices in rotating turbulent convection. *Astrophys. J.*, **742**(1), 34.

Käpylä, P. J., Brandenburg, A., Kleeorin, N., Mantere, M. J., and Rogachevskii, I. 2012. Negative effective magnetic pressure in turbulent convection. *Monthly Notices of the Royal Astronomical Society*, **422**, 2465–2473.

Käpylä, P. J., Brandenburg, A., Kleeorin, N., Käpylä, M. J., and Rogachevskii, I. 2016. Magnetic flux concentrations from turbulent stratified convection. *Astron. Astrophys.*, **588**, A150.

Kazantsev, A. P. 1968. Enhancement of a magnetic field by a conducting fluid. *Sov. Phys. JETP*, **26**(5), 1031–1034.

Kemel, K., Brandenburg, A., Kleeorin, N., and Rogachevskii, I. 2012a. Properties of the negative effective magnetic pressure instability. *Astron. Nachr.*, **333**(2), 95–100.

Kemel, K., Brandenburg, A., Kleeorin, N., Mitra, Dh., and Rogachevskii, I. 2012b. Spontaneous formation of magnetic flux concentrations in stratified turbulence. *Solar Phys.*, **280**(2), 321–333.

Kemel, K., Brandenburg, A., Kleeorin, N., Mitra, Dh., and Rogachevskii, I. 2013a. Active region formation through the negative effective magnetic pressure instability. *Solar Phys.*, **287**(1-2), 293–313.

Kemel, K., Brandenburg, A., Kleeorin, N., and Rogachevskii, I. 2013b. Non-uniformity effects in the negative effective magnetic pressure instability. *Physica Scripta*, **2013**(T155), 014027.

Kersale, Evy., Hughes, D. W., and Tobias, S. M. 2007. The nonlinear evolution of instabilities driven by magnetic buoyancy: a new mechanism for the formation of coherent magnetic structures. *Astrophys. J. Lett.*, **663**(2), L113.

Khain, A., Pinsky, M., Elperin, T., Kleeorin, N., Rogachevskii, I., and Kostinski, A. 2007. Critical comments to results of investigations of drop collisions in turbulent clouds. *Atmosph. Res.*, **86**(1), 1–20.

Khain, A. P., and Pinsky, M. 2018. *Physical Processes in Clouds and Cloud Modeling*. Cambridge: Cambridge University Press.

Kharzeev, D. E. 2014. The chiral magnetic effect and anomaly-induced transport. *Prog. Part. Nucl. Phys.*, **75**, 133–151.

Kharzeev, D. E., Liao, J., Voloshin, S. A., and Wang, G. 2016. Chiral magnetic and vortical effects in high-energy nuclear collisions – A status report. *Prog. Part. Nucl. Phys.*, **88**, 1–28.

Kippenhahn, R. 1963. Differential rotation in stars with convective envelopes. *Astrophys. J.*, **137**, 664.

Kitchatinov, L. L., Pipin, V. V., and Rüdiger, G. 1994. Turbulent viscosity, magnetic diffusivity, and heat conductivity under the influence of rotation and magnetic field. *Astron. Nachr.*, **315**(2), 157–170.

Kleeorin, N., and Rogachevskii, I. 1994a. Effective Ampère force in developed magnetohydrodynamic turbulence. *Phys. Rev. E*, **50**(4), 2716.

Kleeorin, N., and Rogachevskii, I. 1994b. Nonlinear theory of magnetic fluctuations in random flow: the Hall effect. *Phys. Rev. E*, **50**(1), 493.

Kleeorin, N., and Rogachevskii, I. 1999. Magnetic helicity tensor for an anisotropic turbulence. *Phys. Rev. E*, **59**(6), 6724.

Kleeorin, N., and Rogachevskii, I. 2003. Effect of rotation on a developed turbulent stratified convection: the hydrodynamic helicity, the α effect, and the effective drift velocity. *Phys. Rev. E*, **67**(2), 026321.

Kleeorin, N., and Rogachevskii, I. 2006. Effect of heat flux on differential rotation in turbulent convection. *Phys. Rev. E*, **73**(4), 046303.

Kleeorin, N., and Rogachevskii, I. 2007. Nonlinear turbulent magnetic diffusion and effective drift velocity of a large-scale magnetic field in two-dimensional magnetohydrodynamic turbulence. *Phys. Rev. E*, **75**(6), 066315.

Kleeorin, N., and Rogachevskii, I. 2008. Mean-field dynamo in a turbulence with shear and kinetic helicity fluctuations. *Phys. Rev. E*, **77**(3), 036307.

Kleeorin, N., and Rogachevskii, I. 2012. Growth rate of small-scale dynamo at low magnetic Prandtl numbers. *Phys. Scripta*, **86**(1), 018404.

Kleeorin, N., and Rogachevskii, I. 2018. Generation of large-scale vorticity in rotating stratified turbulence with inhomogeneous helicity: mean-field theory. *J. Plasma Phys.*, **84**, 735840303.

Kleeorin, N., and Ruzmaikin, A. 1982. Dynamics of the average turbulent helicity in a magnetic field. *Magnetohydrodynamics*, **18**(3), 116–122.

Kleeorin, N., Rogachevskii, I., and Ruzmaikin, A. A. 1989. Negative magnetic pressure as a trigger of large-scale magnetic instability in the solar convective zone. *Sov. Astron. Lett*, **15**, 274–277.

Kleeorin, N., Rogachevskii, I., and Ruzmaikin, A. A. 1990. Magnetic force reversal and instability in a plasma with advanced magnetohydrodynamic turbulence. *Sov. Phys. JETP*, **70**, 878–883.

Kleeorin, N., Mond, M., and Rogachevskii, I. 1993. Magnetohydrodynamic instabilities in developed small-scale turbulence. *Phys. Fluids B: Plasma Phys.*, **5**(11), 4128–4134.

Kleeorin, N., Rogachevskii, I., and Eviatar, A. 1994a. A mechanism of magnetic flux rope formation in the ionosphere of Venus. *J. Geophys. Res.*, **99**(A4), 6475–6481.

Kleeorin, N., Rogachevskii, I., and Ruzmaikin, A. 1994b. Solar irradiance variations and nonlinear mean field dynamo. *Solar Phys.*, **155**(2), 223–234.

Kleeorin, N., Rogachevskii, I., and Ruzmaikin, A. 1995. Magnitude of the dynamo-generated magnetic field in solar-type convective zones. *Astron. Astrophys.*, **297**, 159–167.

Kleeorin, N., Mond, M., and Rogachevskii, I. 1996. Magnetohydrodynamic turbulence in the solar convective zone as a source of oscillations and sunspots formation. *Astron. Astrophys.*, **307**, 293.

Kleeorin, N., Moss, D., Rogachevskii, I., and Sokoloff, D. 2000. Helicity balance and steady-state strength of the dynamo generated galactic magnetic field. *Astron. Astrophys.*, **361**, L5–L8.

Kleeorin, N., Rogachevskii, I., and Sokoloff, D. 2002a. Magnetic fluctuations with a zero mean field in a random fluid flow with a finite correlation time and a small magnetic diffusion. *Phys. Rev. E*, **65**(3), 036303.

Kleeorin, N., Moss, D., Rogachevskii, I., and Sokoloff, D. 2002b. The role of magnetic helicity transport in nonlinear galactic dynamos. *Astron. Astrophys.*, **387**(2), 453–462.

Kleeorin, N., Kuzanyan, K., Moss, D., Rogachevskii, I., Sokoloff, D., and Zhang, H. 2003a. Magnetic helicity evolution during the solar activity cycle: observations and dynamo theory. *Astron. Astrophys.*, **409**(3), 1097–1105.

Kleeorin, N., Moss, D., Rogachevskii, I., and Sokoloff, D. 2003b. Nonlinear magnetic diffusion and magnetic helicity transport in galactic dynamos. *Astron. Astrophys.*, **400**(1), 9–18.

Kleeorin, N., Rogachevskii, I., Sokoloff, D., and Tomin, D. 2009. Mean-field dynamos in random Arnold-Beltrami-Childress and Roberts flows. *Phys. Rev. E*, **79**(4), 046302.

Kleeorin, N., Rogachevskii, I., Soustova, I. A., Troitskaya, Yu. I., Ermakova, O. S., and Zilitinkevich, S. 2019. Internal gravity waves in the energy and flux budget turbulence-closure theory for shear-free stably stratified flows. *Phys. Rev. E*, **99**(6), 063106.

Kleeorin, N., Safiullin, N., Kuzanyan, K., Rogachevskii, I., Tlatov, A., and Porshnev, S. 2020. The mean tilt of sunspot bipolar regions: theory, simulations and comparison with observations. *Monthly Notices of the Royal Astronomical Society*, **495**(1), 238–248.

Kleeorin, Ya., Safiullin, N., Kleeorin, N., Porshnev, S., Rogachevskii, I., and Sokoloff, D. 2016. The dynamics of Wolf numbers based on nonlinear dynamos with magnetic helicity: comparisons with observations. *Monthly Notices of the Royal Astronomical Society*, **460**(4), 3960–3967.

Kolmogorov, A. N. 1941. Dissipation of energy in the locally isotropic turbulence. *Dokl. Akad. Nauk. SSSR A*, **32**, 16–18.

Kolmogorov, A. N. 1942. The equations of turbulent motion in an incompressible fluid. *Izvestia Akad. Sci., USSR; Phys.*, **6**, 56–58.

Kolmogorov, A. N. 1962. A refinement of previous hypotheses concerning the local structure of turbulence in a viscous incompressible fluid at high Reynolds number. *J. Fluid Mech.*, **13**(1), 82–85.

Kolmogorov, A. N. 1991. The local structure of turbulence in incompressible viscous fluid for very large Reynolds numbers. *Proc. Roy. Soc. London A*, **434**(1890), 9–13.

Kolmogorov, A. N., Petrovsky, I, G., and Piskunov, N. S. 1937. Investigation of the equation of diffusion combined with increasing of the substance and its application to a biology problem. *Bull. Moscow State Univ. Ser. A: Math. Mech.*, **1**(6), 1–25.

Kondo, J., Kanechika, O., and Yasuda, N. 1978. Heat and momentum transfers under strong stability in the atmospheric surface layer. *J. Atmosph. Sci.*, **35**(6), 1012–1021.

Kraichnan, R. H. 1962. Turbulent thermal convection at arbitrary Prandtl number. *Phys. Fluids*, **5**(11), 1374–1389.

Kraichnan, R. H. 1968. Small-scale structure of a scalar field convected by turbulence. *Phys. Fluids*, **11**(5), 945–953.

Kraichnan, R. H. 1994. Anomalous scaling of a randomly advected passive scalar. *Phys. Rev. Lett.*, **72**(7), 1016.

Kraichnan, R. H., and Montgomery, D. 1980. Two-dimensional turbulence. *Rep. Prog. Phys.*, **43**, 547–619.

Krause, F., and Rädler, K.-H. 1980. *Mean-Field Magnetohydrodynamics and Dynamo Theory*. Oxford: Pergamon Press.

Krishnamurti, R., and Howard, L. N. 1981. Large-scale flow generation in turbulent convection. *Proc. Natl. Acad. Sci. USA*, **78**(4), 1981–1985.

Kritsuk, A. G., Norman, M. L., Padoan, P., and Wagner, R. 2007. The statistics of supersonic isothermal turbulence. *Astrophys. J.*, **665**(1), 416.

Kulsrud, R. M., and Anderson, S. W. 1992. The spectrum of random magnetic fields in the mean field dynamo theory of the galactic magnetic field. *Astrophys. J.*, **396**, 606–630.

Lamb, D., and Verlinde, J. 2011. *Physics and Chemistry of Clouds*. Cambridge: Cambridge University Press.

Landau, L. D., and Lifshits, E. M. 1987. *Fluid Mechanics*. Oxford: Pergamon.

Lesieur, M. 2008. *Turbulence in Fluids*. Dordrecht: Springer.

Li, D., Katul, G. G., and Zilitinkevich, S. S. 2016. Closure schemes for stably stratified atmospheric flows without turbulence cutoff. *J. Atmosph. Sci.*, **73**(12), 4817–4832.

Li, X.-Y., Brandenburg, A., Svensson, G., Haugen, N. E. L., Mehlig, B., and Rogachevskii, I. 2018. Effect of turbulence on collisional growth of cloud droplets. *J. Atmosph. Sci.*, **75**(10), 3469–3487.

Li, X.-Y., Brandenburg, A., Svensson, G., Haugen, N. E. L., Mehlig, B., and Rogachevskii, I. 2020. Condensational and collisional growth of cloud droplets in a turbulent environment. *J. Atmosph. Sci.*, **77**(1), 337–353.

Libby, P. A. 1985. Theory of normal premixed turbulent flames revisited. *Prog. Energy Combust. Sci.*, **11**(1), 83–96.

Libby, P. A. 1996. *An Introduction to Turbulence*. Boca Raton: CRC Press.

Liberman, M., Kleeorin, N., Rogachevskii, I., and Haugen, N. E. L. 2017. Mechanism of unconfined dust explosions: turbulent clustering and radiation-induced ignition. *Phys. Rev. E*, **95**(5), 051101.

Liberman, M., Kleeorin, N., Rogachevskii, I., and Haugen, N. E. L. 2018. Multipoint radiation induced ignition of dust explosions: turbulent clustering of particles and increased transparency. *Combust. Theory and Modelling*, **22**(6), 1084–1102.

Liberman, M. A. 2010. *Introduction to Physics and Chemistry of Combustion: Explosion, Flame, Detonation*. New York: Springer-Verlag.

Lighthill, J. 1986. *Informal Introduction to Fluid Mechanics*. Cambridge: Cambridge University Press.

Lindborg, E. 2006. The energy cascade in a strongly stratified fluid. *J. Fluid Mech.*, **550**, 207.

Lipatnikov, A. 2012. *Fundamentals of Premixed Turbulent Combustion*. Boca Raton: CRC Press.

Lipton, A. 2001. *Mathematical Methods for Foreign Exchange: A Financial Engineer's Approach*. New York: World Scientific.

Lohse, D., and Xia, K.-Q. 2010. Small-scale properties of turbulent Rayleigh-Bénard convection. *Annu. Rev. Fluid Mech.*, **42**, 335–363.

Losada, I. R., Brandenburg, A., Kleeorin, N., Mitra, Dh., and Rogachevskii, I. 2012. Rotational effects on the negative magnetic pressure instability. *Astron. Astrophys.*, **548**, A49.

Losada, I. R., Brandenburg, A., Kleeorin, N., and Rogachevskii, I. 2013. Competition of rotation and stratification in flux concentrations. *Astron. Astrophys.*, **556**, A83.

Losada, I. R., Brandenburg, A., Kleeorin, N., and Rogachevskii, I. 2014. Magnetic flux concentrations in a polytropic atmosphere. *Astron. Astrophys.*, **564**, A2.

Losada, I. R., Warnecke, J., Brandenburg, A., Kleeorin, N., and Rogachevskii, I. 2019. Magnetic bipoles in rotating turbulence with coronal envelope. *Astron. Astrophys.*, **621**, A61.

Loureiro, N. F., Uzdensky, D. A., Schekochihin, A. A., Cowley, S. C., and Yousef, T. A. 2009. Turbulent magnetic reconnection in two dimensions. *Monthly Notices of the Royal Astronomical Society: Letters*, **399**(1), L146–L150.

Loureiro, N. F., Samtaney, R., Schekochihin, A. A., and Uzdensky, D. A. 2012. Magnetic reconnection and stochastic plasmoid chains in high-Lundquist-number plasmas. *Phys. Plasmas*, **19**(4), 042303.

Lumley, J. L. 2007. *Stochastic Tools in Turbulence*. New York: Dover.

L'vov, V. S., Procaccia, I., and Rudenko, O. 2008. Turbulent fluxes in stably stratified boundary layers. *Physica Scripta*, **2008**(T132), 014010.

Mahrt, L. 2010. Variability and maintenance of turbulence in the very stable boundary layer. *Boundary-Layer Meteorol.*, **135**(1), 1–18.

Mahrt, L. 2014. Stably stratified atmospheric boundary layers. *Annu. Rev. Fluid Mech.*, **46**, 23–45.

Mahrt, L., and Vickers, D. 2005. Boundary-layer adjustment over small-scale changes of surface heat flux. *Boundary-Layer Meteorol.*, **116**(2), 313–330.

Majda, A. J., and Kramer, P. R. 1999. Simplified models for turbulent diffusion: theory, numerical modelling, and physical phenomena. *Phys. Rep.*, **314**(4–5), 237–574.

Marcus, P. S. 1993. Jupiter's Great Red Spot and other vortices. *Annu. Rev. Astron. Astrophys.*, **31**(1), 523–569.

Mathieu, J., and Scott, J. 2000. *An Introduction to Turbulent Flow*. Cambridge: Cambridge University Press.

Mauritsen, T., Svensson, G., Zilitinkevich, S. S., Esau, I., Enger, L., and Grisogono, B. 2007. A total turbulent energy closure model for neutrally and stably stratified atmospheric boundary layers. *J. Atmos. Sci.*, **64**(11), 4113–4126.

Maxey, M. R. 1987. The gravitational settling of aerosol particles in homogeneous turbulence and random flow fields. *J. Fluid Mech.*, **174**, 441–465.

McComb, W. D. 1990. *The Physics of Fluid Turbulence*. Oxford: Oxford Science Publications.

Mehlig, B., Wilkinson, M., Duncan, K., Weber, T., and Ljunggren, M. 2005. Aggregation of inertial particles in random flows. *Phys. Rev. E*, **72**(5), 051104.

Miropol'sky, Yu. Z. 2001. *Dynamics of Internal Gravity Waves in the Ocean*. Vol. 24. Dordrecht: Springer.

Mitra, Dh., Candelaresi, S., Chatterjee, P., Tavakol, R., and Brandenburg, A. 2010. Equatorial magnetic helicity flux in simulations with different gauges. *Astron. Nachr.*, **331**(1), 130–135.

Mitra, Dh., Brandenburg, A., Kleeorin, N., and Rogachevskii, I. 2014. Intense bipolar structures from stratified helical dynamos. *Monthly Notices of the Royal Astronomical Society*, **445**(1), 761–769.

Mitra, Dh., Haugen, N. E. L., and Rogachevskii, I. 2018. Turbophoresis in forced inhomogeneous turbulence. *Europ. Phys. J. Plus*, **133**(2), 1–8.

Moffatt, H. K. 1961. The amplification of a weak applied magnetic field by turbulence in fluids of moderate conductivity. *J. Fluid Mech.*, **11**(4), 625–635.

Moffatt, H. K. 1963. Magnetic eddies in an incompressible viscous fluid of high electrical conductivity. *J. Fluid Mech.*, **17**, 225–239.

Moffatt, H. K. 1978. *Magnetic Field Generation in Electrically Conducting Fluids*. Cambridge: Cambridge University Press.

Moffatt, H. K. 1981. Some developments in the theory of turbulence. *J. Fluid Mech*, **106**, 27–47.

Moffatt, H. K. 1983. Transport effects associated with turbulence with particular attention to the influence of helicity. *Rep. Prog. Phys.*, **46**(5), 621.

Moffatt, H. K., and Dormy, E. 2019. *Self-Exciting Fluid Dynamos*. Vol. 59. Cambridge: Cambridge University Press.

Monin, A. S., and Yaglom, A. M. 1971. *Statistical Fluid Mechanics*. Vol. 1. Cambridge, Mass: MIT Press.

Monin, A. S., and Yaglom, A. M. 2013. *Statistical Fluid Mechanics*. Vol. 2. New York: Dover.

Nappo, C. J. 2013. *An Introduction to Atmospheric Gravity Waves*. London: Academic Press.

Nazarenko, S. 2011. *Wave Turbulence*. Vol. 825. Berlin: Springer.

Newcomb, W. A. 1961. Convective instability induced by gravity in a plasma with a frozen-in magnetic field. *Phys. Fluids*, **4**(4), 391–396.

Niemela, J. J., Skrbek, L., Sreenivasan, K. R., and Donnelly, R. J. 2001. The wind in confined thermal convection. *J. Fluid Mech.*, **449**, 169.

Obukhov, A. M. 1941. On the distribution of energy in the spectrum of turbulent flow. *Bull. Acad. Sci. USSR, Geog. Geophys.*, **5**, 453–466.

Obukhov, A. M. 1949. Structure of the temperature field in turbulent flows. *Bull. Acad. Sci. USSR, Geog. Geophys.*, **13**, 58–69.

Ohya, Y. 2001. Wind-tunnel study of atmospheric stable boundary layers over a rough surface. *Boundary-Layer Meteorol.*, **98**(1), 57–82.

Olla, P. 2010. Preferential concentration versus clustering in inertial particle transport by random velocity fields. *Phys. Rev. E*, **81**(1), 016305.

Orszag, S. A. 1970. Analytical theories of turbulence. *J. Fluid Mech.*, **41**(2), 363–386.

Ossendrijver, M. 2003. The solar dynamo. *Astron. Astrophys. Rev.*, **11**(4), 287–367.

Ostrovsky, L. A., and Troitskaya, Yu. I. 1987. A model of turbulent transfer and dynamics of turbulence in a stratified shear-flow. *Izv. AN SSSR FAO*, **23**(10), 1031–1040.

Panchev, S. 1971. *Random Functions and Turbulence*. Oxford: Pergamon Press.

Pandey, V., Perlekar, P., and Mitra, Dh. 2019. Clustering and energy spectra in two-dimensional dusty gas turbulence. *Phys. Rev. E*, **100**(1), 013114.

Pandya, R. V. R., and Mashayek, F. 2002. Turbulent thermal diffusion and barodiffusion of passive scalar and dispersed phase of particles in turbulent flows. *Phys. Rev. Lett.*, **88**(4), 044501.

Parker, E. N. 1955a. The formation of sunspots from the solar toroidal field. *Astrophys. J.*, **121**, 491.

Parker, E. N. 1955b. Hydromagnetic dynamo models. *Astrophys. J.*, **122**, 293.

Parker, E. N. 1966. The dynamical state of the interstellar gas and field. *Astrophys. J.*, **145**, 811.

Parker, E. N. 1979. *Cosmical Magnetic Fields: Their Origin and Their Activity*. Oxford: Clarendon Press.

Parodi, A., von Hardenberg, J., Passoni, G., Provenzale, A., and Spiegel, E. A. 2004. Clustering of plumes in turbulent convection. *Phys. Rev. Lett.*, **92**(19), 194503.

Pasquill, F., and Smith, F. B. 1983. *Atmospheric Diffusion*. Chichester: Ellis Horwood.

Pedlosky, J. 1987. *Geophysical Fluid Dynamics*. New York: Springer.

Perlekar, P., Benzi, R., Nelson, D. R., and Toschi, F. 2010. Population dynamics at high Reynolds number. *Phys. Rev. Lett.*, **105**(14), 144501.

Peters, N. 2004. *Turbulent Combustion*. New York: Cambridge University Press.

Pétrélis, F., and Fauve, S. 2006. Inhibition of the dynamo effect by phase fluctuations. *Europhys. Lett.*, **76**(4), 602.

Piquet, J. 2013. *Turbulent Flows: Models and Physics*. Berlin: Springer.

Piterbarg, L., and Ostrovskii, A. 2013. *Advection and Diffusion in Random Media: Implications for Sea Surface Temperature Anomalies*. Berlin: Springer.

Pope, S. B. 2000. *Turbulent Flows*. Cambridge: Cambridge University Press.

Pouquet, A., Frisch, U., and Léorat, J. 1976. Strong MHD helical turbulence and the nonlinear dynamo effect. *J. Fluid Mech.*, **77**(2), 321–354.

Pouquet, A., Rosenberg, D., Marino, R., and Herbert, C. 2018. Scaling laws for mixing and dissipation in unforced rotating stratified turbulence. *J. Fluid Mech.*, **844**, 519–545.

Priest, E. R. 1982. *Solar Magnetohydrodynamics*. Dordrecht: Reidel Publ. Co.

Pruppacher, H. R., Klett, J. D., and Wang, P. K. 1998. *Microphysics of Clouds and Precipitation*. Abingdon: Taylor and Francis.

Rädler, K.-H. 1980. Mean-field approach to spherical dynamo models. *Astron. Nachr.*, **301**(3), 101–129.

Rädler, K.-H. 1982. On dynamo action in the high-conductivity limit. *Geophys. & Astrophys. Fluid Dynamics*, **20**(3-4), 191–211.

Rädler, K.-H., and Brandenburg, A. 2009. Mean-field effects in the Galloway–Proctor flow. *Monthly Notices of the Royal Astronomical Society*, **393**(1), 113–125.

Rädler, K.-H., Kleeorin, N., and Rogachevskii, I. 2003. The mean electromotive force for MHD turbulence: the case of a weak mean magnetic field and slow rotation. *Geophys. Astrophys. Fluid Dyn.*, **97**, 249–274.

Rädler, K.-H., Brandenburg, A., Del Sordo, F., and Rheinhardt, M. 2011. Mean-field diffusivities in passive scalar and magnetic transport in irrotational flows. *Phys. Rev. E*, **84**(4), 046321.

Reeks, M. W. 1983. The transport of discrete particles in inhomogeneous turbulence. *J. Aerosol Sci.*, **14**(6), 729–739.

Reeks, M. W. 2005. On model equations for particle dispersion in inhomogeneous turbulence. *Int. J. Multiph. Flow*, **31**(1), 93–114.

Rehmann, C. R., and Koseff, J. R. 2004. Mean potential energy change in stratified grid turbulence. *Dynamics of Atmospheres and Oceans*, **37**(4), 271–294.

Reynolds, W. C., Parekh, D. E., Juvet, P. J. D., and Lee, M. J. D. 2003. Bifurcating and blooming jets. *Annu. Rev. Fluid Mech.*, **35**(1), 295–315.

Richardson, L. F. 1920. The supply of energy from and to atmospheric eddies. *Proc. Roy. Soc. London A*, **97**(686), 354–373.

Righetti, M., and Romano, G. P. 2004. Particle-fluid interactions in a plane near-wall turbulent flow. *J. Fluid Mech.*, **505**, 93.

Roberts, P. H., and Soward, A. M. 1975. A unified approach to mean field electrodynamics. *Astron. Nachr.*, **296**(2), 49–64.

Roberts, P. H., and Stix, M. 1971. *The Turbulent Dynamo*. Boulder: NCAR.

Rogachevskii, I., and Kleeorin, N. 1997. Intermittency and anomalous scaling for magnetic fluctuations. *Phys. Rev. E*, **56**(1), 417.

Rogachevskii, I., and Kleeorin, N. 1999. Anomalous scalings and dynamics of magnetic helicity. *Phys. Rev. E*, **59**(3), 3008.

Rogachevskii, I., and Kleeorin, N. 2000. Electromotive force for an anisotropic turbulence: intermediate nonlinearity. *Phys. Rev. E*, **61**(5), 5202.

Rogachevskii, I., and Kleeorin, N. 2001. Nonlinear turbulent magnetic diffusion and mean-field dynamo. *Phys. Rev. E*, **64**(5), 056307.

Rogachevskii, I., and Kleeorin, N. 2003. Electromotive force and large-scale magnetic dynamo in a turbulent flow with a mean shear. *Phys. Rev. E*, **68**(3), 036301.

Rogachevskii, I., and Kleeorin, N. 2004. Nonlinear theory of a "shear-current" effect and mean-field magnetic dynamos. *Phys. Rev. E*, **70**(4), 046310.

Rogachevskii, I., and Kleeorin, N. 2006. Small-scale magnetic buoyancy and magnetic pumping effects in a turbulent convection. *Geophys. Astrophys. Fluid Dyn.*, **100**(3), 243–263.

Rogachevskii, I., and Kleeorin, N. 2007. Magnetic fluctuations and formation of large-scale inhomogeneous magnetic structures in a turbulent convection. *Phys. Rev. E*, **76**(5), 056307.

Rogachevskii, I., and Kleeorin, N. 2015. Turbulent fluxes of entropy and internal energy in temperature stratified flows. *J. Plasma Phys.*, **81**, 395810504.

Rogachevskii, I., and Kleeorin, N. 2018. Mean-field theory of differential rotation in density stratified turbulent convection. *J. Plasma Phys.*, **84**, 735840201.

Rogachevskii, I., and Kleeorin, N. 2019. Generation of a large-scale vorticity in a fast-rotating density-stratified turbulence or turbulent convection. *Phys. Rev. E*, **100**(6), 063101.

Rogachevskii, I., and Kleeorin, N. 2021. Compressibility effects in a turbulent transport of temperature field. *Phys. Rev. E*, **103**(1), 013107.

Rogachevskii, I., Kleeorin, N., Käpylä, P. J., and Brandenburg, A. 2011. Pumping velocity in homogeneous helical turbulence with shear. *Phys. Rev. E*, **84**(5), 056314.

Rogachevskii, I., Kleeorin, N., Brandenburg, A., and Eichler, D. 2012. Cosmic-ray current-driven turbulence and mean-field dynamo effect. *Astrophys. J.*, **753**(1), 6.

Rogachevskii, I., Ruchayskiy, O., Boyarsky, A., Fröhlich, J., Kleeorin, N., Brandenburg, A., and Schober, J. 2017. Laminar and turbulent dynamos in chiral magnetohydrodynamics. I. Theory. *Astrophys. J.*, **846**(2), 153.

Rogachevskii, I., Kleeorin, N., and Brandenburg, A. 2018. Compressibility in turbulent MHD and passive scalar transport: mean-field theory. *J. Plasma Phys.*, **84**, 735840502.

Rüdiger, G. 1989. *Differential Rotation and Stellar Convection: Sun and Solar-Type Stars*. New York: Gordon and Breach.

Rüdiger, G., and Kichatinov, L. L. 1993. Alpha-effect and alpha-quenching. *Astron. Astrophys.*, **269**, 581–588.

Rüdiger, G., Hollerbach, R., and Kitchatinov, L. L. 2013. *Magnetic Processes in Astrophysics: Theory, Simulations, Experiments*. Weinheim: John Wiley & Sons.

Ruzmaikin, A., Shukurov, A. M., and Sokoloff, D. D. 1988. *Magnetic Fields of Galaxies*. Dordrecht: Kluwer Academic.

Ruzmaikin, A. A., and Shukurov, A. M. 1982. Spectrum of the galactic magnetic fields. *Astrophys. Space Sci.*, **82**(2), 397–407.

Sabelnikov, V. A., and Lipatnikov, A. N. 2013. Transition from pulled to pushed premixed turbulent flames due to countergradient transport. *Combustion Theory and Modelling*, **17**(6), 1154–1175.

Sabelnikov, V. A., and Lipatnikov, A. N. 2014. Speed selection for traveling-wave solutions to the diffusion-reaction equation with cubic reaction term and Burgers nonlinear convection. *Phys. Rev. E*, **90**(3), 033004.

Sabelnikov, V. A., and Lipatnikov, A. N. 2015. Transition from pulled to pushed fronts in premixed turbulent combustion: theoretical and numerical study. *Combustion and Flame*, **162**(7), 2893–2903.

Sabelnikov, V. A., and Lipatnikov, A. N. 2017. Recent advances in understanding of thermal expansion effects in premixed turbulent flames. *Annu. Rev. Fluid Mech.*, **49**, 91–117.

Safiullin, N., Kleeorin, N., Porshnev, S., Rogachevskii, I., and Ruzmaikin, A. 2018. Nonlinear mean-field dynamo and prediction of solar activity. *J. Plasma Phys.*, **84**, 735840306.

Sagaut, P., and Cambon, C. 2008. *Homogeneous Turbulence Dynamics*. Berlin: Springer.

Salazar, J., De Jong, J., Cao, L., Woodward, S. H., Meng, H., and Collins, L. R. 2008. Experimental and numerical investigation of inertial particle clustering in isotropic turbulence. *J. Fluid Mech.*, **600**, 245.

Salmon, R. 1998. *Lectures on Geophysical Fluid Dynamics*. New York: Oxford University Press.

Sardina, G., Schlatter, P., Brandt, L., Picano, F., and Casciola, C. M. 2012. Wall accumulation and spatial localization in particle-laden wall flows. *J. Fluid Mech.*, **699**(1), 50–78.

Schekochihin, A. A., Cowley, S. C., Taylor, S. F., Maron, J. L., and McWilliams, J. C. 2004. Simulations of the small-scale turbulent dynamo. *Astrophys. J.*, **612**(1), 276.

Schekochihin, A. A., Haugen, N. E. L., Brandenburg, A., Cowley, S. C., Maron, J. L., and McWilliams, J. C. 2005. The onset of a small-scale turbulent dynamo at low magnetic Prandtl numbers. *Astrophys. J.*, **625**(2), L115.

Schekochihin, A. A., Iskakov, A. B., Cowley, S. C., McWilliams, J. C., Proctor, M. R. E., and Yousef, T. A. 2007. Fluctuation dynamo and turbulent induction at low magnetic Prandtl numbers. *New Journal of Physics*, **9**(8), 300.

Schober, J., Rogachevskii, I., Brandenburg, A., Boyarsky, A., Fröhlich, J., Ruchayskiy, O., and Kleeorin, N. 2018. Laminar and turbulent dynamos in chiral magnetohydrodynamics. II. Simulations. *Astrophys. J.*, **858**(2), 124.

Schober, J., Brandenburg, A., Rogachevskii, I., and Kleeorin, N. 2019. Energetics of turbulence generated by chiral MHD dynamos. *Geophys. Astrophys. Fluid Dyn.*, **113**(1–2), 107–130.

Schober, J., Brandenburg, A., and Rogachevskii, I. 2020. Chiral fermion asymmetry in high-energy plasma simulations. *Geophys. Astrophys. Fluid Dyn.*, **114**(1–2), 106–129.

Schrijver, C. J., and Zwaan, C. 2008. *Solar and Stellar Magnetic Activity*. Vol. 34. Cambridge: Cambridge University Press.

Schussler, M., Caligari, P., Ferriz-Mas, A., and Moreno-Insertis, F. 1994. Instability and eruption of magnetic flux tubes in the solar convection zone. *Astron. Astrophys.*, **281**, L69–L72.

Seinfeld, J. H., and Pandis, S. N. 2016. *Atmospheric Chemistry and Physics: From Air Pollution to Climate Change*. New York: John Wiley & Sons.

Shaw, R. A. 2003. Particle-turbulence interactions in atmospheric clouds. *Annu. Rev. Fluid Mech.*, **35**(1), 183–227.

Shraiman, B. I., and Siggia, E. D. 1995. Anomalous scaling of a passive scalar in turbulent flow. *C. R. Acad. Sci.*, **321**(7), 279–284.

Shraiman, B. I., and Siggia, E. D. 2000. Scalar turbulence. *Nature*, **405**(6787), 639–646.

Shukurov, A., Sokoloff, D., Subramanian, K., and Brandenburg, A. 2006. Galactic dynamo and helicity losses through fountain flow. *Astron. Astrophys.*, **448**(2), L33–L36.

Siebert, H., Gerashchenko, S., Gylfason, A., Lehmann, K., Collins, L. R., Shaw, R. A., and Warhaft, Z. 2010. Towards understanding the role of turbulence on droplets in clouds: in situ and laboratory measurements. *Atmosph. Res.*, **97**(4), 426–437.

Silant'ev, N. A. 2000. Magnetic dynamo due to turbulent helicity fluctuations. *Astron. Astrophys.*, **364**, 339–347.

Singh, N. K. 2016. Moffatt-drift-driven large-scale dynamo due to α fluctuations with non-zero correlation times. *J. Fluid Mech.*, **798**, 696–716.

Singh, N. K., and Jingade, N. 2015. Numerical studies of dynamo action in a turbulent shear flow. I. *Astrophys. J.*, **806**(1), 118.

Singh, N. K., and Sridhar, S. 2011. Transport coefficients for the shear dynamo problem at small Reynolds numbers. *Phys. Rev. E*, **83**(5), 056309.

Singh, N. K., Rogachevskii, I., and Brandenburg, A. 2017. Enhancement of small-scale turbulent dynamo by large-scale shear. *Astrophys. J. Lett.*, **850**(1), L8.

Sofiev, M., Sofieva, V., Elperin, T., Kleeorin, N., Rogachevskii, I., and Zilitinkevich, S. S. 2009. Turbulent diffusion and turbulent thermal diffusion of aerosols in stratified atmospheric flows. *J. Geophys. Res.*, **114**(D18).

Sokoloff, D., Bao, S. D., Kleeorin, N., Kuzanyan, K., Moss, D., Rogachevskii, I., Tomin, D., and Zhang, H. 2006. The distribution of current helicity at the solar surface at the beginning of the solar cycle. *Astron. Nachr.*, **327**(9), 876–883.

Sokolov, D. D. 1997. The disk dynamo with fluctuating spirality. *Astron. Rep.*, **41**(1), 68–72.

Spiegel, E. A., and Weiss, N. O. 1980. Magnetic activity and variations in solar luminosity. *Nature*, **287**(5783), 616–617.

Spruit, H. C. 1974. A model of the solar convection zone. *Solar Phys.*, **34**(2), 277–290.

Spruit, H. C. 1981. Motion of magnetic flux tubes in the solar convection zone and chromosphere. *Astron. Astrophys.*, **98**, 155–160.

Spruit, H. C., and Van Ballegooijen, A. A. 1982. Stability of toroidal flux tubes in stars. *Astron. Astrophys.*, **106**, 58–66.

Sreenivasan, K. R., Tavoularis, S., and Corrsin, S. 1982. A test of gradient transport and its generalizations. Pages 96–112 of *Turbulent Shear Flows 3*. Berlin: Springer.

Sridhar, S., and Singh, N. K. 2010. The shear dynamo problem for small magnetic Reynolds numbers. *J. Fluid Mech.*, **664**, 265–285.

Sridhar, S., and Singh, N. K. 2014. Large-scale dynamo action due to α fluctuations in a linear shear flow. *Monthly Notices of the Royal Astronomical Society*, **445**(4), 3770–3787.

Sridhar, S., and Subramanian, K. 2009a. Nonperturbative quasilinear approach to the shear dynamo problem. *Phys. Rev. E*, **80**(6), 066315.

Sridhar, S., and Subramanian, K. 2009b. Shear dynamo problem: quasilinear kinematic theory. *Phys. Rev. E*, **79**(4), 045305.

Steenbeck, M., Krause, F., and Rädler, K.-H. 1966. Berechnung der mittleren Lorentz-Feldstärke für ein elektrisch leitendes Medium in turbulenter, durch Coriolis-Kräfte beeinflußter Bewegung. *Zeitschrift für Naturforschung A*, **21**(4), 369–376.

Stix, M. 2012. *The Sun: An Introduction*. Berlin: Springer.

Strang, E. J., and Fernando, H. J. S. 2001. Vertical mixing and transports through a stratified shear layer. *J. Phys. Oceanogr.*, **31**(8), 2026–2048.

Stretch, D. D., Rottman, J. W., Venayagamoorthy, S. K., Nomura, K. K., and Rehmann, C. R. 2010. Mixing efficiency in decaying stably stratified turbulence. *Dynamics of Atmospheres and Oceans*, **49**(1), 25–36.

Subramanian, K., and Brandenburg, A. 2004. Nonlinear current helicity fluxes in turbulent dynamos and alpha quenching. *Phys. Rev. Lett.*, **93**(20), 205001.

Sukoriansky, S., and Galperin, B. 2008. Anisotropic turbulence and internal waves in stably stratified flows (QNSE theory). *Physica Scripta*, **2008**(T132), 014036.

Sundaram, S., and Collins, L. R. 1997. Collision statistics in an isotropic particle-laden turbulent suspension. Part 1. Direct numerical simulations. *J. Fluid Mech.*, **335**, 75–109.

Sutherland, B. R. 2010. *Internal Gravity Waves*. Cambridge: Cambridge University Press.

Swaminathan, N., and Bray, K. N. C. 2011. *Turbulent Premixed Flames*. New York: Cambridge University Press.

Taylor, G. I. 1922. Diffusion by continuous movements. *Proc. London Math. Soc.*, **2**(1), 196–212.

Taylor, J. R. 1972. *Scattering Theory, the Quantum Theory on Nonrelativistic Collisions*. New York: Dover.

Tennekes, H., and Lumley, J. L. 1972. *A First Course in Turbulence*. Cambridge, MA: MIT Press.

Thomas, J. H., and Weiss, N. O. 2008. *Sunspots and Star Spots*. Cambridge: Cambridge University Press.

Tikhomirov, V. M. (Ed.). 1991. *Selected Works of A. N. Kolmogorov*. Boston: Kluwer.

Tobias, S. M., and Hughes, D. W. 2004. The influence of velocity shear on magnetic buoyancy instability in the solar tachocline. *Astrophys. J.*, **603**(2), 785.

Tobias, S. M., Brummell, N. H., Clune, T. L., and Toomre, J. 2001. Transport and storage of magnetic field by overshooting turbulent compressible convection. *Astrophys. J.*, **549**(2), 1183.

Toppaladoddi, S., Succi, S., and Wettlaufer, J. S. 2017. Roughness as a route to the ultimate regime of thermal convection. *Phys. Rev. Lett.*, **118**(7), 074503.

Toschi, F., and Bodenschatz, E. 2009. Lagrangian properties of particles in turbulence. *Annu. Rev. Fluid Mech.*, **41**, 375–404.

Townsend, A. A. 1980. *The Structure of Turbulent Shear Flow*. Cambridge: Cambridge University Press.

Tritton, D. J. 1988. *Physical Fluid Dynamics*. Cambridge: Cambridge University Press.

Tsinober, A. 2009. *An Informal Conceptual Introduction to Turbulence*. Berlin: Springer.

Tsinober, A. 2014. *The Essence of Turbulence as a Physical Phenomenon*. Dordrecht: Springer.

Turner, J. S. 1973. *Buoyancy Effects in Fluids*. Cambridge: Cambridge University Press.
Twomey, S. 1977. *Atmospheric Aerosols*. Amsterdam: Elsevier.
Vainshtein, S. I., and Cattaneo, F. 1992. Nonlinear restrictions on dynamo action. *Astrophys. J.*, **393**, 165–171.
Van Aartrijk, M., and Clercx, H.J.H. 2008. Preferential concentration of heavy particles in stably stratified turbulence. *Phys. Rev. Lett.*, **100**(25), 254501.
Van Dyke, M. 1982. *Album of Fluid Motion*. Stanford: Parabolic Press.
Van Reeuwijk, M., Jonker, H. J. J., and Hanjalić, K. 2005. Identification of the wind in Rayleigh–Bénard convection. *Phys. Fluids*, **17**(5), 051704.
Vergassola, M. 1996. Anomalous scaling for passively advected magnetic fields. *Phys. Rev. E*, **53**(4), R3021.
Vilenkin, A. 1980. Equilibrium parity-violating current in a magnetic field. *Phys. Rev. D*, **22**(12), 3080.
Vincenzi, D. 2002. The Kraichnan–Kazantsev dynamo. *J. Stat. Phys.*, **106**(5-6), 1073–1091.
Vishniac, E. T., and Brandenburg, A. 1997. An incoherent α–Ω dynamo in accretion disks. *Astrophys. J.*, **475**(1), 263.
Vishniac, E. T., and Cho, J. 2001. Magnetic helicity conservation and astrophysical dynamos. *Astrophys. J.*, **550**(2), 752.
Wang, L.-P., and Maxey, M. R. 1993. Settling velocity and concentration distribution of heavy particles in homogeneous isotropic turbulence. *J. Fluid Mech.*, **256**, 27–68.
Warhaft, Z. 2000. Passive scalars in turbulent flows. *Annu. Rev. Fluid Mech.*, **32**(1), 203–240.
Warhaft, Z. 2008. Laboratory studies of droplets in turbulence: towards understanding the formation of clouds. *Fluid Dyn. Res.*, **41**(1), 011201.
Warnecke, J., Losada, I. R., Brandenburg, A., Kleeorin, N., and Rogachevskii, I. 2013. Bipolar magnetic structures driven by stratified turbulence with a coronal envelope. *Astrophys. J. Lett.*, **777**(2), L37.
Warnecke, J., Losada, I. R., Brandenburg, A., Kleeorin, N., and Rogachevskii, I. 2016. Bipolar region formation in stratified two-layer turbulence. *Astron. Astrophys.*, **589**, A125.
Weiss, N. O. 1966. The expulsion of magnetic flux by eddies. *Proc. R. Soc. Lond. A*, **293**(1434), 310–328.
Weiss, N. O., and Proctor, M. R. E. 2014. *Magnetoconvection*. New York: Cambridge University Press.
Weiss, S., Seiden, G., and Bodenschatz, E. 2014. Resonance patterns in spatially forced Rayleigh-Bénard convection. *J. Fluid Mech.*, **756**, 293.
Wheelon, A. D. 1957. Spectrum of turbulent fluctuations produced by convective mixing of gradients. *Phys. Rev.*, **105**(6), 1706.
Wilkinson, M. 2016. Large deviation analysis of rapid onset of rain showers. *Physical Rev. Lett.*, **116**(1), 018501.
Wilkinson, M., Mehlig, B., and Bezuglyy, V. 2006. Caustic activation of rain showers. *Phys. Rev. Lett.*, **97**(4), 048501.
Wood, A. M., Hwang, W., and Eaton, J. K. 2005. Preferential concentration of particles in homogeneous and isotropic turbulence. *Int. J. Multiphase Flow*, **31**(10–11), 1220–1230.
Xu, H., and Bodenschatz, E. 2008. Motion of inertial particles with size larger than Kolmogorov scale in turbulent flows. *Physica D*, **237**(14-17), 2095–2100.
Yakhot, V., and Orszag, S. A. 1986. Renormalization group analysis of turbulence. I. Basic theory. *J. Sci. Comput.*, **1**(1), 3–51.

Yokoi, N. 2013. Cross helicity and related dynamo. *Geophys. Astrophys. Fluid Dyn.*, **107** (1-2), 114–184.

Yokoi, N. 2018a. Electromotive force in strongly compressible magnetohydrodynamic turbulence. *J. Plasma Phys.*, **84**(5).

Yokoi, N. 2018b. Mass and internal-energy transports in strongly compressible magnetohydrodynamic turbulence. *J. Plasma Phys.*, **84**(6).

Yokoi, N., and Brandenburg, A. 2016. Large-scale flow generation by inhomogeneous helicity. *Phys. Rev. E*, **93**(3), 033125.

Yokoi, N., and Yoshizawa, A. 1993. Statistical analysis of the effects of helicity in inhomogeneous turbulence. *Phys. Fluids A: Fluid Dynamics*, **5**(2), 464–477.

Yoshizawa, A. 1990. Self-consistent turbulent dynamo modeling of reversed field pinches and planetary magnetic fields. *Phys. Fluids B: Plasma Physics*, **2**(7), 1589–1600.

Yousef, T. A., Heinemann, T., Schekochihin, A. A., Kleeorin, N., Rogachevskii, I., Iskakov, A. B., Cowley, S. C., and McWilliams, J. C. 2008a. Generation of magnetic field by combined action of turbulence and shear. *Phys. Rev. Lett.*, **100**(18), 184501.

Yousef, T. A., Heinemann, T., Rincon, F., Schekochihin, A. A., Kleeorin, N., Rogachevskii, I., Cowley, S. C., and McWilliams, J. C. 2008b. Numerical experiments on dynamo action in sheared and rotating turbulence. *Astron. Nachr.*, **329**(7), 737–749.

Zaichik, L. I., Alipchenkov, V. M., and Sinaiski, E. G. 2008. *Magnetoconvection*. New York: John Wiley.

Zakharov, V. E., L'vov, V. S., and Falkovich, G. 2012. *Kolmogorov Spectra of Turbulence: Wave Turbulence*. Berlin: Springer.

Zeitlin, V. 2007. *Nonlinear Dynamics of Rotating Shallow Water: Methods and Advances*. Amsterdam: Elsevier.

Zeldovich, Ya. B. 1957. The magnetic field in the two-dimensional motion of a conducting turbulent fluid. *Sov. Phys. JETP*, **4**, 460–462.

Zeldovich, Ya. B., Ruzmaikin, A. A., and Sokoloff, D. D. 1983. *Magnetic Fields in Astrophysics*. New York: Gordon and Breach.

Zeldovich, Ya. B., Ruzmaikin, A., Molchanov, S. A., and Sokoloff, D. D. 1984. Kinematic dynamo problem in a linear velocity field. *J. Fluid Mech.*, **144**, 1–11.

Zeldovich, Ya. B., Barenblatt, G. I., Librovich, V. B., and Makhviladze, G. M. 1985. *Mathematical Theory of Combustion and Explosion*. New York: Plenum.

Zeldovich, Ya. B., Molchanov, S. A., Ruzmaikin, A. A., and Sokolov, D. D. 1988. Intermittency, diffusion and generation in a nonstationary random medium. *Sov. Sci. Rev. C. Math Phys.*, **7**, 1–110.

Zeldovich, Ya. B., Ruzmaikin, A. A., and Sokoloff, D. D. 1990. *The Almighty Chance*. Singapore: World Scientific.

Zhang, H., Sokoloff, D., Rogachevskii, I., Moss, D., Lamburt, V., Kuzanyan, K., and Kleeorin, N. 2006. The radial distribution of magnetic helicity in the solar convective zone: observations and dynamo theory. *Monthly Notices of the Royal Astronomical Society*, **365**(1), 276–286.

Zhang, H., Moss, D., Kleeorin, N., Kuzanyan, K., Rogachevskii, I., Sokoloff, D., Gao, Y., and Xu, H. 2012. Current helicity of active regions as a tracer of large-scale solar magnetic helicity. *Astrophys. J.*, **751**(1), 47.

Zhu, X., Stevens, R. J., Verzicco, R., and Lohse, D. 2017. Roughness-facilitated local 1/2 scaling does not imply the onset of the ultimate regime of thermal convection. *Phys. Rev. Lett.*, **119**(15), 154501.

Zilitinkevich, S. S. 1991. *Turbulent Penetrative Convection*. Aldershot: Avebury Technical.

Zilitinkevich, S. S., Elperin, T., Kleeorin, N., and Rogachevskii, I. 2007. Energy- and flux budget (EFB) turbulence closure model for stably stratified flows. Part I: steady-state, homogeneous regimes. *Boundary-Layer Meteorol.*, **125**, 167–192.

Zilitinkevich, S. S., Elperin, T., Kleeorin, N., Rogachevskii, I., Esau, I., Mauritsen, T., and Miles, M. W. 2008. Turbulence energetics in stably stratified geophysical flows: strong and weak mixing regimes. *Quarterly J. Roy. Meteorol. Soc.*, **134**(633), 793–799.

Zilitinkevich, S. S., Elperin, T., Kleeorin, N., L'vov, V., and Rogachevskii, I. 2009. Energy- and flux-budget turbulence closure model for stably stratified flows. Part II: the role of internal gravity waves. *Boundary-Layer Meteorol.*, **133**(2), 139–164.

Zilitinkevich, S. S., Esau, I., Kleeorin, N., Rogachevskii, I., and Kouznetsov, R. D. 2010. On the velocity gradient in stably stratified sheared flows. Part 1: asymptotic analysis and applications. *Boundary-Layer Meteorol.*, **135**(3), 505–511.

Zilitinkevich, S. S., Elperin, T., Kleeorin, N., Rogachevskii, I., and Esau, I. 2013. A hierarchy of energy- and flux-budget (EFB) turbulence closure models for stably stratified geophysical flows. *Boundary-Layer Meteorol.*, **146**, 341–373.

Author Index

Subject Index

www.ingramcontent.com/pod-product-compliance
Lightning Source LLC
LaVergne TN
LVHW062208200926
848484LV00006B/1352
9781316518601